8·21·96

CORALS IN SPACE AND TIME

D0878893

John W Wells, Emeritus Professor of Geology, Cornell University, was delighted when I asked him if I might dedicate this book to him in recognition of all he has done for me and for so many others who work on corals. John died on 12 January 1994 after a long and very distinguished career as a coral palaeontologist and taxonomist.

J.V., February 1995

CORALS
IN SPACE AND TIME
THE BIOGEOGRAPHY AND
EVOLUTION OF THE SCLERACTINIA

J.E.N. VERON

AUSTRALIAN INSTITUTE OF MARINE SCIENCE
CAPE FERGUSON, TOWNSVILLE, QUEENSLAND

COMSTOCK/CORNELL
ITHACA AND LONDON

First published 1995 by Cornell University Press.

Library of Congress Cataloging-in-Publication data

Veron, J. E. N. (John Edward Norwood)
    Corals in space and time: biogeography and evolution of the Scleractinia / J.E.N. Veron.
        p. cm.
    Includes bibliographical references and index.
    ISBN 0–8014–8263–1 (alk. paper)
    1. Scleractinia — Geographical distribution. 2. Scleractinia — Evolution. I. Title.
QL377.C7V47 1995
593.6 — dc20   94–33540
                                                                          CIP

Printed in Australia by Southwood Press Pty Ltd

CONTENTS

PART C MODERN DISTRIBUTIONS

PREFACE

This book is designed to bring together a multitude of field observations about the taxonomy, biogeography, palaeontology, and biology of corals, and combine them with explanatory concepts of evolution. In an unavoidable sequence of deductions, this process has revealed significant conflicts between observable geographic variation within coral species and widely held general concepts about species. It questions the common assumption of operational taxonomy, that species remain discrete units over great geographic distances. This in turn draws into question present explanations of biogeographic patterns in corals and major elements of their speciation. The outcome is a rethink of concepts of species in corals and central aspects of the control of their evolution.

In short, when faced with factual realities of coral species on biogeographic scales, the broad spectrum of biological properties we have long associated with 'species' becomes artificial. As biogeographic patterns are the products of change over time, the same observations apply to temporal scales. The search for an explanation begs reconsideration of a spectrum of issues long associated with neo-Darwinian natural selection. Corals, perhaps alone among marine invertebrates, have been studied sufficiently well in essential disciplines — taxonomy, biogeography, palaeontology, reproductive biology and at least some genetics — to be the subject of this kind of inquisition.

The alternative to neo-Darwinian natural selection proposed in this book may initially seem obscure, but ultimately it borders on the obvious. That proposal is that the evolution of corals is not primarily driven by any form of competition between species, but rather by constantly changing patterns of ocean circulation creating, over geological time, constantly changing patterns of genetic connectivity. This mechanism, termed *surface circulation vicariance*, drives both speciation and hybridisation. The outcome — constantly changing patterns of affinities, termed *reticulate evolution* — has long been recognised in plants, but has not been considered as a general explanatory evolutionary concept in animals.

For corals, reticulate evolution underpins almost all that is known about species, and this has far-reaching consequences for almost all studies of corals based on species. It is stressed, however, that this book is not intended to construct a case for reticulate evolution in corals, but rather to highlight the data and the issues involved in evolutionary concepts, and to interrelate these across traditional discipline boundaries. The case for and against the concluding concepts of the nature of coral species, and the mechanisms controlling their evolution (the concluding two chapters) is the hypothetical outcome of the book as a whole, not isolated aspects of it.

ACKNOWLEDGMENTS

It was not my original intention to write a book that demanded a distillation of almost everything I have ever learned about corals, but that is what this book eventually became. The writing was a long and sometimes tortuous process, made so by the convoluted nature of the book's complex of subjects, but also compelled by criticisms of the manuscript by the many colleagues who took the time and trouble to help me with it. This help has been essential: I have not been able to write this book in the way I have written, or co-authored, others. I did not foresee this book's ultimate boundaries, nor the range of subjects that needed to be included, nor the wealth of points of view that had to be accommodated.

My original aim was to write a detailed review about the distribution of the corals of the central Indo-Pacific. This task grew ever longer and more complicated. A study of the corals of Japan, undertaken specifically for this book, had to be published separately as a monograph. Then the central database had to be published as another monograph. However, when the manuscript was eventually completed, I asked Mary Stafford-Smith (James Cook University) to review it for me. The undeniable logic of her criticisms convinced me that I had not only to rewrite the whole book (starting from the title), but also to reconsider key aspects of it. The outcome, she insisted, had to highlight my concepts of species and their evolution, make them clear to non-specialists, support them with evidence, and explain them 'unambiguously and in the light of alternatives'.

Critical readers will judge for themselves the successes and failures of this demanding but all-engrossing venture, but whatever the book's shortfalls, they do not occur for want of helping hands. At first I had no intention of seeking widespread criticisms of the (rewritten) manuscript, but it became clear, from an initial three reviews, that if I was ever able to make the complexities of the book clear and accurate for a wide readership, I would have to have input from experts in many different biological and geological disciplines. And so at different times, and with different requests, I sent ever-altered parts of the text to an increasing number of very busy people. I shudder to think of the collective time and effort these people have put into it.

And so my thanks and gratitude go to many who have given me the precious gift of their knowledge, time, and experience: Peter Glynn

(University of Miami) for his considerations of biogeographic issues, Rick Grigg (University of Hawaii) for his editorial advice and criticisms of palaeo-oceanographic issues, Terry Hughes (James Cook University) for his particularly detailed criticisms of ecological issues, Alan Dartnall (Townsville) for his biological overviews, Russell Kelley (Coral Sea Images) for his geological perspectives, John Benzie (Australian Institute of Marine Science) for his comments on coral genetics, John Chappell (Australian National University) for his help with palaeoenvironments, Dave Miller (James Cook University) for his help with molecular techniques, Don Potts (University of Southern California) for many discussions of coral taxonomy and evolution, Robert van Syoc (California Academy of Sciences) for his comments on systematics, Nancy Budd (University of Iowa) for her help with the Cenozoic fossil record, Bob Buddemeier (University of Kansas) for his help with geochemistry and calcification, and Bette Willis (James Cook University) for her help with coral reproduction.

Having then become convinced that any amount of criticism of a book such as this is never enough, I asked Mark Stafford-Smith (terrestrial ecologist with the Australian Commonwealth Scientific and Industrial Research Organisation) to review the whole (and by then much revised) manuscript. Review it he did, in detail and in concept, and in the process prompted me to go ahead with further revisions and rethinks, some of which I had been contemplating, others that I hadn't. Mary and Mark Stafford-Smith between them have been essential to this book's existence, and I am indeed grateful for their efforts.

Several colleagues, including many of those noted above, have provided me with unpublished data and reference material. I especially thank Steve Cairns (Smithsonian Institution), for his compilation of genera of extant azooxanthellate corals, Francesca Bosellini (University of Modena) for data on the fossil genera of the Tethys Sea, Brian Rosen (Natural History Museum, London) and George Stanley (University of Montana) for palaeontological references and discussions about them, Jim Maragos (University of Hawaii) for our several discussions about coral biogeography, and Mark Spalding (World Conservation Monitoring Centre) for his computer compilation of global reef distributions. Acknowledgments for taxonomic and biogeographic information are given in the monographs referred to above: I especially thank the volunteers cited therein who worked long and hard for many years on the data compilations this book has required.

Illustrations for the book were drawn by Marty Eden, with additions by Steve Clarke. Zolley Florian of James Cook University prepared photomicrographs. Several staff members of the Australian Institute of Marine Science have assisted with the book's production: Liz Howlett straightened out the literature and index; Karen Handley did the darkroom photography; Christine Cansfield-Smith assisted with proof reading; and the library staff were, as ever, invaluable. However, very special acknowledgments are due to Kim Navin who, over years, has compiled, corrected, sifted and analysed biogeographic data for me, and whose computer skills have been vital.

Finally, it is a pleasure to acknowledge the influence on all my work of the late John Wells of Cornell University. Our association began in the mid-1970s when, one hot summer's day while sitting on a Devonian reef in the Australian bush, I asked him why corals occurred where they do. 'Well ... that's a good question,' he said. And so it turned out to be.

1

ABOUT
THIS
BOOK

This book is based on an extensive taxonomic and biogeographic database of scleractinian corals that has been twenty years in the making and now underpins most recent research on Indo-Pacific zooxanthellate species. It reveals a great deal about the nature of coral species, the mechanisms that control their diversity, how they are distributed, and why they occur where they do.

THE BOOK AT A GLANCE

The following chapters are a fusion of the aforementioned database with explanations of evolution from the biological and geological literature. The outcome incorporates many previously expressed points of view, but challenges principal concepts of the major schools of biogeographic and evolutionary theory. It also raises issues that are potentially relevant to the understanding of all marine life.

The book describes how coral species most probably change in space and time, and seeks to explain the nature of species. These matters are not just academic. All research that is based on species is, to some extent, undermined by uncertainties about concepts of them. Even where there are no apparent problems with taxonomy or identification, there are always pervasive doubts about their geographic, environmental, reproductive or genetic boundaries. These doubts affect the interpretation of all studies, no matter what the field of research.

Coral 'species' are very differently interpreted in palaeontology, taxonomy, biogeography, population ecology, genetics and molecular biology. These differences create confusion and barriers to understanding, barriers that are identifiable and often avoidable. This book examines the nature of coral species and argues that they do not have a special status in the taxonomic hierarchy. Species are not units in which evolutionary change invariably occurs. They are not single lineages in evolutionary time. They do not have centres of evolutionary origin. They show no evidence of having been formed by discrete 'events' in time, and there is no association between their present place of occurrence and any apparent place of origin.

At lower taxonomic levels, corals are perpetually fusing and separating in space and in time. Species commonly form evolutionary networks, not branching trees. Most individual species are a part of a network. They may have various internal isolating barriers between their component races or populations, as well as external genetic and reproductive links to other species within the same genus, and perhaps beyond. At any one location each species will appear to be relatively distinct; these distinctions break down with geographic distance and therefore also break down in evolutionary time.

The process just described is sometimes known as 'reticulate' evolution, a concept almost completely restricted to studies of terrestrial plants. In the case of corals, evolutionary change is ultimately driven by palaeoclimatic cycles. These cycles are transferred to the genetic composition of species through fluctuations in ocean circulation patterns, which inevitably create ever-changing fluctuations in genetic connectivity. This process simultaneously drives both speciation and hybridisation. Geological intervals of weak ocean circulation reduce genetic connectivity and separate the components of the reticulate network; intervals of strong circulation have the reverse effect.

Fluctuations in ocean circulation vectors affect all coral species simultaneously. Intervals of mass convergence (hybridisation) of species occur when circulation is strong; intervals of mass divergence of species (speciation), and also of local extinction, occur when circulation is weak. Species diversity thus changes continuously and mostly in a gradual, benign way. This is a process of non-biologically-driven mass synchronous evolution.

The separate species diversities of the Indo-Pacific and of the Atlantic are, initially, outcomes of their separate Cenozoic histories. In the Indo-Pacific the geographic source of diversity is a taxonomically uniform equatorial band extending from the Red Sea to the western Pacific. Regional species diversity outside that band is the outcome of broad-scale availability of pan-equatorial species and regional survival. Local species diversity is the outcome of regional availability of species and local survival. This is a progression of source/sink relationships where the ultimate source of species diversity is limited by niche diversity, genetic connectivity and species interactions. The last (in contrast to neo-Darwinian theories of natural selection) is the only point of biological control of species diversity and of evolutionary (hybridisation, speciation and extinction) rates.

The nature of coral species, their diversity, their distribution and their evolution, are all interwoven subjects. These subjects cannot be viewed in isolation, yet are exceedingly complex as a whole. Of necessity, they are considered together in this book.

USE OF THE TERM 'SPECIES' IN THIS BOOK
The term 'species' can legitimately have a wide range of meanings. In this book (as in most others) species are operational morphotaxonomic units, that is, are units that are recognised by taxonomists and users of taxonomy. (What is 'operational', ie in current use, changes with taxonomic revisions and acceptance of those revisions by users.) The main points about operational species of corals in this book are as follows.

♦   Within a single region, they are:
    1) morphologically distinguishable from other species, and
    2) genetically semi-isolated from other species.

♦   Over their full range, this book concludes that they:
    1) continuously vary, morphologically and genetically, and
    2) are not necessarily morphologically nor genetically isolated from other species.

♦   Spatial patterns of species interact with temporal patterns to produce networks of genetic links. These links are not observable at single points in space or time.

♦   There are many kinds of species: all are human-defined units of identification and are not discrete units in a systematic hierarchy.

Chapter 13 and figure 74, p 228 give further information about species as operational taxonomic units.

GEOGRAPHIC VARIATION IN SPECIES: FIRST OBSERVATIONS

This book has its origins in the very simple observation that reef corals from one tropical central Indo-Pacific region look much like those from adjacent regions, and that differences between regions reflect geographic separation, latitude, size of the region and isolation.

To the discriminating observer, there are few differences among the corals of the separate countries bordering the Coral Sea, or among the corals of the islands of the Indonesia/Philippines Archipelago, or among the corals of the islands of the Caribbean. Differences in corals within each of these regional groups of countries are mainly ones of presence or absence, not of variation *within* species. However, between any two of these groups of countries, differences become combinations of presence/absence and *within*-species variation. The degrees of these differences are broadly correlated with geographic distance.

Geographic distance Using the Great Barrier Reef as a reference point and other countries as examples (figure 66, p 203), these observations can be expanded as follows. The species complements of each country bordering the Coral Sea are similar to that of the Great Barrier Reef and there are few significant differences *within* those species. The species complements of more northerly central Indo-Pacific countries (eg the Ryukyu Islands) are significantly different, and colour, morphological and ecological differences occur *within* those species. Further afield (eg Thailand or the Marshall Islands), species complements are substantially different and doubts arise as to whether some differences are *within* species or *between* species: boundaries between species may be uncertain. Further afield again (eg Hawaii, or Tahiti or the Red Sea), distinctions *between* species and *within* species are mostly uncertain. In the Caribbean all differences are clearly *between* species (there being no species in common between the Great Barrier Reef and the Caribbean).

Latitude Looking a little more closely, the observer will find that the differences within and between species are not just correlated to distance. Within groups of contiguous countries or along continental coastlines, changes with latitude are disproportionately greater than changes with longitude. Thus, for example, the corals of the Great Barrier Reef resemble those of tropical western Australia much more than they do those of adjacent temperate southeastern Australia. Along both Australian continental coastlines, as well as along the offshore islands of southeast Asia, clear latitude-correlated changes occur in most aspects of the coral fauna.

Region size and isolation Cocos (Keeling) Atoll, for example, has a depauperate and distinctive coral fauna in comparison with that of Indonesia to the north and western Australia to the east. The atoll is small and the ocean surrounding it appears to be an isolating barrier.

Simple observations such as those above beg some not-so-simple questions. Are corals where they are because that is where they evolved? Are corals where they are because that is where coral reefs are? Are corals where they are because of physical environment: temperature or ocean circulation or bathymetry or plate tectonics? Or because of biological parameters: ecology or competition? Or because of survival of the fittest? Are there different answers for different taxa? What are coral species: are they singular units or are they real? If so, for whom: palaeontologists or coral taxonomists or biogeographers or geneticists? Do they change with space and time? How can this change be studied? What have genetics and evolutionary theory to say about it? How do distributions relate to the fossil record? How do these questions relate to Darwinian evolution? Does it matter?

The last question can be answered now: yes, it does matter. It matters because the answers to such questions underlie our ability to understand the real world, and not just that of corals. It matters because taxonomy is an empty subject unless it provides us with a framework that reflects what actually exists in nature. Conceptually, that framework is only as good as its faithfulness to the genetic codes that allow it to exist. If we have taxonomic units, then those units are only as good as their capacity to describe relationships that occur in reality. If we have species, then those species are only useful if they can be meaningfully separated in some way from other species.

Editorial note: the use of the term 'coral' in this book is explained in the glossary.

TAXONOMY AND BIOGEOGRAPHY: A BACKGROUND

For reasons of logistics or opportunity, original studies of coral taxonomy have mostly been undertaken within single regions. In contrast, biogeography, by its very nature, must be based on different studies of a very large number of regions or the whole world. These simple historical necessities have forced the two disciplines apart, making them largely independent, especially at species level.

Geographic variation   Historically, intra-regional taxonomic studies have suggested greater differences between regions than actually exist. For example, original studies in the Philippines or Japan or the Marshall Islands have deceptively implied that these countries have few species in common. This problem has long been recognised. In theory the historical proliferation of regional names has been countered by creating 'synonymies'. These are intended to correct nomenclature by separating *within*-species names (synonyms) from *between*-species names, combining the former under one name, and separating the latter under different names. Unfortunately this process (which has been used extensively by almost all coral taxonomists) has seldom been based on any study of inter-regional variation within species: it has been guesswork.

Environment-correlated variation   Environment-correlated morphological variations within species have been a much worse problem in coral taxonomy. The issues are described in chapter 2. Twenty years ago a coral 'species' was simply a description of morphologically similar specimens and included no reference to geographic range, geographic variation, or to variation correlated with physical environment. There was no established basis for distinguishing between species and thus, at that time, *Porites compressa* in Hawaii was the only Indo-Pacific species of coral that had ever been distinguished from any other species by specific criteria such as are used for most major groups of plants and animals today.

The importance of *in situ* observation The advent of scuba diving opened the underwater world to scientific scrutiny. It allowed taxonomists to observe variation in corals more or less as botanists had long observed variation in terrestrial plants. It allowed differences between co-occurring species, and variation within species, to be observed directly. It allowed particular specimens to be selected for study and details of physical environment to be recorded with them. *In situ* studies have thus given enormous impetus to coral taxonomy, as it has to most other fields of whole-organism coral biology. Most research is now based on species, and thus most research contributes to our pool of knowledge about those species. Through this process taxonomy has become strongly tied to all aspects of biology.

The biogeographic scale The impact of *in situ* studies on coral biogeography has been of a different nature because geographic variation, unlike environment-correlated variation, cannot be observed at single locations. Knowledge of geographic variation within species must be accumulated from separate studies in different regions and interpreted through the medium of taxonomic detail. Thus, although biogeography is dependent on taxonomy, the broad-scale data it demands have not, until recently, been available.

If coral taxonomy has been driven by need, coral biogeography has been driven by questions. Fundamental questions, such as those asked above, have been asked of corals since the time of Darwin. It has largely been the biogeographer's domain to address them, and most biogeographers who have done so have been palaeontologists, not biologists. Thus evolutionary aspects of the biogeography of corals have not only had an inadequate taxonomic background, biogeography has not been integrated with biological details, especially those of population genetics, reproductive biology and community ecology. Biogeography of corals has, in effect, bypassed species-oriented taxonomy and the biology that goes with it; instead it has been based on generic data. This may be adequate for some purposes, but coral biogeography is also steeped in concepts of species and genetic mechanisms of speciation. The result is that concepts have either been 'general', or borrowed from other phyla (especially the Mollusca) and proposed mechanisms of evolution have not been derived from data about species. The subject of coral biogeography has therefore become theoretical and remote from biology.

BACKGROUND OF THIS BOOK

Editorial note: Geographic terminology used in this book is explained in the glossary.

I have followed a path that many others have taken: from taxonomy, to biogeography, to concepts of species and their evolution. This journey began when my first efforts to identify corals failed. They failed because names and descriptions based on collections of museum specimens could not be applied to what could be seen on reefs. The investigation of variation within and between species based on *in situ* studies thus became a research priority. The essential requirement was a blend of taxonomy and field identification, in effect an amalgam of historical review and new observation. This task, commenced in 1974, was undertaken on the Great Barrier Reef by a small group of taxonomists. It was completed after eight years and published in the five volumes of *Scleractinia of Eastern Australia* (Veron et al 1976–84).

This protracted taxonomic work on the Great Barrier Reef started to merge with biogeographic issues when we extended it south of the Great Barrier Reef. For most corals, there are enormous morphological variations between (for example) colonies on a reef flat and colonies of the same species on an adjacent lower reef slope. These variations usually intergrade: the continuum they form is clearly correlated with depth (p 17). South of the Great Barrier Reef, the number of species decreases, but species also show change in morphology, colour and habitats. These are correlated with latitude and isolation, not just local physical environment.

Taxonomic work along the whole east Australian coast was followed, over several years, by a series of companion studies along the west Australian coast. In the tropical northwest the number of species was found to be fewer than on the Great Barrier Reef, but their identities were mostly the same. They showed similar, but often not identical, environment-correlated variations; their relative abundances were often different; and often there were minor differences in spatial distribution, morphology and colour. Further south, the number of species was found to further decrease, and again there were intra-specific latitude-correlated changes similar to, but often not the same as, those of the Pacific east coast. In short, geographically defined subspecies became recognisable.

Studies of corals in other countries followed. In the central Indo-Pacific, the total number of species distinguished gradually expanded, and so did their known ranges of morphological variation. Most species were seen to be a complex of geographic intergradations not unlike the depth-correlated intergradations seen down individual reef slopes. Nevertheless most species retained a more-or-less definable individual identity. This identity was usually clear when geographic variation could be traced from one adjacent region to the next, but was usually less clear when there were major geographic disjunctures.

Studies beyond the central Indo-Pacific almost always involved increasing problems of intra-specific variation. Most species (within the Indo-Pacific) exhibit increasing variation with distance (p 203): some remain recognisable, others exhibit patterns of variation well beyond tax-onomically meaningful boundaries. Taxonomic issues under these cir-cumstances become, in effect, biogeographic issues.

The outcome, for this book, is two sets of data. The first covers the central Indo-Pacific, has a relatively well-defined taxonomic framework and is suitable for numerical analysis. The second covers all regions beyond the central Indo-Pacific and (except for the Gulf of Mexico and Caribbean) has a taxonomic framework that is anything but well-defined and is only suitable for numerical analysis at generic level. The biogeo-graphic database of this book (Veron 1993), therefore, clearly distin-guishes between comprehensive and (relatively) robust data for the central Indo-Pacific, and data from other regions of the Indo-Pacific where taxonomic issues progressively blend into biogeographic ones.

The timing of this book is not arbitrary. The above-described studies along the latitudinally contiguous Pacific and Indian Ocean coasts of Australia in the southern hemisphere and its mirror image along the Asian coast in the northern hemisphere have been completed. A number of less-detailed companion studies across the Indo-Pacific and in the Caribbean region have also been completed. All available records of coral occurrences have been extracted from the literature, as well as from most major museum collections (Veron 1993). A global picture has emerged that is certainly nowhere near complete, but is sufficient to draw the prin-cipal conclusions outlined at the beginning of this chapter and to high-light issues for future study.

DATA QUALITY

Data quality varies primarily according to comprehensiveness and taxo-nomic reliability. Comprehensiveness is critical to numerical analyses of species diversity, determination of distribution ranges and regional affini-ties. Taxonomic reliability affects all biogeography but is far more critical to species-level studies than generic-level studies because genera, being mostly well-defined, are much more readily identified than species. Most conclusions about species in coral biogeography have been built on extrapolations of generic data. This occurs most commonly when generic distributions are used as proxy-indicators of species distributions. Generic data are especially inappropriate if used to derive hypotheses about mechanisms of speciation.

TYPES AND USES OF CORAL BIOGEOGRAPHIC DATA

The principal database of this book (Veron 1993) links data of many types and sources. The principal divisions are as described below.

Primary data These are from original field studies by the author and his colleagues in the central Indo-Pacific, specifically for taxonomic and bio-geographic purposes. They are based on a uniform taxonomy, facilitating direct and detailed comparisons between different biogeographic regions. Taxonomic and identification strengths, weaknesses and errors are uniform. The data include estimates of abundance and observations of geographic variation in colony formation, skeletal morphology, ecology and colour. The data provide comprehensive presence/absence records for specific sites along the three continental coastlines of the central Indo-Pacific (eastern and western Australian and eastern Asia). Only primary data are used in examining details about the global centre of coral diversity (chapter 9) and in dealing with biogeographic variation in species (chapter 11).

Reference data   These are from studies in the central Indo-Pacific undertaken by many other taxonomists. They are used for distribution records, but not records of species attributes. Taxonomic and identification weaknesses and errors are not equal, and reliability varies greatly from one author to another. Data may not be comprehensive for any one region and are thus not used in this book for quantitative analysis.

THE PRIMARY GEOGRAPHIC DATABASE OF THIS BOOK

Taxonomic strengths and weaknesses for any given region of the data presented in Veron (1993) can be summarised as follows:

STRENGTHS

♦ a uniform taxonomic framework of recognisable species with known ranges of variation;
♦ detailed correlation between morphological variation and physical environment;
♦ uniform *in situ* taxonomic methodologies for all taxa;
♦ similar levels of comprehensiveness for all locations studied;
♦ exclusion of human-created taxonomic and identification problems through uniformity of nomenclature and direct transfer of original taxonomy to identification.

WEAKNESSES

♦ incomplete data due to undetected species (p 160);
♦ non-resolution of known species complexes (p 204);
♦ omission of species thought to exist (from unidentified specimens or *in situ* sightings), but that have remained unstudied.

Reference data that appeared doubtful in original sources have been re-evaluated by the author. Those from distant countries are often erroneously biased in favour of species described in *Scleractinia of Eastern Australia* (Veron et al 1976–84) and *Corals of Australia and the Indo-Pacific* (Veron 1986), making the corals of these regions appear to have greater affinity with those of the central Indo-Pacific than is actually the case.

STRUCTURE OF THE BOOK

To aid in dealing with the complexities of the multifarious views expressed about the many subjects covered by this book, numerous quotations, asides (in boxed text) and internal cross-references have been employed. Different chapters will have varying value for different readers. Each chapter is designed to stand alone as much as possible, to allow readers to skip introductions and topics of peripheral interest.

To bypass irrelevancies of nomenclature, names of all taxa have been unified to that of Veron (1993). Unless there is a particular reason for doing so, original names used in references are not cited.

Much of the geographic and taxonomic data used in this book is too voluminous to be reproduced in print, especially if specific source references are included. Present intentions are to do this through a computer-readable electronic biogeographic database linked to taxonomic information.

The book is divided into four parts, details of which are listed in the table of contents and summarised in an introduction to each part.

Part A Introductions and reviews (chapters 2 to 6) Each chapter starts with a list of the topics covered. These are treated in detail, or not, according to relevance to species-level biology. More comprehensive or specialised accounts of most topics can be found in other publications, which are cited. These chapters are written for non-specialists. Unless otherwise indicated, they are not intended to promote particular points of view, nor to debate issues.

Part B Fossils and palaeoclimates (chapters 7 and 8) Both chapters start with a list of the topics covered. Chapter 7 summarises those aspects of geological change in palaeoenvironments and continental positions that are relevant to the evolution of corals. Chapter 8 presents the main points of the author's (unpublished) compilation of fossil genera, combined with extant genera (Veron, 1993). This chapter is intended to bridge the gap between palaeontology and the biogeography of extant corals.

Part C Modern distributions: biogeographic data and data analyses (chapters 9 to 11) These chapters are about data, not concepts or conclusions. Each chapter ends with a summary. The biogeographic database described above (Veron 1993) links these chapters to original taxonomic and biogeographic data sources: the main points are summarised in the appendix.

Part D Evolution: the nature and origin of species (chapters 12 and 13) These chapters are about concepts of coral species and conclusions about their taxonomy, systematics, biogeography and evolution. As such, they are a hypothetical synthesis of the information presented throughout the book.

PART A

INTRODUCTIONS
AND REVIEWS

The main topics of chapters 2 to 6 are listed at the beginning of each chapter. The following is a summary of these topics:

THE MAIN POINTS

Some of the following points (identified by cross references) anticipate conclusions made in subsequent parts of this book. These points are not intended to be self-explanatory.

◆ Systematics, taxonomy and identification are very different subjects, requiring different data and interpretations. Systematics is linked to the non-observable genetic organisation of life; taxonomy is linked to observation, and therefore underpins identification. Species in corals are human-created units of identification, not systematics (p 228).

◆ Most coral species, being colonial, display wide ranges of morphological variation. These are correlated with both physical environment and geographic range. It is the extent of geographic variation that makes coral species distinctions arbitrary (p 204).

◆ Morphological taxonomy necessarily depends on *in situ* comparisons between colonies. These comparisons underpin most operational coral taxonomy. Molecular techniques will create the need to revise many decisions based on morphological taxonomy. Molecular techniques are likely to identify genetic units that have no identifying morphological characters (p 229).

◆ Current biogeographic concepts of evolution (Darwinian

centres of origin, panbiogeography, classical vicariance, phylo-genetic biogeography and dispersal and ecological biogeo-graphy) do not adequately explain observed distribution pat-terns of corals. Place of origin is not limited to, nor necessarily correlated with, present-day distribution.

♦ Rapid long-distance larval dispersal creates ever-changing dis-tribution patterns and obscures any trace of place of origin.

♦ Neither the biological species concept, and derivations of it, nor most current views of the process of natural selection, are sup-ported by operational coral taxonomy. Nor are they supported by key aspects of coral reproductive biology.

♦ Coral species diversity is the morphological (taxonomic) prod-uct of genetic (systematic) diversity. Geographic patterns of species diversity are created by geographic pattens of genetic diversity. The latter is controlled by balances between regional diversity sources and local diversity sinks (p 234).

♦ Species diversity within communities is primarily controlled by physical environment and competitive interactions, moderated by the frequency, and the amount, of environmental change. This limits the total diversity communities can support. Communities are thus the ultimate sinks in biodiversity-regulating patterns of source/sink relationships.

♦ Coral species can be artificially hybridised, as can species of other animal phyla. Given evolutionary scales of space and time (figure 17, p 70), almost unlimited combinations of hybridisa-tion and divergence at different taxonomic levels are likely to occur if they can occur.

♦ Temperature and continental boundary currents (p 171), control most *latitudinal* (north–south) biogeographic patterns. A combi-nation of fluctuating sea surface currents and evolution creates most *longitudinal* (east–west) patterns (p 155). Latitudinal and longitudinal biogeographic changes operate in very different time-frames.

♦ Reefs are an evolutionary 'reward' of coral/algal symbiosis, but symbiosis is not a controlling microevolutionary mechanism: most corals can exist without reefs.

2

THE
TAXONOMIC
FRAMEWORK

'Virtually every writer on historical biogeography
remarks that biogeography is subordinate to systematics
... biogeography can be no better than the taxonomy it
must use to describe distributions.'
(Paterson 1981)

The purpose of this chapter is to describe intra-specific morphological variation in corals where relevant to concepts of species, their diversity and their evolution; outline the taxonomic methods behind the primary geographic database of this book; and outline and compare all methods of coral taxonomy. This chapter is introductory to taxonomic aspects of biogeography (chapter 3), general concepts of species, speciation and species diversity (chapter 4), and reproduction (chapter 5). It also provides background to the data analyses of chapters 9 to 11. Chapter 13 focuses on the issues raised.

THE MAIN TOPICS

♦ Coral systematics, taxonomy and identification.
♦ Intra-specific variation in corals.
♦ Human-made issues of coral taxonomy: the 'name game'.
♦ Co-occurrence in the methodology of coral taxonomy and its role in the compilation of the primary geographic database of this book.
♦ Taxonomic methods and data sources: numerical methods; cladograms and cladistics; *in situ* experimental studies; studies of reproduction; physiological studies; molecular and genetic studies.
♦ Bridging, ranking and testing of taxonomic methods.

SYSTEMATICS, TAXONOMY AND IDENTIFICATION

Systematic order is genetically based and conceptual. It represents the true relationship between taxa. Taxonomic order is morphologically based and operational. Systematic order underpins taxonomic order, but the two are not the same (p 228). Species are taxonomic units of identification, not systematics. These distinctions, which are critical to concepts of coral species and mechanisms of speciation, are the subject of chapters 12 and 13.

The primary object of taxonomy is to define the morphological limits of species, separate species, name them and describe them in a form meaningful to other taxonomists. The primary object of 'identification' is to apply the results of taxonomy to other studies, that is, to make them useful to non-taxonomists.

With corals, as (for example) birds, identification depends on the appearance of the living organism and the applicability of guide books. Unlike birds, the appearance of living coral varies greatly with physical environment. For Indo-Pacific corals, any adequate field guide would need, perhaps, fifty photographs per species to be reliable. Such guides do not exist, and this creates a major information barrier. As a consequence, reliable identification requires considerable field knowledge, memory and experience. Many keys have been written for coral identification, but except in limited circumstances (which computerised Expert Systems have the means of overcoming), most are too simplistic to be effective.

Taxonomy is repeatedly blamed for identification difficulties. Some of this is justified, but much is not. Even in situations where all taxonomic issues have been solved, environment-correlated variations remain; and many species continue to be difficult to separate. In fact, if, at some imaginary point in time, all taxonomic issues with corals were resolved, most of the present identification difficulties would remain.

MORPHOLOGICAL VARIATION

Almost all species of zooxanthellate corals display wide ranges of morphological variation. This makes species identification a task for specialists and gives rise to much of the confusion in the taxonomic literature. The nature of morphological variation is best appreciated by dissecting it into its principal components: most species do not show wide variation in all the categories described below, but have most of their variability restricted to one or two categories. Examples of different types of intraspecific variation are indicated in the endnotes of chapter 11. Veron (1993) gives relevant notes and references and many are illustrated in *Scleractinia of Eastern Australia* (Veron et al 1976–84).

Corallite variation within a colony Almost all colonial corals show variation in skeletal structure between different corallites of the same colony. Such variations reach an extreme in *Acropora* where differentiation into axial and radial corallites is a complete dimorphism, and where

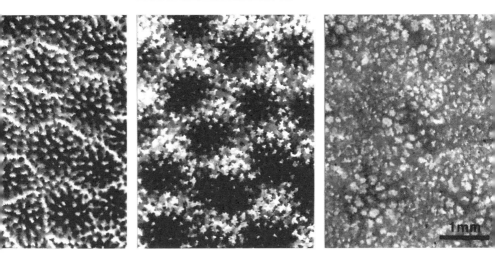

Figure 1 Corallite variation within a colony of *Porites lutea*. The corallites occur within 300 mm of each other, around the lip of the base of a helmet-shaped colony.

Figure 2 A small faviid colony, illustrating intra-colony variation approaching an inter-generic level. Corallites on the top of the colony are *Favites*-like, those on the sides of the same colony are *Favia*-like.

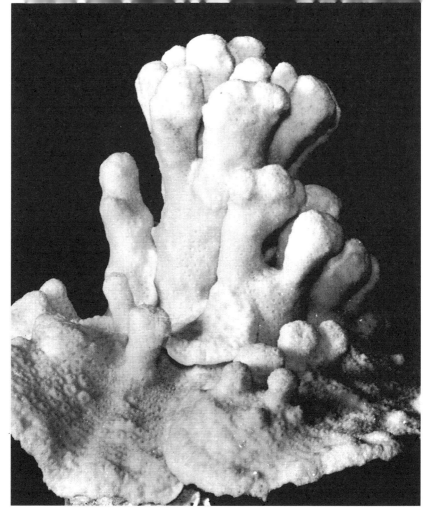

Figure 3 A small colony of *Pavona maldivensis* showing explanate and columnar growth forms combined. These two morphs, which are usually found in different colonies, were originally described as *Siderastrea maldivensis* Gardiner, 1905 and *Pavona (Pseudocolumnastrea) policata* Wells, 1954 and have been placed in different genera or sub-genera by several taxonomists.

radial corallites on proximal and distal parts of the same branch, or on central and peripheral branches of the same colony, may have substantially different characters. Corallite variations in other genera are usually related to the particular location of the corallite within the colony (figure 1) and are due to factors such as space availability, age, predation or microenvironment. In some cases these differences cross normal species boundaries. In extreme cases, different parts of the same colony have corallites with the characteristics of different genera (figure 2).

Morphological variation within different parts of the same colony
A small proportion of coral species show major growth form modification with age or some other factor, so that one part of the colony becomes

quite unlike another part (eg colonies that have both plates and branches or have different types of branches). In almost all cases, whole colonies can be found with only one of the possible growth forms expressed.

Different parts of very large colonies may have substantially different microenvironments (eg different levels of light availability or exposure to wave action), producing a wide range of intra-colony morphologies. Small colonies from equivalent microhabitats usually mirror each part of this variation.

The majority of nominal species descriptions reflect only a small part of the full variation of operational species. There are several instances where variation that is commonly expressed within a single colony has been the basis for different nominal species of the same operational species being placed in different genera (figures 2 and 3).

Variation between colonies within the same biotope Different colonies of the same species growing in close proximity under uniform physical environmental conditions or compound colonies made up of several original colonies that have grown together may show major variations in growth form, polyp or corallite structure. Differences between

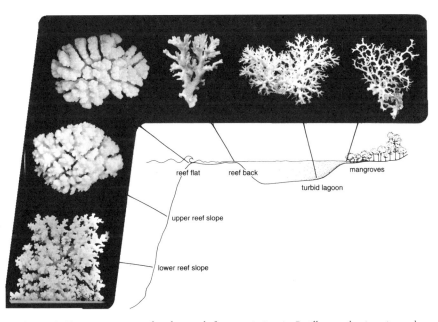

Figure 4 Environment-correlated growth-form variation in *Pocillopora damicornis* on the Great Barrier Reef. This is the most widely studied of all coral species. All ecomorphs (p 229) (indicated here by individual photographs) are linked by a continuum of intermediate growth forms. This amount of intra-specific variation creates major taxonomic problems in corals: for example the colony from the reef flat (top left), has more in common with colonies of other *Pocillopora* species than with colonies of *P. damicornis* from other environments. Although corallite characters are of little use in *Pocillopora* species, this species is readily recognisable throughout the Indo-Pacific.

colonies appear to be primarily genetic except for the effects of size or age or some aspect of individual colony history.

Contiguous variation between physical environmental zones Almost all coral species exhibit major morphological change over a series of contiguous biotopes in response to gradients in one or more physical environmental parameters (figure 4). Over a wide depth range, such as between a reef flat and a lower slope, morphological variation in most species is enormous, so much so that a colony from a reef flat and another of the same species from a nearby lower slope may have less in common morphologically than colonies of different species from the same reef flat.

Variation within regions The full range of variation of a species along an extended latitudinal range (such as along the eastern Australian coast (p 173)) is a product of the combined effects of multiple local environments, isolation, and factors associated with latitude. Intra-specific variation within the Great Barrier Reef as a whole is considerably greater than that found on any individual reef because the range of physical environments is greater. Most species with distributions extending to high latitudes undergo skeletal and other changes that can be correlated with decreasing temperature (p 90) or genetic isolation or both (p 178).

Non-skeletal variation A glance at any field guide shows that the appearance of the living coral is of great importance in species recognition. Soft tissue morphologies are also used as taxonomic characters, especially with large-polyped genera that have polyps extended by day (notably *Goniopora*, *Alveopora* and caryophylliids). So far, in only six species of the caryophylliid genus *Euphyllia* are soft tissues essential to species identification (Veron 1990).

TAXONOMIC METHODS AND CRITERIA

HISTORICAL BACKGROUND

When any taxonomist erects a taxon, what he or she actually does is propose a hypothesis about that taxon that can be accepted or rejected by others. The taxon is only as good as the taxonomy from which it was created; the taxonomy is only as good as the taxonomist who created it, and the methods and material that were available at the time and place of study.

The methods that led to two thousand descriptions of nominal species of extant zooxanthellate corals from around the globe throughout the last century and most of this one are the hand lens, workbench, pen and paper. Throughout most of this time the only concept of 'species' was that they are unchanging products of divine creation. A good illustration of the absence of a species concept is the debate in

PROBLEMS OF HUMAN ORIGIN

Today the practical process of describing a coral species involves two types of research. The first is usually an intra-regional detailed field and laboratory study of morphological variation, and establishing criteria for separating species from their nearest neighbours. The second involves study of museum collections and type specimens, supposedly allowing construction of synonymies, and determination of nomenclatorial priorities and distribution ranges. I add 'supposedly' because this process involves comparative evaluations of type specimens from many different countries, an exercise which presupposes a knowledge of biogeographic variation that few taxonomists have the opportunity to gain (p 5). An additional problem is that the oldest type specimens usually have nomenclatorial priority. These are often as not just pieces of skeleton, with no accompanying detail about locality or physical environment. Taxonomists are still changing species names according to re-evaluations (even rediscovery) of these old museum bits and pieces. A commonly held view is 'better now by me than later by someone else'. For many genera that have a high proportion of type specimens of doubtful status, this process can go on forever. There are many human-created reasons for changing generally accepted species names or generic designations; these have been, and still remain, a major source of instability in all coral nomenclature. Fortunately, they don't affect the delineation of species, and thus can be discounted for present purposes.

'Bernard's Symposium' (Cock 1977): a debate over Bernard's (1903–06) rejection of binomial Latin names in favour of a morphologically and geographically based numbering system[1]. Subsequent to Bernard, all coral taxonomists of note have had at least some experience of living reefs and therefore have had at least some appreciation of the appearance of corals in different reef environments. However, with the single exception of Vaughan's (1907) study of the enigmatic *Porites compressa* in Hawaii, no study of intra-specific variation in any species of coral was published until the early 1970s. Worse, there was not a single case where a taxonomist actually explained how *any* species could be diagnosed. The implication was that a coral species was something defined by a taxonomic description. A further implication was that these descriptions and diagnoses were ends in themselves, absolved from the need to provide an identification base for other research disciplines.

Many methods have been used, or are currently being used, to separate coral species. Some are of historical interest only. Others are superficial or frivolous (Lang 1984 gives many examples). In reality, the first need of any coral taxonomist, no matter what methods are employed, is to gain an understanding of how species differ from each other *in situ* and how they vary intra-specifically *in situ*.

BIG BORING VOLUMES

The volumes of *Scleractinia of Eastern Australia* and many other such taxonomic monographs justify this heading. They suffer a common fault in containing information that is not easily accessed by non-taxonomists. This is partly because of technical terminology but partly because they go relentlessly from one taxon to the next without highlighting issues or points of general interest. Nevertheless, the often-imagined view that they are anachronistic is usually incorrect. Most modern taxonomic monographs are based on an extensive knowledge and understanding of the taxa they deal with. The 'discoveries' of later years are often buried deeply within them.

STUDIES OF CO-OCCURRING COLONIES

In situ study of co-occurring colonies is a simple but very powerful technique in coral taxonomy. It is widely used and is the basis of much of the primary taxonomy of this book (p 8).

Some species are taxonomically straightforward, being reliably distinguished from their nearest neighbours by well-defined attributes. These require only simple *in situ* studies to determine environment-correlated variation. However, this is not the case for any species-rich colonial genus and is certainly not the case for species groups that are both widespread and polymorphic. The basis of study of these species groups must be understood; the procedure varies according to need, but usually follows these steps:

1   Co-occurring colonies of what may appear to be separate species (eg figure 5) are separated *in situ* and representative colonies collected.
2   The morphological basis of this separation, whatever its nature, is determined in the laboratory by any method, morphological or molecular.
3   The same process is repeated in different biotopes and in all available physical environments until laboratory criteria and field criteria separating species are in agreement, and repeatably so in all physical environments.
4   The species is described in terms of its full range of environment-correlated variation, and criteria separating it from all other species.

This procedure allows the gathering of results of a virtually infinite number of 'natural experiments' on almost any combination of species, or species attributes in any combination of natural physical environments. It requires close co-occurrence of colonies of related species and recognisable field characters. It can be repeated in a manner that focuses on ever-finer detail, almost without limit. It is particularly effective for studying problematic species, for separating groups of species that are morphologically similar, and for separating highly polymorphic species.

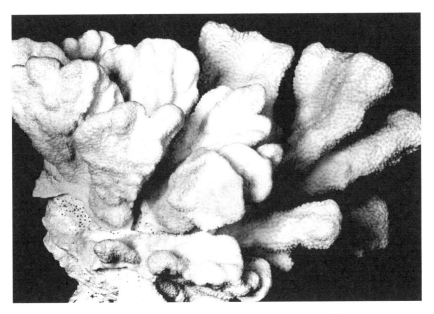

Figure 5 The effectiveness of co-occurrence in determining species differences. The two species illustrated, *Acropora cuneata* (left) and *Acropora palifera* (right), each exhibit wide environment-correlated growth-form variations, but they can be readily separated where they co-occur. Electrophoresis supports this species separation (p 28).

Figure 6 A rare example of where co-occurrence apparently fails to separate species. The two colonies illustrated are near opposite extremes of variation in *Pocillopora* at Clipperton Atoll in the far eastern Pacific. These colonies may represent different species, but their morphologies and skeletal details almost completely intergrade. This intergradation occurs within single large colonies, and also among series of different colonies within the same biotope.

The process has weaknesses. First, because of its reliance on both co-occurrence and *in situ* recognition, very rare species (notably those with narrow physical environmental tolerances) and species that do not have reliably recognisable *in situ* characters (notably massive *Porites*), are prone to being masked by other more common species (p 28). Second, as with all taxonomic methodologies, it can be defeated by species complexes (p 227). In rare instances, for example, with *Pocillopora* at Clipperton Atoll (figure 6), *Porites compressa* at Hawaii and *Lobophyllia hemprichii* on the Great Barrier Reef, co-occurrence may fail to yield intuitively meaningful species separations.

These are not the only morphology-based procedures available to discriminate between species and determine intra-specific variation. The first two types of morphological variation within a colony described above (p 14) can be adequately determined without *in situ* studies. A few species can be discriminated (or confirmed) from inter-regional studies, as described below. Others can be initially detected because of some non-morphological (reproductive or genetic) attribute, then studied by co-occurrence.

THE BIOGEOGRAPHIC SCALE

Variation within species in a single region can be correlated with physical environmental gradients that occur down reef slopes (figure 4), or occur between the windward and leeward sides of reefs, or occur between inshore and offshore reefs. These variations take the form of a range of arbitrarily delineated ecomorphs[2], each ecomorph being a product of genotype and microenvironment. The number of ecomorphs of a species increases as genetic and environment-correlated variations increase, but can usually be satisfactorily determined within a given region if the species is common enough.

If there are no major distribution disjunctures between regions, inter-regional studies can clarify taxonomic issues because taxonomic problems encountered in one region may be readily resolved in another. If two species are confused in one region, but one of those species is absent from another region, the characteristics of the isolated species in the second region will be relatively clear. If what was considered to be one species in one region was found to have had an unaccountably narrowed range of variation in another region, a mistake in the original species diagnosis must be suspected.

There are often major differences in the relative abundances of species in different regions (p 199); a species whose presence may have been masked by a much more abundant species in one region may be readily observable in another.

Geographic subspecies The linking of a given species of one region (with its determined range of ecomorphs) with what is believed to be the same species in another region often involves taxonomic decisions that are less easily substantiated than those made within single regions. If there are no continuous gradations in all taxonomic characters from one region to another, distinct geographic subspecies may be distinguished. These may only be arbitrarily distinguishable (in theory and practice (p 204)) from geographically separate species.

Taxonomic uncertainty also occurs when making intra-specific comparisons between colonies from tropical regions and those from high-latitude regions. The number and distinctiveness of geographic subspecies (p 229) is much higher in temperate regions than in the tropics. High latitude regions are often isolated and are usually environmentally discrete. Thus, the total intra-specific variation of corals within the reefs of tropical regions tends to be less than that within the non-reef environments of high-latitude regions (p 201).

NUMERICAL METHODS

Numerical methods have been successfully used to analyse morphological data, to relate morphological data to non-morphological data (especially reproductive and physiological data), to order species, to create keys and to create cladograms from morphological data. Numerical classifications based on molecular taxonomy have only recently been used and are, as yet, very much in the development stage.

Multivariate techniques (such as canonical variate, discriminate function, principal component and principal coordinate analyses) have often been used to ordinate morphometric data so that underlying biological parameters may be more readily discovered. As yet few studies have yielded useful results in coral taxonomy. Budd (1984) and Cuffey and Pachut (1991) used canonical variate analysis to compare the phenotypic plasticity of extant *Montastrea annularis* and *Favia pallida*, respectively, with fossils of the same genera. Combined with other data (such as from thin sections of *Porites*, *Goniopora* and *Stephanocoenia*; Budd 1986, 1987), this methodology appears to be the most effective way of exploring phylogenies in the fossil record (Budd 1993) (p 143). Comparisons between morphometric results in *Porites* and electrophoretic and co-occurrence studies of the same material (Garthwaite et al 1994) greatly enhance the taxonomy of that genus, but whether morphometric methods on their own can provide a definitive basis for taxonomic decisions remains yet to be established. Perhaps the matter will rest with the rapidly developing technology of automated three-dimensional image analysis.

Gaudin (1988) took her methods of multivariate analyses of skeletal data (including principal component, cluster and a host of regressions) to an honest conclusion in her study of Red Sea *Goniastrea*: 'it may be concluded from this numerical study that instead of *Goniastrea* being subdivided into two species (*G. pectinata* and *G. retiformis*) it is in fact one species with a wide morphological variation'. These two species are, in fact, near opposite ends of the *Goniastrea* spectrum and can be separated *in situ* at a glance. The issue is not with the methods as such, but with the usefulness of simple morphometric data as descriptors of anything that occurs in reality. In-depth morphometric studies (Budd 1983; Willis 1985) of extant corals have not yet produced independent taxonomic revisions.

CLADOGRAMS AND CLADISTICS

Cladistics, or phylogenetic systematics (*sensu* Hennig 1966), is a numerical method of phylogenetic reconstruction in which taxa are grouped into phylogenetic trees or cladograms. From this procedure, sister taxa are those that have shared derived characters inferred to have originated in the latest common ancestor. Shared primitive characters, inferred to have originated in a more remote ancestry, are excluded from the classification subsequent to that ancestor. The terminal nodes (most distal branches) of the cladogram are the most recent taxa; the internal nodes are inferred ancestral taxa, also defined by shared derived characters. The cladogram is intended to reflect the phylogeny of clades and relative distances between them in two dimensions. A number of computer programs are available to generate the most parsimonious solution to sets of characters or character states (Platnick and Funk 1983; Platnick 1987).

While cladistics is a very effective technique in many areas of taxonomy and biogeography (p 41), there are both conceptual and operational objections to its use with corals. The conceptual objections are that it will give misleading results with reticulate phylogenies; that corals do not readily exhibit primitive or derived characters or character states, but rather exhibit unending homoplasy; and that cladistics cannot accommodate information derived from co-occurrence. Operational objections are not criticisms of cladistics as such, but of its use with corals. Cladistics programs can produce visually impressive results that are not based on in-depth knowledge or original field experience. Like so many other methods of numerical analyses, cladistics is prone to the 'black-box' syndrome, in which neither the output, nor the process that generated it, is understood by the user, but rather is something that the user assumes is correctly understood by whoever wrote the algorithm that generated it.

To date, well-conceived cladistic analyses of morphological data have been undertaken in Scleractinia only in the Fungiidae (Hoeksema 1989), where species are mostly taxonomically straightforward (as most are acolonial, with little intra-specific variation). The results, even in this most suitable of coral families, were not accepted by Hoeksema and are unhelpful. Cladistic analysis may be useful in morphological coral taxonomy to illustrate relationships, to examine the logic behind decisions, and to review the comprehensiveness of data: that is, it is useful only as an adjunct to other taxonomic methods (Hull 1979).

FIELD EXPERIMENTAL STUDIES

Some phenotypic indicators of genotype variability can be observed directly (eg observable variation among colonies of the same species in a uniform physical environment indicates something of the plasticity of phenotypic expression of genotypes). Likewise, growth-form changes in response to transplantation provide a clear indication of physical-environmental influence on genotype expression. There are many such methods for separating genetic control of morphological variation (or lack of it), from purely phenotypic expression of physical environment: these are the source of our knowledge of the control of intra-specific variation of almost all corals. Transplants are not needed to determine taxonomic levels of genetic plasticity, but they have contributed to our understanding of it in some species (reviewed, Willis 1990).

No taxonomic decisions have so far been based on field experiments. However, it is desirable that field experimentation be used to study intra-specific variation, especially in combination with genetic methods (such as electrophoresis) or morphometric methods (Willis and Ayre 1985; Hunter 1985).

Self-recognition bioassays, based on the assumption that only corals with identical genotypes will fuse when tied together (eg Reising and Ayre 1985; Hunter 1985), have been used in several studies to determine genotype *in situ*. Stoddart et al (1985) and Heyward and Stoddart (1985) found that this assumption is not necessarily so, and the technique must now be considered to be of limited use.

STUDIES OF REPRODUCTION

Reproductive mechanisms and behaviour have a special relationship with many subjects of this book, because reproduction is intimately linked with evolutionary mechanisms, species concepts and population dynamics. All these subjects have consequences for taxonomy and biogeography. Taxonomic issues raised by studies of reproduction in general and hybridisation in particular have become confused in recent literature. These issues are addressed elsewhere in this book as follows:

- ◆ Concepts of species and of speciation: mechanisms of hybridisation between species and of genetic isolation within species are essential elements of reticulate evolution proposed in chapter 12.
- ◆ Population dynamics: relevant aspects of abundance, demography and dispersion are discussed in chapter 5.
- ◆ Biogeography: reproductive aspects are raised in chapter 3.
- ◆ Taxonomy: intra-specific and inter-specific variation in reproduction is discussed in chapter 5.

PHYSIOLOGICAL STUDIES

In some instances (eg Gattuso et al 1991) physiological criteria have been claimed to overturn morphological coral taxonomy (as in the 'whatever works' concept of Lang 1984). Physiological attributes of species are very likely to vary with physical environment just as morphological attributes do. Physiological attributes are also likely to vary with geographic location, and this variation is likely to be under some form of genetic control. Very little of this is relevant to taxonomy.

MOLECULAR AND GENETIC STUDIES

Taxonomy is only one field of application of molecular studies of corals. Figure 7 summarises the present applicability of the various techniques currently available for the study of coral systematics and phylogeny.

To be suitable for addressing both taxonomic and evolutionary issues, a given molecular sequence must be relatively free of convergence and should change at a uniform rate, depending on the question being addressed[3]. If these criteria are met, molecular data can indicate both the degree of similarity between species or populations, and the time of their divergence. Molecular data are particularly useful in studying polymorphic organisms (because results are not obscured by morphological polymorphisms) and are also valuable in studying populations over wide geographic ranges.

The range and applicability of molecular techniques available to coral taxonomy have changed substantially over the past decade. No single methodology can reasonably be claimed to be a panacea for all problems, although that claim has at times been made for most. The most relevant molecular technique depends on the question being addressed, the time-frame in which the event being investigated occurred, and the applicability of the results to something observable in nature[4]. Gene pools can be studied at many systematic levels: clone, population, subspecies, 'species' and higher (p 228). Each of these levels is associated with a particular range of time-frames, spatial patterns and questions, the last varying according to the differing needs of population genetics, taxonomy, biogeography and systematics.

All commonly used molecular techniques (figure 7 and explained below) are applicable to the most common time-frame of divergence of taxonomic interest (5 to 50 million years). Of these, isozyme electrophoresis and RFLP (restriction fragment length polymorphism) analyses of mitochondrial DNA (see below) have been the two most generally used. Sequence comparison is by far the most powerful technique, but it has only become practical with the advent of PCR (polymerase chain reaction) technology. Fast-evolving isozyme loci or mitochondrial RFLPs have been used to study closely related species (diverging within the last 5 million years or so), but very recent divergences require sequence data

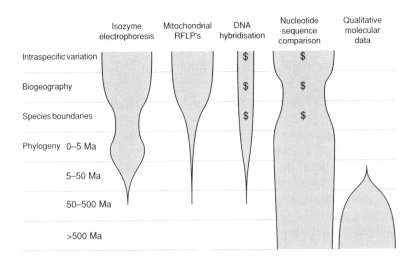

	Isozyme electrophoresis	Mitochondrial RFLP's	DNA hybridisation	Nucleotide sequence comparison	Qualitative molecular data
Intraspecific variation			$	$	
Biogeography			$	$	
Species boundaries			$	$	
Phylogeny 0–5 Ma					
5–50 Ma					
50–500 Ma					
>500 Ma					

Figure 7 Principal techniques currently used in molecular systematics and their general applicability to different objectives (partly after Hillis and Moritz 1990; Moritz 1992). $ signs indicate major cost inhibition. This is a generalised guide only; applicability to corals is highly dependent on specific purpose and taxon selection. (Ma = million years ago)

because of the difficulty of separating uniquely derived character states from random fixation of ancestral genetic polymorphisms (Neigel and Avise 1986). Further back in time (50 to 500 million years), only sequence data for very conserved genes and qualitative molecular data are informative.

Molecular techniques and their applications to coral taxonomy can be only briefly reviewed here.

KARYOTYPING

The morphology of metaphase chromosomes, which are visible by light microscopy, can provide systematic information that is independent of morphological, biochemical and other characteristics. Whilst there have been few attempts to karyotype corals (Wijsman and Wijsman-Best 1973; Heyward 1985), there are no indications of taxonomically significant variations (p 79).

PROTEIN TECHNIQUES

Isozyme electrophoresis The separation of protein isoforms by starch gel electrophoresis has provided the most cost effective and readily used method of determining departures from Hardy-Weinberg predictions and other aspects of genetic variation. The method does not directly provide character-state information and is open to a considerable degree of interpretation, especially concerning the effects of sexual versus asexual reproduction in between- and within-population samples. However, given informed interpretation, the method applied to corals has a great deal of promise, both in taxonomy and in the understanding of species dynamics. Its greatest resolving power with corals is at the level of species or closely related genera (Nei 1987).

Results to date are patchy, but have established the many uses of the technique. Heyward and Stoddart (1985) used electrophoresis of Hawaiian *Montipora* to show that tissue fusion (generally considered indicative of coloniality) could occur between different genotypes. In a first application to Indo-Pacific coral taxonomy, Ayre and Willis (1988) found strong associations between genotype and growth form in *Pavona cactus,* but supported the integrity of this species. Similarly, electrophoresis confirmed the validity of two very similar species of *Acropora* (*A. palifera* and *A. cuneata*) and was used to investigate their local distribution (Ayre et al 1991). It has also been used to study the population structure of *Seriatopora hystrix* (Ayre and Duffy in press) and to separate a 'sibling species' from *Montipora digitata* (Stobart and Benzie 1993).

The technique is presently being used in Australia to study various taxonomic questions in *Montipora* and faviids, especially in relation to reproduction (p 83). It is not, however, always dependable, having failed to recognise morphologically clear distinctions between species of *Platygyra* (Miller 1994).

Probably the most informative application of electrophoresis has been for the genus *Porites* (Garthwaite et al 1994; Potts et al in press). To date, for example, the technique has demonstrated the existence of a 'sibling species' pair initially observed *in situ* (on the basis of colour, correlated with depth), but traditionally combined as a single species, *Porites astreoides.* As with non-molecular taxonomy, the collection of samples was guided by the *in situ* co-occurrence of visually distinct colonies.

Immunology Immunological techniques have been used for most of this century to infer phylogenies by determining the degree of reactivity between antibodies and antigens. Of the several primary methods available, radioimmunoassay is probably the most suitable for taxonomic purposes as it can be used with very small amounts of protein and (like PCR), may be applicable to museum specimens and fossils with residual protein (Lowenstein 1985, 1986). These techniques are, however, technically demanding and (unlike PCR technology) cannot provide unambiguous quantitative data. Radioimmunoassay is currently being used to investigate the origins of Scleractinia (Fautin and Lowenstein 1994) (p 112).

DNA TECHNIQUES

Restriction fragment length polymorphisms (RFLPs) Currently the second most commonly used family of techniques in molecular systematics, these methods indirectly estimate DNA sequence variation by determination of the length of fragments generated by restriction endonucleases. The fragments obtained vary in length and number owing to loss or gain of restriction sites, or insertion or deletion of sequences between sites. RFLPs provide information on the nature as well as the extent of differences between DNA sequences. Analyses can be undertaken at several different levels of resolution; however, interpretation of RFLP data is generally more complex than sequence data (see below).

For other animal groups, mitochondrial DNA has proved particularly informative in RFLP (and sequence) analyses because of its relatively high mutation rate and its maternal (non-recombining) inheritance. Mitochondrial RFLP analysis can be applied to a wide range of problems, notably the 'fingerprinting' of parentage, intra-specific variation, and determination of geographic distribution of phylogenies. Avise et al (1987) coined the term *phylogeography* for the last. Whilst it is highly likely that the methods that have been so successfully used on vertebrates and higher invertebrates can be successfully applied to corals and other lower invertebrates, at the time of writing there are no published applications of these techniques in coral taxonomy.

DNA hybridisation DNA hybridisation techniques permit the estimation of divergence between specific DNA sequences or between whole genomes (although in the latter case the data are not quantitative). Generally specific DNA fragments from one species are radioactively labelled and made single-stranded before hybridising with single-stranded DNA samples from other species immobilised on a membrane. Few attempts have been made to apply DNA hybridisation techniques to coral taxonomy. McMillan and Miller (1990) investigated relationships within *Acropora* using cloned highly repetitive DNA from *A. formosa* as a probe. DNA hybridisation data are, at best, semi-quantitative; PCR technology has greatly facilitated obtaining qualitative sequence data, making DNA hybridisation obsolete as a technique.

Nucleotide sequence comparison We are now entering a golden era of DNA sequence comparison. Of all the molecular techniques, the direct comparison of homologous DNA sequences is the most informative. Because of the requirement for cloning and screening, obtaining nucleotide sequences was such a laborious process that obtaining novel data sets, or extending existing ones, was simply not an option open to taxonomists. However, PCR has changed that. PCR-based procedures are currently being extensively applied to corals and, at the time of writing, are yielding results of family- and higher-level systematic interest (Chen et al in press) from ribosomal DNA (Hillis and Dixon 1991). One of the major advantages of PCR is that coral sequences can now be specifically amplified even in the presence of contaminating DNA from zooxanthellae, therefore fresh tissue (as opposed to gametes) can be used as the source of DNA. Furthermore, the technique works reliably on DNA from museum specimens and has yielded data from fossils in other animal groups. Thus the range of potential applications is enormous.

CATEGORISATION OF SPECIES IN TERMS OF EASE OF DETECTION AND IDENTIFICATION
The following list ranges from the most difficult to the easiest species to detect, taxonomically define, and identify:

1   Non-morphological 'genetic species' that are detectable only by molecular methods (ie have no distinguishing morphological characters, and thus exist in concept only). There are no known examples but this is likely to be a common occurrence (p 229).

2   Species that are detectable only by molecular methods, after which morphological methods can be used for identification. Examples are likely to initially come from *Porites* research.

3   Species that are detectable by co-occurrence methods and require co-occurrence of similar species for confident identification. The *Acropora palifera/cuneata* species pair (figure 5) is an example; there are many such groups in most major genera.

4   Species that, once detected by co-occurrence methods, are identifiable without reference to other species, provided that basic physical-environmental data are available. This includes the majority of species.

5   Species that can be identified without physical-environmental data if the region of origin is known. Less than 20 per cent of the species of major genera can ever be thus identified with confidence. It is noteworthy that most older museum collections and most type specimens fit this category (p 19).

6   Species that can be identified without site or physical environmental data. Species of paucispecific genera and a few other very distinctive species can be included here.

7   Species that can be identified from specimens without any accompanying data. Most Caribbean species, and some species of larger genera that have very distinctive taxonomic characters only, are included here.

Some species require different sorts of additional information (eg tentacle characters in the case of *Euphyllia*).

BRIDGING TAXONOMIC METHODS

Over the past two decades, there has been an explosion of faunal, ecological and physiological research on corals, funded in response to the needs of conservation and management. Almost all of this research is based on 'species'; almost all requires taxonomic support in some form or other. Coral taxonomy is thus 'operational', and it provides support for many hundreds of scientific publications every year. The essential question is how closely does this taxonomy reflect the natural world, and how much is it a human construct? The issues are ones of broad concept as well as of specific detail. They concern: concepts of the nature of species (chapter 13); issues of identification (p 9); human-created nomenclatorial problems (p 19); mistakes and incomplete studies; unresolved taxonomic problems, especially over biogeographic scales; actual and potential conflicts between taxonomic methods.

> Some have claimed that molecular characters are relatively weak...whereas others have claimed that morphological characters are likely to be misleading... Closer examination shows this to be an empty argument (Hillis 1987).

> Comparative studies have shown that morphological change and molecular divergence are quite independent, responding to different evolutionary pressures and following different rules. (Moritz and Hillis 1990)

This book concludes that species boundaries are arbitrary and that the systematic status of one species may be different from that of another. Different taxonomic methods will resolve different morphological and/or geographic boundaries, creating the conflict Moritz and Hillis refer to. No matter what methods are used, operational taxonomy must incorporate morphological distinctions between species. To date, the methods that have been used have created little conflict in results, but there will certainly be widespread 'conflict' in the future unless differences between systematic and taxonomic organisations are understood[5].

Cladistics   Cladistics based on composites of morphological, environmental and ecological data may point to faults or insufficiencies in data and data collecting, and impose rigour in both analysis and description. However, cladistic analysis as an endpoint will never be self-supporting with corals (p 24) and will always involve major loss of information derived from studies of co-occurring colonies (p 20). Cladistic parsimony analyses of molecular data are another matter; these have the potential to test the validity of other taxonomic methods within the context described in chapter 13.

Colony transplants   Results of transplants may indicate faulty original taxonomy or identification or both. Transplants are effective in separating physical environment-correlated polymorphism from genetic variation. To date, such studies have been carried out on only a few species: *Montastrea annularis* and *Siderastrea radians* (Budd 1979) in the Caribbean and *Acropora formosa* (Oliver et al 1983), *Acropora palifera* (Potts 1984a) and *Turbinaria mesenterina* and *Pavona cactus* (Willis 1985) on the Great Barrier Reef. There has been little or no conflict of results, although investigation of such conflicts was not the objective of these studies.

Physiological and behavioural studies   Physiological and behavioural differences between populations, races, geographic subspecies and species do not provide an independent test of taxonomy, but they do add to knowledge of variation.

Figure 8 *Montipora digitata* at Magnetic Island, Great Barrier Reef, showing the *in situ* appearance of a pair of sibling species. The two morphologies have colour differences, can be separated electrophoretically, and are reproductively isolated. Although recognisable by co-occurrence, morphologies and skeletal structure of these two species almost merge, making them indistinguishable from genetic variants of a single species. (Photo: B Stobart)

Reproduction Studies of reproduction have begun to present an array of interesting conceptual problems and practical issues. Of greatest interest are barriers to cross-fertilisation between co-occurring colonies of the same species or groups of 'sibling species' and artificial hybridisation between species (p 83) (figure 8). The spectrum of issues that arise has long existed in botany and must now be addressed in corals (chapters 5 and 13).

Molecular methods Molecular techniques present a powerful array of tests of morphological taxonomy and, in the long-term, may displace morphology as the primary basis for separating species. Hillis (1987) provides a general review of the subject, concluding that disagreements among morphological and molecular systematists over 'species' definitions usually represent a disagreement of concept without due reference to biological realities. This is likely to increasingly be the case with corals.

The main source of conflict will come from the arbitrary nature of species boundaries, and hence debate as to what are species and what are subspecific taxa. The first such debate concerns the well-known 'species' *Montastrea annularis,* which has recently been claimed to be 'at least three species' by Knowlton et al (1992). Van Veghel and Bak (1993) give an opposite view, based on comparable data. The difference of opinion can be considered an example of a general situation, as most 'species' (whether divisible into smaller 'species' or not) have at least the systematic complexity of the M. annularis complex. The relevant points are that there may be no 'correct' answer because species are arbitrary; that different information sources and purposes target different taxonomic levels or units; and that taxonomy must accommodate information from multiple sources.

Molecular techniques, if used and interpreted expertly and not as 'black boxes', have the potential to overcome most weaknesses of morphological taxonomy and vice versa. If not used and interpreted expertly, they can create meaningless divisions in taxa (at any level) in a manner that becomes progressively more hypothetical and remote from biological reality. Of all molecular techniques available, isozyme electrophoresis has provided information most *directly* relevant to the taxonomic level of species. DNA sequence comparisons may, in time, displace electrophoresis and give a third perspective of species boundaries.

Molecular tests of the morphological taxonomy underlying the primary geographic database of this book are few. Willis and Ayre (1985) gave electrophoretic data compatible with the very wide intra-specific variation of *Pavona cactus* described by Veron and Pichon (1979) and showed that this was under a substantial level of genetic control. Ayre et al (1991) used electrophoresis to support and extend the distinction between *Acropora palifera* and *A. cuneata* reported by Veron and Wallace (1984). Many other electrophoretic studies have been conducted on other Indo-Pacific corals (p 27). These do not 'test' morphological taxonomy because of the intermediate step of identification (p 14), but all are based on species distinctions proposed in *Scleractinia of Eastern Australia* and are not in conflict with them.

To date, there have been no 'tests' of molecular techniques against each other. As their applicability varies with systematic level, such tests would always be qualified, but they are an essential step to the understanding and application of what different methodologies have to offer.

3

INTRODUCING CORAL BIOGEOGRAPHY

Coral biogeography and hypotheses about mechanisms of coral speciation, have always been closely linked subjects and for this reason are treated together in this chapter. Genetic concepts of speciation, historically an almost separate subject, are introduced in chapter 4. This chapter is divided into three parts, as follows:

Biogeographic concepts: an introduction for readers new to biogeographic theory.

Observations about coral distributions: a history of observations about the global distributions of coral.

Explanatory coral biogeography: a summary of hypotheses that have been proposed to explain coral distributions.

THE MAIN TOPICS

- Pre-continental-drift biogeography: Darwinian centres of evolution.
- The challenge to Darwinian centres: Croizat's 'panbiogeography' and from it, 'vicariance' biogeography.
- The relationship between vicariance biogeography and cladistics.
- Problems for vicariance: dispersion and the founder principle.
- Ecological biogeography and the equilibrium theory of island biogeography.
- Observations about coral distributions: generic and species levels.
- Explanatory coral biogeography in relation to: centres of origin concepts, vicariance concepts and cladistics, dispersion biogeography, glacio-eustatic biogeography, ecological biogeography, palaeobiogeography and the biogeography of specific geographic situations.

> Biogeography, as a topic for discourse or discussion, is in some ways like religion: both topics lend themselves to ever more complicated treatment in the abstract, which is apt to border even on the miraculous, but which is apt to crumble in confrontation with concrete facts of life. (Nelson and Platnick 1981)

Many previously expressed points of view about coral biogeography are compatible with what we now know of the nature and distribution of species, but the principal schools of explanation are not. There are two good historical reasons for this: the first is the absence of a species-level taxonomy adequate for inter-regional comparisons, and the second is the implicit assumption that places of occurrences of species are somehow correlated with places of their origin. Other reasons are current elements of 'follow-the-leader', encouraging displacement of original thought by computer programs which, because of their conceptual basis, predetermine much of the outcome. It is little wonder that so many cladistic studies have similar end products, and that so much debate concerns end-product detail rather than original questions.

Previous authors have not (for any biota) identified surface circulation vicariance as an evolutionary mechanism, or reticulate evolution as being the outcome of this mechanism, nor have the ramifications of physical, as opposed to biological, control of evolution been explored. This chapter is thus mostly about theories that, combined evidence suggests, should be rejected, in part or in whole, in the light of increasing knowledge about the nature of coral species. This is not to say that past theories are without present value. On the contrary, many previously expressed concepts of coral biogeography are strongly supported by present knowledge of species. Also, evolutionary mechanisms will vary with different biotas and geographic situations, and thus no single general concept may be even remotely applicable to specific evolutionary processes or events.

BIOGEOGRAPHIC CONCEPTS

'OBSERVATION' AND 'EXPLANATION'

These two word concepts, which are continually used by biogeographers, warrant comment at the outset. Observing distribution patterns, and explaining those *patterns* by a *process*, are two very different undertakings. Most literature deals with either or both, together or alone. Where used together in the context of numerical or conceptual correlation, the reader should be aware of precisely what is being correlated with what, as well as how and why, for this is a repeated source of error in much biogeography. Observations (patterns) may contribute nothing to explanations (processes): for example, the observed distribution patterns of genera do not help explain any process of speciation.

> In biogeography perhaps more than any other subject, a holistic approach is
> necessary. The solution of any biogeographic problem, no matter how small,
> requires the following of a great many varied lines of inquiry. (Pielou 1979)

Historically biogeography is split into two approaches. The first (which
includes this book) is 'grassroots up' and is observation-driven, the
second is 'top down' and is concept-driven. Pielou's (1979) book
Biogeography remains an excellent general introduction to the former.
At a conceptual level, biogeography is an extraordinarily divided and
divisive subject, the principal reasons being its complexity, its close
relationship to evolutionary theory, and the dogma of reductionist
thinking (the process of reducing natural complexity to simple explanation).

The following introduction is selectively brief; historical detail can
be found in the reviews by Nelson and Platnick (1981) and Humphries
and Parenti (1986), and the subject as a whole is introduced in several
chapters of Myers and Giller's (1988) *Analytical Biogeography*.

CENTRES OF ORIGIN

Darwin's centres-of-origin theory, the first of the major theories of biogeography, is strongly refuted as a whole by present knowledge of coral
species, but the theory has enormous historical interest, especially as
most of its many detractors have fared less well.

> Work...in the Museum, has already extended to comparisons...with the view
> of ascertaining whether there is any probability of tracing a genetic connec-
> tion between the animals of...different geographical areas, and how far geo-
> graphical distribution and specific distinction are primary facts in the plan
> of creation. It must be obvious that the question of the origin of species is
> not likely to be discussed successfully before the laws of geographical distri-
> bution of organised beings have been satisfactorily ascertained. (Agassiz
> 1865)

The creationist world of Agassiz, and so many of his contemporaries,
was brought to a close by Darwin, and it is with that great thinker that
coral biogeography, along with so much else, was born. Darwin proposed that species were not created, but evolved from centres of creation, whence they dispersed. His theory that species evolve has long
since ceased to be rationally disputed, but not so the mechanisms he
proposed. These, to this day, have remained controversial. The basic
reason is that we can readily observe the results of evolution, but very
seldom its mechanisms. Observation and explanation are juxtaposed
from the very beginning.

Darwin (1859) proposed that dispersion was primarily driven by
the evolution of new species by processes of natural selection[1], which
displaced older and less competitive species to more distant places (figure 9). This is but one aspect of Darwin's theory of the origin of species,

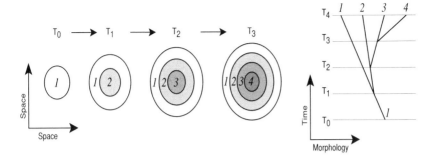

Figure 9 Darwinian concept of centre of origin of species. Circles represent the distribution ranges of individual species. Species 1 is the oldest species, species 4 the most derived. Species are progressively displaced from the centre of origin at times T_1–T_4 and have concentric, not overlapping ranges. Place of occurrence (left) and phylogeny (right) are correlated.

yet it is one of his most thoughtful: it integrated the known distribution of many plant and animal groups, the known fossil record and palaeoclimates, as well as barriers and vehicles controlling dispersion. Darwin entered the fields of marine and island biogeography and he (1842) founded the now well-known theory of the origin of atolls by island subsidence.

Darwin's centres of origin theory, or refinements of it, dominated all coral biogeographical thinking until the theory of continental drift appeared to make a fundamental aspect of it — geographic centres — untenable[2]. However, it is well-remembered history that the theory of continental drift was responsible for one of the greatest upheavals in biogeographic and evolutionary thinking since Darwin, an upheaval that came belatedly in the 1960s, led by the leaders of the time, notably Darlington (1965). We tend to forget that the coral systematic bible of Wells (1956) (p 108), with its never-repeated array of palaeobiogeographic data, predates this time. Even in Stehli and Wells (1971), the most quoted paper on coral biogeography to this day, no causal relationship between coral distributions and continental positions is suggested. As late as 1974 Briggs enlightened many a marine biologist with his exposé of plate tectonics in his *Marine Zoogeography*.

Briggs (1984) gives a modern synopsis of centres of origin, clarifying his concept as being synonymous with 'evolutionary centre' or 'centre of evolutionary radiation'. He documents the central Indo-Pacific as being the world's principal centre of marine species diversity. There are, presumably, other centres, 'the characteristics of which are: (1) large geographical size, (2) heterogeneous topography, (3) warm and relatively steady temperatures, (4) maximum species diversity for the general part of the world in which they are located, and (5) possession of the most advanced species and genera of those groups of organisms that are well represented'. Briggs's last point, which follows Darwin, is the very opposite of Hennig's (1966) 'rule' (which underpins the cladistic biogeography), that the most primitive characters are found within the earliest-occupied part of an area. Briggs notes the role of barriers in the Darwinian sense, and observes that species that cross these barriers may well undergo allopatric speciation. Thus, the centres act as primary nurseries for larger regions into which their products are gradually dispersed.

Briggs's primary example in the marine world, following Stehli and Wells (1971), is zooxanthellate corals. The crucial point (re-examined in chapter 8) is that the central Indo-Pacific is inhabited by the youngest genera and that the average age of genera progressively increases with distance from the centre. The strength of Briggs's central argument is clear: that at least as far as many or most shallow-water tropical biota are concerned, the central Indo-Pacific is indeed a centre, from which biota are, indeed, able to disperse to neighbouring regions. What is not clear is the difference between a 'centre of diversity' and a 'centre of origin', and that is not just a matter of semantics.

The centre of origin theory of biogeography is the only major theory that predicts a *specific* geographic pattern in the occurrences of species. The pattern has seldom been supposed to be expressed in terms of concentric circles as in figure 9, but it has predicted the occurrence of *amphitropical (amphiequatorial, bipolar* or *relict) disjunct distributions*, where isolated populations of the same species occur either side of the equator.

AMPHITROPICAL DISTRIBUTIONS

These are disjunct distributions where the population or species occurs subtropically both sides of the equator. They have been most commonly recorded in fish, and many explanations have been offered for them. These have involved centre of origin theory where amphitropical distributions are relicts of equatorial extinctions (Newman 1986; Briggs 1987), Pleistocene displacement of tropical faunas by temperate faunas (Randall 1981), isothermic submergence (primarily of fish), Miocene temperature increase displacing tropical species (White 1986) and various associations with plate tectonics in general or with island integration (p 58) in particular (Springer 1982).

Disjunct distributions in central Indo-Pacific corals show no significant amphitropical component (p 193). On a larger scale, there are similarities between the corals of Hawaii and the southeast Pacific (p 164).

PANBIOGEOGRAPHY

Either the Darwinians bury me, or I them. (Croizat 1981)

I will, directly challenged this time by Brian Rosen from the British Museum under my name, deal separately in coming pages with my own (pan)bio-geography, vicariance/vicariism, and alien 'vicariance biogeography', in order that the reader may form, once for ever, a solid understanding of these matters, ready to detect for himself the countless false-hoods, mis-constructions etc., that now foul this history of their origin, growth, etc. (Croizat 1982, in one of the less well edited articles of *Systematic Zoology* of the time)

Panbiogeography is very much the subject of Leon Croizat, biogeogra-phy's most prolific writer and the most outspoken opponent of both cen-tres of origin and vicariance biogeography. Panbiogeography (Seberg 1986 gives a useful review; Craw 1988a) is a method in which many 'road maps' of taxa can be superimposed to produce 'the dispersion high-ways of life'. This involves the plotting of the distributions of taxa on maps of present-day geography and joining adjacent ('disjunct') points with lines forming 'tracks' (Croizat 1974). Where many tracks coincide, 'generalised tracks' are formed. Where many generalised tracks intersect, 'gates' or 'nodes' occur. Gates are the initial points (centres of origin?) from which life dispersed. The joining of gates finally produces common baselines, and these tend to encircle the far southern hemisphere (pre-sumably why most panbiogeographers are New Zealanders). In principle, the method highlights the massing of distribution, not the boundaries of distribution.

Croizat's principal work (in the 1950s and 1960s) predated con-tinental-drift theory, although he proposed, and now others propose, considerable interaction with global palaeobiogeography (Craw 1988b). There are many issues associated with panbiogeography (most raised by Craw and Weston 1984) that space and relevance pre-clude from discussion in this book, but it may be noted that the sub-ject has much more relevance to terrestrial than marine biogeography, that as a method it does not cope well with either a strong fossil record or long-distance dispersal, and that as a concept (which it inevitably becomes), it is not *the* radical alternative to Darwinian cen-tres of origin that Croizat claimed, nor is it necessarily over-ridden by its offspring, vicariance biogeography, as is widely supposed. Gates and generalised tracks (and vicariant distributions) eventually become meaningless in a marine physical environment where the base map effectively changes with the vagaries of palaeoclimates, ocean currents, long-distance dispersal and, above all, sea level changes[3].

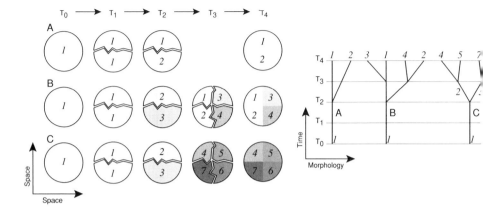

Figure 10 Classical vicariance mechanism of speciation. Circles represent the distribution ranges of individual species. In the simplest possible example (A), the distribution range of species 1 is divided by a barrier (the zig-zag) which is maintained until a second species has formed. In example B, there are two barriers and two derived species have formed (in addition to the two original species). In example C, there are the same two barriers, but four derived species have formed. Place of occurrence (left) and phylogeny (right) are correlated.

VICARIANCE BIOGEOGRAPHY

Despite an extraordinary level of conflict in the literature of vicariance biogeography, classical vicariance processes[4], recognised from the time of Darwin himself, are common sense. If one or more barriers form across a species' distribution, the divided population might diverge in time, forming two or more distinct populations, which, if able to remain reproductively isolated after the barrier(s) are removed, may then be two or more different species. This process on a larger scale leads to patterns of species that have no centre of origin. There is no requirement for dispersal to occur (figure 10).

Vicariance speciation requires that an ancestral species occupies the geographic area of its descendants (as illustrated in figure 10) and that this area becomes divided up as those descendants progressively evolve. The process, therefore, produces ever-increasing numbers of species, each occupying an ever-decreasing geographic range. The result is ever-increasing endemicity, and hence a close association between the concepts of endemicity and vicariance.

Vicariance speciation can only work in the absence of dispersion across the aforementioned barrier(s). Organisms do, however, disperse (or migrate) from their place of origin. Thus dispersion (or migration), which increases distribution range, must alternate in time with the absence of dispersion, which allows vicariance to take place.

As a general concept thus far, vicariance as a mechanism of speciation may be applied to a wide range of biota, including corals: the process of species formation through fluctuations in ocean surface circulation (chapter 12) is a vicariant one. However, it is from here on that coral biogeography and classical vicariance theory must part company. The reasons, which are brought out in subsequent pages of this book, are as follows:

♦ Dispersion (migration), the Achilles heel of vicariance, is of overwhelming importance in most corals (p 86).

♦ The place of origin of coral species is not limited to, nor can it be correlated with, their present place of occurrence (p 233).

♦ The process of evolutionary change neither commences nor ceases in any particular point in time, although intervals of species formation *are* separated from intervals of dispersion (p 212).

♦ There is a dynamic relationship between the increase in number of species by vicariance and the decrease in number of species by hybridisation (p 216).

Concepts aside, there are two problems with the *application* of vicariance to corals: first, it does not help the present knowledge of species-level taxonomy or distribution, and, second, it is closely associated with the restrictive assumptions of cladistics.

CLADOGRAMS IN CORAL BIOGEOGRAPHY

The cladistic method requires cladograms to be constructed for different groups of taxa. The taxa in each separate cladogram are then replaced by the areas in which they occur. Common patterns among *area cladograms* are then generated by parsimony analysis (recently, Kluge 1988; Brooks 1990) or component/consensus analysis (Page 1988, 1990a).

A wide range of assumptions is necessarily involved at different levels in this process:

♦ Cladograms assume vertical transmission of characters by inheritance. They exclude horizontal transmission of characters by hybridisation and dispersion. These two processes are analogous, but they are not the same (Sober 1988).

♦ In molecular systematics, where cladistic methods are most useful, the terminal taxa are genomes. Cladograms indicate patterns of association, but not degree of association, between genomes. Theoretically this can be avoided, but in practice it seldom is. The pattern may thus be fragile, that is, changeable by a change in fusion strategy or an increase in information.

♦ The substitution of genomes for species and of species for area both require assumptions that may be unjustified (discussed by Doyle 1992). The question, for both genomes and species, is one of data quality and quantity, and also of interpretation.

♦ Similarities among independent gene trees are, it might be assumed, due to a common underlying species tree. However, this need not be the case, especially where there is incomplete divergence, as in reticulate phylogenies (p 216).

For many (perhaps most) major groups of organisms, these negative aspects of biogeographic cladistics are outweighed by the rigour it demands of data collection and analysis. Not so with corals, for reasons given below (p 52).

The subject of biogeography can be divided up in different ways and thus vicariance and/or cladistic biogeography or variants of them have been given other names. The oldest is 'phylogenetic biogeography', methodologically defined by Brundin (1981) as 'the study of the causal connections between phylogenesis (*development of the hierarchy in time and space*), anagenesis (*transformation of characters in time and space*), allopatry (*vicariance*), sympatry (*dispersal*) and paleographical events'. The various brands of biogeography in the vicariance school are all reductionist concepts that have varying relevance to different taxa.

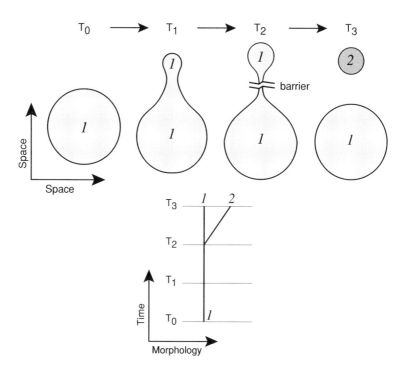

Figure 11 'Dispersal' and 'founder' mechanism of evolution. Shapes represent the distribution ranges of individual species. Species 1 migrates to a previously unoccupied place and speciates after a barrier is formed. Place of occurrence (left) and phylogeny (right) are correlated.

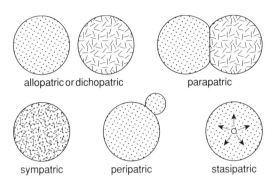

Figure 12 Common terms used to describe both mechanisms of speciation and, except for stasipatric populations, geographic relationships between populations. These geographic relationships are only arbitrarily distinguishable: the crucial variable — the fraction of a breeding group composed of immigrants from another group — may be very small and impossible to determine (p 69). Theoretically, only a small amount of migration between populations may be necessary to maintain genetic cohesion in the absence of differential selection.

DISPERSION BIOGEOGRAPHY AND THE FOUNDER PRINCIPLE

The basic concept of dispersion biogeography, proposed by Mayr (1942) and expanded by Carson (1971), is again simple (figure 11). A species comes to occupy a previously unoccupied place by dispersing (migrating) there (eg a 'founder' event), a barrier breaks genetic communication between the two parts of the new species range, and speciation takes place. As with vicariance biogeography, speciation is allopatric (figure 12) and the outcome is seen in patterns of endemism. The most essential difference between this and vicariance is that this process requires dispersion to occur, while vicariance requires dispersion *not* to occur. If species co-occur after speciation, additional dispersion is equally necessary in both concepts.

Dispersion biogeography is generally linked with centre-of-origin biogeography as the alternative to vicariance (Simberloff et al 1981) but, as with so much biogeography, this highlights extremes rather than common ground. The dynamics of a hypothetical case involve a great deal of variation in the relative size and composition of the parent and daughter gene pools; their degree of genetic communication; their distance apart; and their dynamics through time. All non-allopatric distribution patterns illustrated in figure 12 intergrade. Allopatric patterns can also intergrade with non-allopatric patterns, and may do so even between very disjunct populations (such as between the central and far eastern Pacific, p 165). Most of the evolutionary issues concerning spatial separation between populations of the same species are irrelevant to corals because of their capacity for long-distance dispersion.

DIFFERENCES *IN PRINCIPLE* BETWEEN THE MAIN SCHOOLS OF BIOGEOGRAPHY

♦ Centre-of-origin biogeography alone requires sympatric speciation.

♦ Centre-of-origin biogeography requires that dispersion occurs *after* speciation. Vicariance biogeography requires that it does not occur at all *during* speciation. Dispersion biogeography requires that dispersion occurs *before* speciation. Panbiogeography as a method makes no assumptions.

♦ Vicariance biogeography leads to increased species diversity. The other concepts do not necessarily alter species diversity *at place of origin*.

♦ All concepts have a causal link between place (or pattern) of evolution and place (or pattern) of occurrence. The patterns are different in each case.

ECOLOGICAL BIOGEOGRAPHY

Biogeography obviously interfaces with ecology and has often been viewed as reductionist ecology on big temporal and spatial scales. In the spectrum of subjects from geology to genetics, there is a window where this is clearly true. In the context of this book, however, the demarcation between biogeography and ecology is adequately clear, the most basic distinction being that the former addresses evolutionary processes in evolutionary time-frames, whereas these time-frames are mostly irrelevant to the latter. Ecological observation, however, can be relevant to biogeography, for example, in the statement 'long-lived, stress-tolerant species are evolutionarily conservative and relatively immune from extinction' (Vermeij 1978).

The *equilibrium theory of island biogeography* (Preston 1962; MacArthur and Wilson 1963, 1967), an essentially ecological theory, had the sort of impact on biogeographic concepts twenty years ago that vicariance biogeography has today. When rates of immigration to an island are balanced by local extinction, the number of species, but not necessarily the identity of those species, remains approximately constant. For any particular island there is a dynamic balance between immigration, local extinction, proximity to a continental mainland and the area of the island. There are several assumptions or predictions in this (speciation is irrelevant; an underlying state of equilibrium exists; species are equal; islands are only isolated by distance; islands are ecologically equal, etc) that have attracted a substantial critical literature. 'Equilibrium theory' seems to have gone out of fashion, but as far as zooxanthellate corals are concerned, it is relevant in a regional context.

OBSERVATIONS ABOUT
CORAL DISTRIBUTIONS

Historically, study of the distributions of corals has been approached from observations and analysis of global generic distribution data, and through geographically restricted observations about species (p 3). There has been very little cross-fertilisation between these groups of studies, reflecting differences in data availability, timing, aims and scope.

GENERIC DISTRIBUTIONS

Interest in coral distribution originated in the early observations of Dana (1843) and Darwin (1859) that coral reefs have a temperature-correlated tropical distribution. The subject was raised many times (eg by Vaughan 1918, 1919; Davis 1928; Yonge 1940; Vaughan and Wells 1943) before any really useful global distribution data were compiled. Evolutionists following Darwin, even to the present day, have given impetus to the accumulation of coral biogeographic data by stimulating thought about evolutionary mechanisms and the relationships between species.

Wells was the first coral taxonomist to collect comprehensive distribution data for broad-scale biogeographic purposes. These generic-level data, covering the whole Indo-Pacific, were initially published in 1954a. This data set, repeatedly updated and expanded, has contributed greatly to biogeographic studies, including those of the present as well as other authors. Wells observed that:

♦ Most Indo-Pacific genera are widely and uniformly distributed, but some have restricted distributions, and others are widely distributed but rare.
♦ Some genera are widely distributed but are essentially non-reefal.
♦ There are only minor provinces in the Indo-Pacific, leaving vast areas of it undivided by unique generic compositions.
♦ There is a clear correlation between generic contours of diversity[5] and sea surface temperature.
♦ Genera 'drop out' in the same sequence from regions of high to regions of low diversity, leaving peripheral faunas of similar composition[6].

The first complete and best-known global generic contour map is that of Stehli and Wells (1971), which can be compared for historical interest with the maps of Rosen (1971a), Coudray and Montaggioni (1982) and Veron (1985) in figure 13. These differ in detail rather than essence: the shape of Wells's Indo-Pacific centre of diversity is retained, and his regional patterns of longitudinal and latitudinal attenuation remain similar in principle. The existence of a secondary

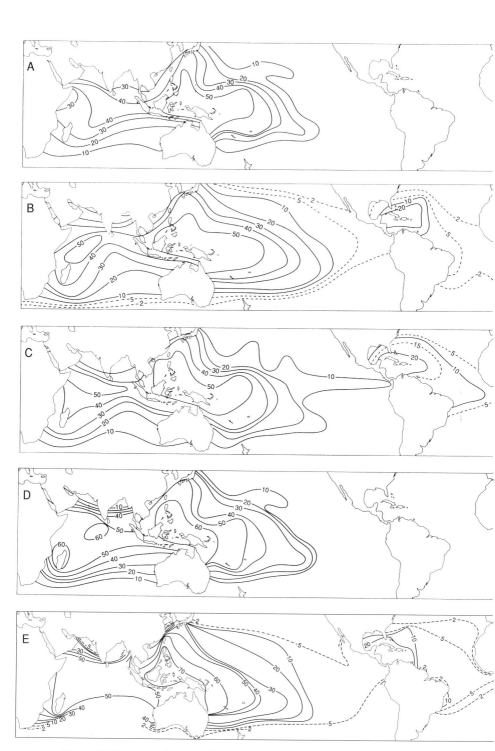

Figure 13 Contours of generic diversity, re-drawn from (A) Wells (1954a), (B) Stehli and Wells (1971a), (C) Rosen (1971), (D) Coudray and Montaggioni (1982) and (E) Veron (1985) (see text). Figure 48 (p 158) shows contours based on generic *ranges* generated from the database of the present study.

centre of diversity in the western Indian Ocean and/or the Red Sea is less clear: relatively high diversity in the Red Sea is indicated by the first and last maps (of figure 13), less so by the others. In reality some of the differences between these maps are artefacts of contour-drawing across the empty southern Indian Ocean. In a recent, but anachronistic, review of coral biogeography, Achituv and Dubinsky (1990) reproduce Stehli and Wells's original map and largely ignore subsequent studies.

General observations made from global analyses of genera (principally Wells 1955a, 1955b, 1956, 1969; Stehli and Wells 1971; Rosen 1971a, 1975, 1981, 1984, 1988a, b; Veron 1974, 1985, 1986, 1988; Scheer 1984) can be summarised as follows:

- ◆ There are areas of high diversity in the tropical western margins of the world's three great ocean basins, that of the Indian Ocean being relatively ill-defined.
- ◆ Within the Indo-Pacific, there is little generic variation within the centre of high diversity.
- ◆ There is attenuation of species diversity latitudinally and longitudinally from these centres.
- ◆ Regions distant from centres of species diversity tend to have similar genera.
- ◆ Latitudinal attenuation occurs in the same or similar 'drop-out' sequence in the northern and southern hemispheres.
- ◆ Latitudinal attenuation is highly correlated with sea surface temperature.
- ◆ The mean generic age of Caribbean corals is twice that of Indo-Pacific corals and centres of diversity have relatively young generic ages (specifically the outcome of Stehli and Wells 1971, figures 14 and 15).

These are the general observations. Explanations of distributions have also been offered in many of these studies.

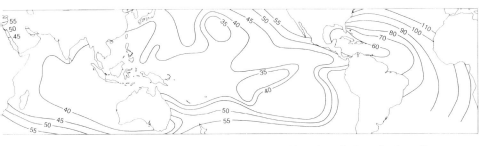

Figure 14 Contours of average generic age plotted by Stehli and Wells (1971) when all genera with no fossil record are deleted. Figure 42 (p 140) shows contours re-computed from the biogeographic database of the present study.

THE USEFULNESS OF GENERIC DATA IN EXPLANATORY
CORAL BIOGEOGRAPHY

Compared with species, genera have few problems in coral taxonomy and biogeography, and immeasurably fewer problems in palaeontology. This makes generic data very useful, but the uses and limitations need to be identified.

Generic names, so widely used in corals, usually serve the practical purpose of identification labels, in which case their phylogenetic position is usually irrelevant. In biogeography, genera are very useful in this sense: for example, the total number of genera is a useful measure of total *heterogeneity* (p 156). The number of species represented by a genus, however, may vary enormously, thus generic data are much less useful indicators of *diversity*.

Species composition of genera becomes important when generic labels are used to specify groups of species, for example, in generic distribution maps. In these cases, the genus will have meaning only if it represents a group of species that are genuinely related. The issue here is that generic units may have doubtful boundaries (contain species that have ecomorphs that cross generic boundaries (p 14), or have species that are intermediate between two genera, p 264).

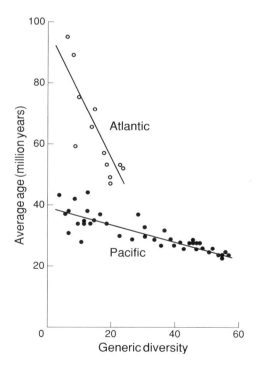

Figure 15 Relationships between diversity and average generic age for a series of Indo-Pacific and Atlantic locations, plotted by Stehli and Wells (1971).

In general, species of uncertain genera are not sufficiently numerous to have a significant *quantitative* affect on the value of the genus as a taxonomic unit, but they do reflect the unsatisfactory *qualitative* status of some, giving these value only as names without status.

SPECIES DISTRIBUTIONS

The change of subject now is not just a change from the genus to the species: it is a change from the museum and library to the reef and involves a major change in taxonomic complexity.

Species-level coral taxonomy in historical perspective has been a complicated, uncertain subject. Perhaps for this reason, most early studies of species-level distributions have remained hidden from, or avoided by, coral biogeographers. Yet links between taxonomy and biogeography have always existed, at least in theory[6]. Stronger links were forged when field observations started to have a role in taxonomic decisions; phenetic species with discrete morphologies are far more easily associated with museum collections from a particular country than are polymorphic species with extensive synonymies from many different countries. Yet, even with its history of faulted conclusions, it is surprising that virtually none of the species-level distribution records from early taxonomy were married with concurrent ideas of biogeography and evolution.

The studies of Yabe and his pre-war Japanese colleagues (summarised in Yabe and Sugiyama 1935) are relevant to biogeography. Although they give a false impression of regional isolation of Japan, they show that species (according to the concept of these authors) can be widely distributed, that they have continuous ranges from the tropics to high latitudes, that they do not replace one another along these sequences, and that species diversity attenuates in a manner correlated with sea surface temperature.

In his study of the corals of the Marshall Islands, Wells (1954a) analysed species-level distribution data, but (like his Japanese predecessors) confined his analyses to the atolls themselves, reserving broader biogeographic analyses for generic data. The evolutionary and biogeographic study of the Fungiidae by Wells (1966) remained the only species-level biogeographic account of any Scleractinia up to those of the author and his colleagues. Since then Hoeksema (1989) has produced a detailed revision of Wells's work, with a discussion of biogeographic aspects. Other studies up to the present one are confined to either small areas, or small species groups.

EXPLANATORY CORAL BIOGEOGRAPHY

> It has always been easier in biogeography to devise explanations than to test them. There can be few subjects that have offered so much scope for so many people to speculate at such length on so little evidence. Whole schools have grown up around favoured groups of theories and methods, often with little real exchange between them. (Rosen 1988b)

As Rosen points out, a very wide range of explanations have been offered for coral distributions. For most, this involves the implicit assumption that, at species level, the Indo-Pacific is divisible into different provinces, each with high levels of endemism. It is that assumption, together with an absence of species-level distribution data, that has allowed speculative explanations to reign as freely as they have.

The most useful review of explanatory coral biogeography is Rosen's (1988b). He lists thirteen theories (incorporated among those below) that have been put forward to explain the origin of Cenozoic to Recent coral distributions. The perspective is geological and conceptual, thus most region-specific and taxon-specific studies are not included. Each theory is analysed for speculations about speciation, extinctions and distribution change. From this, Rosen proposed a 'biogeographical system' in which there are three groups of processes: maintenance, distributional change and origination. Some of the points Rosen and others make are examined in the following chapters of this book; suffice to note here that this author has no argument with what, for him, is a central point of Rosen's review: that all coral biogeographers (the present author excepted) have assumed that place of present occurrence is linked, somehow, with place of origin.

The following notes about hypotheses of other authors are necessarily brief and, in many cases, do not do them justice. They are not criticisms: some should be seen in their historical context; others are viable alternatives to the views presented in this book. The coverage is not complete: the very general and the very specific are omitted. Each hypothesis is numbered and summarised in italics. Reflections added by the present author are separated in a subsequent paragraph.

CENTRE-OF-ORIGIN CONCEPTS

Briggs, as described above (p 37), has singularly promoted neo-Darwinian biogeography, but not as single-mindedly as his vicariance-promoting detractors have claimed. Although the crux of the debate is the difference between centre of origin and centre of diversity, dispersion is a key aspect. As Briggs (1987) points out, dispersion is an 'everyday' event, while classical vicariance is relatively rare: 'dispersion may be looked upon as a continuing inexorable process while vicariance, when it occurs in one habitat usually stimulates dispersal in another'.

Hypothesis 1 *There is an Indo–West Pacific centre of diversity from which species have spread by adapting to less favourable conditions in marginal regions.* (Stehli and Wells 1971)

This classical view cannot now be supported: distributions of species display no evidence of geographic displacement and very little evidence of amphitropical distributions (chapter 10). Several of the studies listed below, notably at hypothesis 10 (McManus 1985), have components supportive of centres-of-origin concepts.

Hypothesis 2 *The trend for older genera to be more widely distributed, as concluded by Stehli and Wells (1971), may be a statistical artefact.* (Vermeij 1978; Jokiel and Martinelli, 1992)

This correlation (figures 14 and 15) is re-examined below (p 139).

VICARIANCE CONCEPTS AND CLADISTIC BIOGEOGRAPHY

Concepts and methodologies that can be put under this heading are as varied and muddled as the rest of biogeography combined.

IMPORTED CONCEPTS AND GENERALISATIONS IN BIOGEOGRAPHY

The interface between biogeography and taxonomy is inevitably the biogeographer's weak point, for empirical biogeography is always subservient to taxonomy. There are two avenues of escape open to the biogeographer without taxonomy: *conceptualisation* and *generalisation*. Many biogeographers try varying amounts of both, and do so successfully because, with practice, the subject can be made so fuzzy that it becomes immune to the barrage of little facts that are the stock-in-trade of taxonomists.

The relevance of imported concepts. Most biogeographic concepts are primarily related to terrestrial animals, where physical environmental barriers are frequent, and active dispersion occurs by way of the adult phase of the life cycle. Sea surface currents provide a relatively benign and passive dispersion vehicle, and most dispersion in the ocean is carried out during a larval phase of the life cycle. Biogeographic concepts developed from terrestrial biota have very doubtful relevance to the ocean.

The relevance of generalities. Biogeographically, zooxanthellate corals have more in common with other zooxanthellate organisms (eg reef molluscs) and organisms with similar mechanisms of larval dispersion (eg many groups of annelids, molluscs and echinoderms) — even with seagrasses and mangroves — than they have with their azooxanthellate coral relatives. Zooxanthellate corals are not a taxon: they are an ecophysiologically defined group of taxa. Generalities made by lumping disparate groups of organisms such as zooxanthellate and azooxanthellate corals seldom retain intended relevance.

CORAL BIOGEOGRAPHY FROM CLADISTICS?

This could become a popular field of endeavour as (apart from other sources) distribution data can be obtained from the database of the present study[7] and morphometric data directly applicable to it can be obtained from the taxonomic monographs and specimens upon which these distributions are based. Cladistic analysis of these data *will* yield results that, given the spectrum of issues presented in this book, *will* suggest attractive so-called 'Just-so' stories. I do not support cladistic analysis in coral biogeography for the following reasons:

◆ Most coral taxa lack diagnostic characters and also lack character states which are, in any sense, primitive or derived. Cladistic analysis will readily construe artificial character states that have no meaningful status (p 24).

◆ At the present state of development, cladistics does not meaningfully cope with reticulate evolution (p 216), or distributions that are unrelated to place of origin, or with complex intra-specific geographic variations.

◆ The use of morphometric data (the common but not obligatory source of quantitative non-molecular data) as proxy-indicators of a taxonomy that is based on a great deal more than morphometrics is unacceptable. The quality control of any taxonomic data, whatever its form or composition, is that it be self-supporting (ie contains all the information on which the taxonomy is built). Only cladistic data that incorporate the results of studies of co-occurrences (p 20) could overcome this objection.

◆ Only the distribution range of (most) Caribbean species, and that part of the range of most Indo-Pacific species that falls within the central Indo-Pacific region, is sufficiently well known for numeric treatment (p 8).

Hypothesis 3 *Far eastern Pacific corals are derived from a previously widespread, pan-Tethyan coral biota that has since been modified by tectonic events, speciation and extinctions.* (McCoy and Heck 1976; Heck and McCoy 1978)

Heck and McCoy (1978), who purport to incorporate 'current biogeographic thinking' into a 'coherent hypothesis', primarily devote their paper to denigrating the study of Dana (1975) on the origin of far eastern Pacific corals by long-distance dispersal. Their widely quoted study contains a high level of factual error, considering the information sources claimed. Today it stands discredited in favour of Dana's alternative (hypothesis 30 below).

Hypothesis 4 *The major features of coral distribution are latitudinal and are primarily controlled by temperature and climate; regional features are primarily longitudinal and are due to geo-tectonic events, enhanced by glacio-eustatic* *change concentrating speciation in outlying islands.* (Rosen 1984)

Rosen (1984) presents the reader with a formidable array of vicariance ideas from which he concludes that historical factors have been relatively neglected in accounting for diversity patterns. He suggests that the main longitudinal features of these patterns (differences in numbers of genera in the Atlantic and Indo-Pacific centres of diversity and the diversification of Indo-Pacific corals since the middle Miocene) can be explained in terms of glacio-eustatic changes superimposed on 'a geographical template of island patterns'. Pandolfi (1994) gives a similar view within the Papua New Guinea region.

While Rosen's ideas are not generally compatible with the dispersalist school, he proposes the existence of island refuges, a suggestion that he notes 'is attractive because it decouples evolutionary aspects from subsequent distribution patterns...implying no causal connection'. His paper, however, is an extensive discussion of evolutionary links with distribution patterns of extant corals, as are those of the other authors with whom he associates his ideas. Clearly many of Rosen's views (especially the different basis of latitudinal and longitudinal patterns, and the role of temperature in determining the former) are supported by the present study, while those that relate to endemicity and absence of long-distance dispersal are not.

Hypothesis 5 *Indo-Pacific distribution of the* selago *group of* Acropora *species has been determined by sequential west-to-east vicariance events. Depending on assumptions accepted in generating area cladograms, vicariance barriers were determined to have occurred at the mouth of the Red Sea, then west of Western Australia, then at the Indonesian arc. Speciation events may or may not have been followed by dispersal, but no dispersal is needed to explain area cladograms* (Wallace et al 1991). *These conclusions can be developed into an historical hypothesis for the world-wide distribution of* Acropora.

The author's comments on the use of cladistics (p 41) are relevant to this study. The distribution of species across the Indian Ocean is only superficially known and does not provide an adequate basis for this study.

Hypothesis 6 *Indo-Pacific distributions of* Symphyllia *and* Coscinaraea *species result from peripheral vicariance speciation. A cladistic analysis of the author's taxonomic and distribution data of* Symphyllia *and* Coscinaraea *show that species endemic to high-latitude Australia are the most derived (ie the most recent). Species originations occur in peripheral regions in response to geologic vicariance events.* (Pandolfi 1992)

There is no evidence to support correlation between present place of occurrence and mechanisms of speciation, nor the suggestion that the endemic species are the most derived.

DISPERSION BIOGEOGRAPHY

Unlike coral vicariance biogeography, studies claiming long-distance dispersion as the basis of distribution have not been dependent on generalised preconceptions (p 51).

Hypothesis 7 *El Niño Southern Oscillation (ENSO) events change surface circulation vectors and may thus be responsible for extending distribution ranges. Species immigrations, range extensions, establishment of 'expatriate' populations and the setting of the stage for speciation may all follow from ENSO events.* (Richmond 1990)

The present study supports this hypothesis and extends it to a wider scale.

Hypothesis 8 *Westward-flowing ocean currents lead to accumulation of species in the western side of the major oceanic basins. This is demonstrated by computer simulation. Results accord with observations about long-distance dispersion by rafting, and reservations about the validity of generic age patterns.* (Jokiel and Martinelli, 1992)

This is a clear statement of the role of surface circulation in producing patterns of species diversity in the Pacific. It is not generally applicable to Indian or Atlantic Ocean distributions.

GLACIO-EUSTATIC BIOGEOGRAPHY

Several authors have placed much importance on Plio-Pleistocene sea level changes as the primary driving force of evolutionary change. Special significance is placed by Rosen (1988a) on differences of opinion between himself, McManus (1985) and Potts (1985) as to whether the different processes of speciation and distribution change occur at high sea levels or low sea levels.

Hypothesis 9 *Plio-Pleistocene glaciation had little ecological or evolutionary effect. Coral communities repositioned themselves up and down reef slopes as sea level rose and fell, without major long-term consequences.* (Vaughan and Wells 1943; Goreau 1969; Newell 1971)

Lack of Plio-Pleistocene extinction is strong evidence supporting this view. It would appear especially true in regions like the Caribbean and central Pacific where continental shelves are narrow or absent (p 124).

Hypothesis 10 *Plio-Pleistocene glaciation has enhanced speciation. The high number of species of some coral genera in southeast Asia, the centrality of the region within the ranges of many species and the high level of regional endemicity are the result of regional vicariance speciation. The centre of diversity is also the centre of origin and speciation occurs at low sea levels.* (McManus 1985)

Potts (1985) discusses this view in relation to Rosen (1984) and his own (1983–85) model. There is, again, no evidence to suggest any correlation between present place of occurrence and place of origin.

Hypothesis 11 *Plio-Pleistocene glaciation has retarded speciation. Evaluation of genetic continuity over ecologically meaningful time-frames indicates that glacio-eustatic change would be more likely to inhibit speciation than enhance it. Polymorphism, generation times and ecological distributions allow species to survive multiple glacio-eustatic changes, which serve to enhance genetic communication rather than create barriers.* (Potts 1983, 1984a, b, 1985)

Veron (1985) supported this view; it is the precursor to the hypothesis of surface circulation vicariance presented in this book.

ECOLOGICAL BIOGEOGRAPHY

There are a large number of ecological explanations of coral distributions in the literature. Only those on biogeographic scales are included here; those on ecological scales are referred to in chapters 4 and 6.

Hypothesis 12 *Cold coastal upwelling in the equatorial central Indo-Pacific caused destruction of shallow-water habitats at low sea levels. Extinctions due to low temperature, rather than barriers, were created during low sea levels and these explain plate-endemic fish faunas.* (Springer and Williams 1990)

Available data on sea surface temperature during glacial intervals do not support this hypothesis, at least as it might be applied to corals.

Hypothesis 13 *Evolution proceeds most rapidly in regions of warmest water.* (Stehli and Wells 1971)

There is no evidence for or against this view, which is also an aspect of centre-of-origin biogeography (refer to hypothesis 1).

Hypothesis 14 *Corals had a cosmopolitan Indo-Pacific distribution, being constrained primarily by physiological limitations. Climatic changes that affect those limitations are primarily responsible for biogeographic variation. The most recently evolved species have the least ecophysiological tolerance and are relatively susceptible to climatic change, and thus tend to accumulate in centres of diversity where conditions are relatively benign.* (Newell 1971)

Various extensions or modification of physiological aspects of this hypothesis are found in Vermeij (1978). There is scant evidence that physiological factors limit coral distributions in this way.

Hypothesis 15 *Greater species diversity in the tropics is due to greater 'effective' evolutionary time (evolutionary speed), resulting from shorter generation times, faster mutation rates and faster selection at higher temperatures. Temperature determines evolutionary speed, not species numbers.* (Rhode 1992)

Although no specific reference is made to corals, this study is included here because of its intrinsic interest and (distant) relevance to the control of diversity by source/sink mechanisms.

Hypothesis 16 *Inorganic nutrient availability creates latitudinal attenuation.* (W. Wiebe conference presentation)

This proposal awaits supporting study. It is empirically refuted by correlations between temperature and latitudinal attenuations in the central Indo-Pacific (p 92).

Hypothesis 17 *Calcium carbonate saturation, along with temperature, controls distribution. Carbon dioxide solubility increases with decreasing temperature and hence aragonite saturation decreases.* (Buddemeier and Fautin in press)

This proposal also awaits supporting study. It is a viable addition (rather than a simple alternative) to ecological explanations of temperature-correlated latitudinal attenuation because it operates semi-permanently rather than in annual cycles.

PALAEOBIOGEOGRAPHY

There are a wealth of palaeobiogeographic observations in the literature concerning past distributions and areas of endemism, explained in terms of plate tectonics and palaeoclimates (chapter 7).

Hypothesis 18 *Climatic change during the Neogene has led to adaptation and speciation in high latitudes. Climatic warming during the late Cenozoic has created a surge of vicariance speciation in equatorial regions followed by dispersal to higher latitudes as these became warmer because they were pre-adapted to warmer conditions.* (Valentine 1984a,b)

The place of corals in this general scheme (if any) is uncertain, but the concept is a popular one. Distribution ranges of corals have fluctuated greatly with relatively minor temperature fluctuations.

Hypothesis 19 *Environmental conditions partitioned Mesozoic corals into small isolated basins, which allowed sympatric speciation to produce a large number of local endemics.* (Beauvais 1992)

This is a passing attempt to relate a modern theory of speciation to a long-extinct fauna. There are several other such occurrences in the palaeontological literature. There is no supportive evidence other than by analogy with extant corals.

BIOGEOGRAPHIC STUDIES OF SPECIFIC GEOGRAPHIC REGIONS

Once focus moves from the global to the specific by restricting space, time, or taxon, conceptual breadth may be lost, but relevance and accuracy are usually gained[8]. The following briefly lists explanatory biogeographic studies that are broadly region-specific.

LITHOSPHERIC PLATE BOUNDARIES

Several authors (notably Springer 1982; Springer and Williams 1990) have observed distribution discontinuities corresponding to lithospheric plate boundaries. These are often correlated with the proximity of land mass and thus may be associated with change in substrates, nutrients, coastal upwelling and surface circulation patterns.

Hypothesis 20 *Patterns of species diversity are related to tectonic patterns as well as climatic patterns. Tectonic patterns underpin Wells's (1954a) generic contours* (figure 14, p 47). (Valentine 1973)

This article followed soon after general acceptance of the theory of continental drift. The general idea has been formative for many other authors, notably Rosen (above).

Hypothesis 21 *The region of northern Papua New Guinea is particularly diverse environmentally and faunally.* (Pandolfi 1994; Hoeksema 1994)

This region, extending to eastern Indonesia, has the world's highest species diversity and is therefore a focus of hypotheses about distributions.

HIGH-LATITUDE CORAL REEFS

Many explanations have been proposed for latitudinal limitations to coral and coral reef distribution. Those relevant to the central Indo-Pacific are discussed in chapter 10.

Hypothesis 22 *Competition with macroalgae at the Houtman Abrolhos Islands primarily limits the latitudinal distribution of corals and coral reefs.* (Crossland 1981, 1984; Johannes et al 1983)

Latitudinal limits are ultimately determined by temperature, but this indeed appears to be via ecological mechanisms, especially annual competition with macroalgae (p 96).

INDIAN OCEAN

Hypothesis 23 *Biotas of the Seychelles are remnants of continental crusts that have retained their original biogeographic affinities.* (Briggs 1987)

Although not necessarily relevant to corals, the importance of regional geological events is appropriately emphasised.

HAWAII

Hawaii, with its very high level of faunal and floral endemicity (eg Kay and Palumbi 1987), has been a creative centre of thought about historical biogeography.

Hypothesis 24 *Hawaiian and Canton Atoll corals are the product of local extinction and subsequent immigration.* (Maragos and Jokiel 1978; Grigg 1981)

Both regions have been depopulated in geological time and subsequently repopulated from the west. Although one of the oldest explanations of the origin of corals in the central Pacific, this theory remains strongly supported today. Corals have been present in Hawaii since the early Oligocene (Grigg 1988).

Hypothesis 25 *Relict endemic biotas of the Hawaiian Ridge are the product of horizontal lithospheric plate movements during the Tertiary.* (Pielou 1979; Rotondo et al 1981)

Time scales of plate movement are too great to explain observed distribution patterns in the central Pacific, but juxtaposition of the Line Islands and the Equatorial Counter Current during the Miocene may have been an important source of propagules for the far eastern Pacific (Vermeij 1978; Newman 1986).

Hypothesis 26 *The Hawaiian and Polynesian foci of endemism are the source of present endemics as well as species that have dispersed westward.* (Kay 1984)

This hypothesis (centre-of-origin biogeography in reverse) is based on distribution data of higher invertebrates and fish. It is not possible to prove that Hawaiian endemics evolved in Hawaii; affinities between Hawaiian corals and those of the southeast Pacific indicate outward (eastward) dispersion from the Indo-Pacific centre of diversity (p 100).

Hypothesis 27 *No endemics, and only species with broad geographic distributions are found at Johnston Atoll.* (Maragos and Jokiel 1986)

This observation has been made of several remote places, but not Hawaii, where endemism reaches the highest known level for the Pacific. As the above authors point out, the reason for this pattern of endemicity is probably a function of habitat diversity and relative species abundance. Johnston Atoll, being reefal, would have a much lesser range of habitats than Hawaii, where habitat diversity, combined with marginal climatic conditions, may have allowed the continued existence of a number of species in refuge situations.

Hypothesis 28 *Hawaiian shallow-water fauna, including corals, are relatively young and must be accounted for by long-distance dispersal. Faunas of Hawaii and the southeast Pacific are relicts following extinctions further west.* (Newman 1986)

This is a proposal for amphitropical distributions and is an alternative to the 'island integration theory' of Rotondo et al (1981) (hypothesis 25). As Newman points out, time-frames exclude island integration as a basis of origination of Hawaiian fauna.

CENTRAL AND FAR EASTERN PACIFIC

Hypothesis 29 *The fossil record indicates a high species turnover in the Pitcairn Islands.* (Paulay and Spencer 1988)

This observation is similar to the conclusions of Grigg (1981) for Hawaii (hypothesis 24).

Hypothesis 30 *Far eastern Pacific corals are comparatively recent immigrants from the central Pacific.* (Dana 1975)

In one of the seminal papers of coral biogeography, Dana traced the history of far eastern Pacific corals from the Cretaceous to the extant, concluding that the latter had a western Pacific origin after the Pliocene closure of the Isthmus of Panama, and Pleistocene extinction of all species thus isolated. This view of the origination of far eastern Pacific species was held by some earlier authors (Newell 1971) and has since been supported by others (eg Vermeij 1978; Glynn and Wellington 1983; Cortés 1986; Grigg and Hey 1992) (p 144).

Hypothesis 31 *Plate tectonics is largely responsible for the distribution of Indo-Pacific corals. Vertical movements (bathymetric change) of lithospheric plates, especially in the central and eastern Pacific, have created present distributions within climatic constraints.* (Coudray and Montaggioni 1982; Rosen 1984)

Species distributions over the eastern Pacific cannot be explained in the time-frame of short-distance dispersal and vertical movements of terrains; climate does not comprehensively explain longitudinal distribution in the Indo-Pacific (p 235).

Hypothesis 32 *Corals in remote marginal regions have unstable species compositions.*

This has been observed in many places, including southeast and southwest Australia (p 175), Japan (p 177), Hawaii (p 164) and southeast Polynesia (Paulay and Spencer 1988), and the far eastern Pacific (Glynn 1976; Glynn and Wellington 1983; Wilson 1990, 1991; Cortés et al 1994).

WEST AFRICA AND MEDITERRANEAN

Hypothesis 33 *Corals have colonised West African islands by long-distance dispersal (a) from the Caribbean via Bermuda and the Azores* (Chevalier 1966), *(b) from Brazil* (Laborel 1974), *(c) from the Mediterranean Tethys.*

Laborel (1974) discusses the first two alternatives, but comes to no conclusions (p 168). The Eocene epi-continental sea, which crossed western Africa from the Tethys to the eastern Atlantic (figure 29, p 120), could have been a dispersal corridor.

Hypothesis 34 *Corals do not occur in the Mediterranean because of exclusion due to physico-chemical conditions.* (Zabala and Ballesteros 1989)

Geological history, not physico-chemical conditions, demonstrates why there are no zooxanthellate corals in the Mediterranean (p 122).

 If viewed as a whole, the strengths of the above studies include information exchange over traditional taxonomic boundaries and across widely disparate research fields. This may have encouraged objectivity and testability. Several computer-based packages have been designed to try to enhance this process and extract the subject of coral biogeography from the domination of its dusty taxonomic masters. Weaknesses tend to have similar sources, in particular, pre-packaged, pre-conceived numerical solutions, linked to dogma.

Many and varied though the explanations of coral distributions are, they must be blended with general genetic explanations of speciation (p 62), and with ecological theories of diversity (p 69) before it is possible to gain a balanced appreciation of the range of concepts the literature has to offer. Key questions that arise concern, first, the nature of physical environmental limitations; second, the importance of historical factors; and, third, the nature of distributional and evolutionary change. One all-important question that has not been asked concerns the relationship between species distributions, biodiversity and intra-specific variation.

Be that as it may, the subject of coral biogeography, with its strong links with biodiversity and conservation, seems to have a future that has not been so assured since the monumental upheaval that followed Darwin.

4

SPECIES CONCEPTS AND SPECIES DIVERSITY

As with all concepts of species and evolution, the two subjects of this chapter are inherently interlinked in complex ways (figure 16). Considering that they are about the very organisation of life itself, this is hardly surprising; what is surprising is that the explanatory literature contains so few real milestones since, perhaps, the biological species concept of Mayr (1942).

The role of species concepts in coral taxonomy and the mechanisms of speciation that various authors have proposed in order to explain coral distributions have been introduced in the two preceding chapters. The present chapter follows on from these, outlining general points of view, then turning to aspects of the literature specifically relevant to corals.

THE MAIN TOPICS

◆ General concepts of species and speciation: an introduction for readers new to evolutionary theory.

◆ General species concepts applied to corals: the evolutionary mechanisms which underpin explanatory coral biogeography; singular versus plural species; the status of the taxon species; non-allopatric speciation; divergence and genetic transilience.

◆ General concepts of species diversity and an introduction to the mechanisms which limit biodiversity and regulate the rate of evolution (p 233).

Editorial note: Terminology issues with concepts of species are noted in the glossary. Figure 12, p 43 illustrates some common terms.

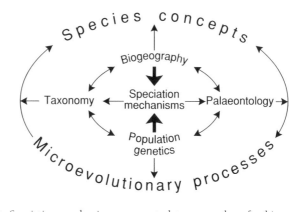

Figure 16 Speciation mechanisms are central to a complex of subjects, concepts and processes, all acting over enormously variable spatial and temporal scales.

GENERAL CONCEPTS OF SPECIES AND SPECIATION

> Concepts of species are found in two very distinct fields of biological endeavour, taxonomy (the science of classification) and population genetics. In both these fields distinct concepts of species are used, and their conflation is a major source of confusion. When an author glides imperceptibly from...one concept to another distinct concept, a subtle kind of nonsense is generated, which is exceedingly difficult to detect. (Paterson 1981)

The steps forward in 'the species debate' (below), though important, are few in number and all seem fated to become submerged in the confusion to which Paterson refers. The crux of the confusion — the distance between taxonomic and genetic concepts — has been highlighted by many authors, but is frequently forgotten. In the context of this book, the 'debate' has three significant areas of impact: on concepts that underlie operational coral taxonomy; on the theoretical basis and practical issues of coral biogeography; and on mechanisms of evolution.

An important issue underlies the first two of these subjects. Most coral biologists have a working familiarity with what they recognise and understand as species, yet a large part of the general 'what are species' literature is devoted to telling us that there is a wide spectrum of uncertainty, or interpretive faults, with these species that we recognise. Perhaps those who are very familiar with coral species could be forgiven for feeling dismissive of these issues as being of either peripheral importance or the result of the accumulation of theoretical abstractions. However, a wide range of *real* issues regarding coral species are very much the concern of coral taxonomy and biogeography.

THE 'SPECIES DEBATE'

Are general concepts of species and the process of speciation inseparable? Do operational species *have* to conform to a concept? Or should it be vice versa? Are species individual entities, defined by their own attributes (ie self-defining), or are they just human-constructs from assemblages of populations united by descent, just as genera are assemblages of species united by descent? If species are real and individual, must they be self-defining, irrespective of the mechanisms and/or forces which created and which maintain that definition?

The role of taxonomy is to delineate, define and describe species, and as such, it is a science unto itself as well as a servant of other disciplines. There are no constraints on methods, but there are definite constraints on how results are interpreted, for the taxonomist cannot afford the luxury of vague and often changing concepts. The end products of taxonomy must be taxa that are recognisable in nature, or the closest possible approximation to them. Some form of species concept must underlie any taxonomic study, but practice must come first. There is thus a functional distinction between taxonomic (ie operational) species and genetic (ie conceptual) species as 'units of evolution'. Much of the early 'species debate' has this seemingly obvious origin.

Concepts of species These mostly apply to all species and are usually composites of information obtained from many different sources. They are best derived from detailed observation and comprehensive knowledge, and must allow for exceptions.

Processes of speciation These are mostly governed by genetic mechanisms and may be significant for some taxa and not others.

Concept and process are not necessarily linked.

THE 'NEO-DARWINIAN SYNTHESIS'

> Species are units fixed in the plan of creation...divine appointments which cannot be obliterated. (Dana 1857)

Dana, geologist and coral taxonomist, described 18 per cent of all coral species in currently operational taxonomy. Nearly half of all today's operational coral species were described by contemporaries of Darwin.

The neo-Darwinian synthesis that resulted from the integration of Mendelian genetics with evolutionary theory was long based on the broad premise that evolutionary change within species resulted from natural selection acting on variations within populations, variations that ultimately arose from random mutations (Charlesworth et al 1982 and many earlier reviews). For decades there has been a consensus that biological species, unlike other taxa, are units within which genes are, or can be, freely exchanged, but between which gene flow does not occur, at least under normal circumstances (Dobzhansky 1935, 1937; Mayr 1942, 1963; Lewontin 1974, among many others). Species were thus considered to be reproductively isolated from other species. When the dimension of time is added, this concept was expanded to include the 'evolutionary species' (Simpson 1944, Grant 1981; Wiley 1978), one that exists through time with an ancestor–descendant lineage different from that of every other species. In both cases the 'species' was considered to be the most clearly defined taxonomic and evolutionary unit with a special reality in nature that generally incorporated observable morphological, ecological, reproductive and/or geographic discontinuities.

For decades there has also been a widely held view that speciation in animals usually proceeded by divergence of allopatric populations, with other modes of speciation being largely insignificant. This view has been challenged back and forth several times (reviewed by Futuyma and Mayer 1980), yet it remains a central issue in concepts of species and speciation.

CURRENT SPECIES CONCEPTS

Leaving aside a wealth of historical debate, five developments encapsulate current concepts of species. Only a brief sketch of them (mostly after Templeton 1989, to which readers are referred) is given here because they have little bearing on operational coral taxonomy, or on coral biogeography, or on present conclusions about the nature and evolution of coral species (chapter 13).

The evolutionary species concept (Simpson 1944) is, by default, the concept of most taxonomists and palaeontologists where species are envisaged as populations, or groups of populations that have had a common evolutionary history. Species are held together by developmental, genetic and ecological constraints, not just heredity. This is the nearest any general concept comes to operational coral taxonomy. As a concept, however, it is empty, contributing no guidance as to what are acceptable criteria for taxonomic decisions.

The isolation species concept has had a long and much debated history (from Dobzhansky 1935), whence it became the popular, intuitive, 'biological species concept'. Mayr (1963) defined isolation species as 'groups of actually or potentially interbreeding populations which are reproductively isolated from other such groups'. Species are thus defined in terms of isolating mechanisms (see the box). This concept has very variable relevance to different faunal groups. For corals, which are believed to have great potential for hybridisation (p 83), the only isolating mechanism available is hybrid inviability or sterility, which remains, as yet, unstudied. The isolation concept must thus be offset against the notions that, given enough space and time, if hybridisation *can* occur, it probably *will* occur, and that corals form identifiable species because of a combination of ecological interactions (in the form of cascades of source/sink relationships) and temporary patterns (in evolutionary time-frames) of genetic connectivity (which are ultimately environmentally, not biologically, controlled, p 216).

ISOLATING AND COHESION MECHANISMS IN ANIMALS (AFTER MAYR 1963 AND TEMPLETON 1989, RESPECTIVELY)

ISOLATING MECHANISMS:

1) Mechanisms that prevent inter-specific crosses (premating mechanisms):
   a) Potential mates do not meet (seasonal and habitat isolation).
   b) Potential mates meet but do not mate (ethological isolation).
   c) Copulation attempted but no transfer of sperm takes place (mechanical isolation).
2) Mechanisms that reduce full success of inter-specific crosses (postmating mechanisms):
   a) Sperm transfer takes place but egg is not fertilised (gametic mortality).
   b) Egg is fertilised but zygote dies (zygote mortality).
   c) Zygote produces an F_1 hybrid of reduced viability (hybrid inviability).
   d) F_1 hybrid is fully viable but partly or completely sterile, or produces deficient F_2 (hybrid sterility).

COHESION MECHANISMS:

1) Genetic exchangeability. The factors that define the limits of gene flow:
   a) Mechanisms promoting genetic identity through gene flow.
      i) Fertilisation systems: the exchange of gametes.
      ii) Developmental systems: the products of fertilisation giving rise to fertile adults.
   b) Isolating mechanisms preventing gene flow.
2) Demographic exchangeability. The factors that limit the spread of genetic variants through genetic drift and natural selection:
   a) Replaceability: genetic drift promotes genetic identity.
   b) Displaceability:
      i) Selective fixation: natural selection promotes genetic identity by favouring the fixation of genetic variants.
      ii) Adaptive transitions: natural selection favours adaptations that alter demographic exchangeability.

The recognition species concept follows Paterson's (1981, 1985) argument that the process of allopatric speciation has nothing to do with intrinsic isolating mechanisms (above) because the speciating faunas are already isolated (eg by classical vicariance, figure 10). Species may be maintained sympatrically by such mechanisms as courtship behaviour, coordination of reproductive effort, self-fertilisation and asexual reproduction and thus exist without reference to other species, that is they are self-recognising through common fertilisation. In a sense this concept is the inverse of the isolation concept, because the isolating barriers (above) that prevent inter-specific hybridisation are also mechanisms that facilitate intra-specific diversity. It is thus the opposite side of the same coin, and clearly is more relevant to corals. However, it is not a concept that is useful in explaining, let alone guiding, taxonomic decisions.

The cohesion species concept proposed by Templeton (1989), is 'inclusive of populations of individuals having the potential for phenotypic cohesion through intrinsic cohesion mechanisms' (see box). The concept is an amalgam of its predecessors, designed to apply to all taxa and identifying the specific mechanisms of evolution by speciation. To this extent it is a useful summary of the best of its predecessors, but, as with all proposed concepts of species, it presupposes a primarily biological control of the composition of species and, the emphasis being on genetic mechanisms, takes little account of spatial variation within species and none of temporal variation.

The phylogenetic species concept (Cracraft 1987) emphasises common origin and is a methodological concept where species are 'the smallest aggregation of populations (sexual) or lineages (asexual) diagnosable by a unique combination of character states in comparable individuals' (Nixon and Wheeler 1990). This concept, being character-based, requires the assumption of a hierarchical taxonomic pattern. Being cladistic, it is not, in the author's view, applicable to corals (p 41).

CURRENT CONCEPTS AS APPLIED TO CORALS

The spectrum of issues about species in general cannot meaningfully be introduced in this book. The following are some general points that appear to have direct relevance to corals.

Is there one kind of species or many?

Biological variation is distributed phylogenetically, geographically and chronologically. What parts of...'the tree of life' contain species? Where along phylogenies, over geography and in time, are species' boundaries located? Should biologists recognise chronospecies, morphospecies, taxonomic species, microspecies, asexual species, and so on; or is there only one kind of species? (Vrba 1985)

We must resist at all costs the tendency to superimpose a false simplicity on the exterior of science to hide incompletely formulated scientific foundations. (Hull 1970)

Cladists, among others, have long sought to make operational a universal or singular notion of species. Although this view is accepted by some taxonomists in principle, few have accepted as a guide any single, universally applicable, theory, concept or method (p 24). Specific genetic mechanisms alone can be considered within a 'single concept' approach, that is, can be considered potentially applicable to all corals. If specific mechanisms are extended, even to local populations, some common ground is unavoidably lost; if mechanisms are turned into concepts and made applicable to whole species, then whole genera, and thence all life, all common ground is progressively lost. Given the diversity of life on earth, and the spectrum of genetic coding that underlies it, this is hardly surprising.

> Species do not need to be explained any more than genera or phyla do...the common-sense-notion of species-taxa as unique entities is mainly an artefact of a perception restricted in space and time. (Linden 1990)

A case is presented throughout this book that the different kinds of species Vrba refers to exist independently and may have nothing in common; that species are individual and do not necessarily have common properties; and that species concepts applicable to one group of organisms (in the present case, zooxanthellate corals) may have nothing in common with those applicable to another taxa. Mishler and Donoghue (1982), Rieppel (1986) and Hoffman (1989), among many others, debate the issues.

A special status for species? Are populations, subspecies and species different? Levinton (1988), among others, says they are not and that the bulk of evidence from variability in chromosomes, allozymes and morphology indicates that the sorts of variation found within species is qualitatively similar to that found between closely related species. Levinton further notes that in the case of allozymes, there is a smooth increase in degree of differentiation from intra-population, to closely related inter-specific, to more distantly related inter-specific comparisons. It follows that processes of speciation are not fundamentally different from those that lead to genetic divisions of populations.

A philosophical or biological question?

> The species problem has to do with biology, but it is fundamentally a philosophical problem. (Ghiselin 1974)

This much-debated statement (Hull 1976; Gayon 1990) is undeniable, and there is now a significant literature in support of it. Species cannot now, and never will, be definable, because they are ultimately human-created concepts or 'philosophical' points of view (p 238).

Speciation and morphological change

> We have argued...that speciation is not a driving force in morphological evolution, in the sense that it is the principal process that generates variation or provides the necessary conditions for adaptive evolution. (Levinton 1988)

Speciation is almost never observable[1] in nature; it occurs on a time scale beyond the reach of humans and may well be independent of any morphological change, karyotypic change, significant DNA sequence divergence, significant isozyme differentiation or shift in niche or habitat (Templeton 1981 and many other authors).

Reticulate or dichotomous evolution?

> The evolutionary importance of hybridization seems small in the better-known groups of animals. (Mayr 1963)

> We will not consider reticulate evolution arising from hybridization between related species or the occasional transfer of genetic material between taxa by viruses. (Futuyma 1986)

> In contrast [to animals], examples of natural hybridization are common among plants. (Harrison 1993)

Despite Harrison's acceptance of reticulate evolution in animals and his appreciation of its hypothetical consequences, he (and it seems, all other authors) consider it to be a phenomenon restricted to geographically restricted 'hybrid' or 'suture' zones.

Reticulate evolution in animals, if considered by evolutionists or geneticists at all, is almost always discounted on the basis that it may well occur within species, but (by definition) does not occur between species. The conflict between evolutionary mechanisms and concepts of species is only too obvious. This is pursued in the context of coral evolution in chapter 13; the issue, however, is a general one and is not just restricted to corals.

Monophyly or polyphyly? The conflict between monophyly and interbreeding between populations (perhaps another way of putting the 'reticulate versus branching evolution' question) is debated by de Queiroz and Donoghue (1988). The outcome is a 'disjunctive' definition where species are either populations (monophyletic or not) or monophyletic groups of populations. These authors conclude that what some may see as 'insurmountable difficulties' with species, others will see as 'simple facts of life' and that 'this is the species problem'. With corals 'the species problem' is a complex one: there can be no assumptions of monophyly at any taxonomic level.

Isolating mechanisms

> All authors who have written on isolating mechanisms...have stressed the enormous diversity of such devices. (Mayr 1988)

This, more than any other subject, is at the core of any species concept of corals. Given time and opportunity (both abundant in coral evolution), it is reasonable to assume that if species can hybridise, they probably have done so in the past, and will do so in the future. Two types of isolating mechanisms are important; the first is genetic and involves reproductive strategies; the second is ecological. Very little is known about either, but, given the high levels of genetic plasticity in corals, it is likely that the latter is the more important. Thus, although different species may hybridise (and they do, p 83) and produce a fertile F_1, ecological factors are likely to be at the core of the hybrid's survival. The number of hybrids in natural populations is unknown, but again raises the hypothetical question, what is it that is hybridising: what is a species?

Punctuated evolution: genetic revolutions, founder events, genetic transilience and stasipatric speciation These are all different aspects of related phenomena, where a small population displaces a parent population, precipitating rapid,'punctuated', evolutionary change. Terms are easily confused: 'punctuated evolution' refers to a macroevolutionary event (p 148) while the other terms principally underlie microevolutionary processes. In Mayr's original concept 'genetic revolutions' can follow founder events (p 42), by the displacement of parent genotypes by outcrossing from a peripheral 'founder' population. Founder speciation can also be induced by 'genetic transilience' (Templeton 1980), involving fixation of a small part of the founder genome by the overcoming of isolating barriers between populations (eg Carson and Templeton 1984)[2]. In both cases the outcome could be a sudden speciation event. Stasipatric speciation (figure 12, p 43) may follow from a chromosome rearrangement or mutation in a very small population, which can spread through the parent population if hybrids have reduced fitness, and thus form hybridisation barriers (White 1968, 1978).

Editorial note: Descriptors of space and time are explained in the glossary.

SPECIES DIVERSITY IN CORALS

CORAL BIOGEOGRAPHY, BIODIVERSITY AND COMMUNITY ECOLOGY
In principle, these are three fields that differ only in emphasis. Biogeography emphasises the historical past (hence the term 'historical biogeography'), and thus reaches back into evolutionary time-frames. The distribution of biodiversity is the biogeography of the present, usually with emphasis on small spatial scales. Community ecology is a more evolved term, with emphasis on short-term dynamics and environment. In practice these fields are a long way apart, perhaps because the partitions that separate them are less those of familiar subject matter, and more those of hard-to-manage concepts of spatial and temporal scales.

Recent interest in biodiversity (as indicated by species diversity (p 234)) has inevitably resulted in a broadening in the use of this term. The primary division of the field — organism-based measures of diversity versus molecular-based measures of diversity — applies equally to taxonomy and biogeography. With both taxonomy and biogeography, there are cladistic measures of biodiversity that can be based on either type of data. Indices of diversity, which reflect relative abundance, bring these fields into the realm of ecology.

Ecologists are beginning to realise that local diversity bears the imprint of such global processes as dispersal and species production and of unique historical circumstance. (Ricklefs 1987, who gives a particularly good introduction to local versus regional processes in the maintenance of biodiversity)

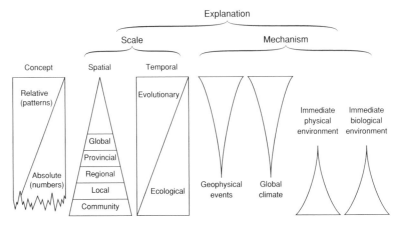

Figure 17 Concepts of biodiversity can be relative or absolute. The former usually involves patterns, the latter numbers. Explanations of these patterns and numbers vary according to spatial and temporal scale and also with the mechanism of control. Global patterns on an evolutionary time-scale are primarily controlled by geophysical events and global climate, yet these intergrade with the smallest spatial and temporal scale where control is with the immediate physical and biological environment.

In chapter 13, it is argued that, within the Indo-Pacific, total coral species diversity within the equatorial band of high diversity (figure 75, p 235) is the primary parameter limiting the rate of evolutionary change. Other geographic regions are connected to this band in a cascade of source/sink relationships (p 75). These source/sink relationships can be introduced by exploring the proposition that coral species diversity, if generated on a biogeographic scale, is maintained, and thus limited, on a local one.

Almost all theories of coral biogeography have something to say about *relative* species diversity (eg the location of centres of diversity, or geographic patterns of diversity) and much to say about the mechanisms by which that diversity *increases* (ie mechanisms of speciation). They have less to say about *absolute* species diversity (eg the total number of species in the world, or the total number in a particular region) or the mechanisms by which that diversity is *maintained*. This last subject wears the label 'ecology', but in terms of explanation of species diversity, ecological and biogeographic concepts are equally important: the two form a causal continuum from the smallest scale to the largest.

For coral reefs, as for many other ecosystems, the biogeographer's explanations of relative species diversity meet the ecologist's explanations of absolute species diversity in the context of almost limitless independent variations in space and time (figure 17). This seldom fails to generate confusion when explanations are proposed. Explanations become compounded when combinations of space and time are viewed in the context of their individual controlling mechanisms and become impossibly compounded if each of these mechanisms is considered to have some sort of interactive relationship with each other mechanism. This complexity, if imagined in the context of full scales of space and time, and if viewed in the light of a wealth of recent ecological literature, is the scenario of the following discussion.

The long held general view is that absolute species diversity is locally maintained by habitat or niche diversity[3]. This appears generally true; the highest coral species diversity within the Great Barrier Reef, for example, is found around complex coastlines of high islands (figure 60, p 182), because these contain a greater range of habitat types than platform reefs. However, this says little about mechanisms of maintenance of diversity, merely that, as coral species have preferred habitats, the greater the range of habitats the greater the number of species (if niche widths are assumed to be constant).

THE COMMUNITY SCALE

The focus of the community scale is at the very bottom of figure 17, and on mechanisms of maintenance of coral species diversity, not its local total, and certainly not its origins. The mechanisms proposed in figure 17 are, simply, *immediate physical environment* and *immediate biological environment*, the latter essentially being species–species interactions. These alternatives can be considered in three propositions (below), the last (disturbance) highlighting the distinction between *equilibrium* and *non-equilibrium* communities, which respectively encapsulates the notion of stable climaxes as opposed to perpetual succession created by perpetual disturbance.

These alternatives are currently not just aspects of the history of ecology: they represent a debate that is currently very much alive and hence not answered to widespread satisfaction. This debate, as with all aspects of coral species dynamics, is made complex by the ranges of spatial and temporal scales involved. However, much supposed complexity that repeatedly arises in the literature (reviewed Karlson and Hurd 1993) is irrelevant, being based on the addition of historical non-issues and current real ones, as well as on the addition of taxon-specific or location-specific issues and general ones.

Proposition 1 *Coral species diversity is the outcome of environmental conditions and community history.*
Physical environment creates boundaries, defines habitats and mediates the habitat occupancy of coral species. In this context communities can be regarded as essentially temporary aggregates of available coral species in response to prevailing physical environment, and coral species diversity is thus governed through environment–coral-species interactions. The principal conclusion of Huston's (1985) review of patterns of coral species diversity was that it is maintained by 'reduced rates of competitive displacement resulting from decreased light

availability and growth rates with depth'. It is indeed true that coral species numbers change with depth, and that coral species occupy limited ranges of habitats, the physical-environmental parameters of which are mostly depth-correlated. But there is very little evidence of displacement, competitive or otherwise, over depth ranges that are wider than individual zones.

Proposition 2 *Coral species diversity is the outcome of ecological succession of species interactions.*

Species interactions, through conflict or competition for space or other resources, demonstrably cause local changes in area occupancy, and hence relative abundance. That change might be in the form of never-ending cycles of occupancy by the same group of species or transitory sets of species, might be random or stochastic in nature, or might be directional, involving successional displacement of species. Cyclical changes due to species interactions, like changes controlled by physical environment, require no starting point. Successional changes do, and that starting point in an ecological time-frame must be some form of disturbance. Rainforests and coral reefs have long been the target of this subject, presumably because they are well-defined ecologically, abundant and conspicuous.

Coral communities of unchanging species diversity may, presumably, continue to exist indefinitely in unchanging physical environments. These communities may have constant species compositions or changing ones, and may (depending on definition) be described as *stable, equilibrium* or *climax* communities. Whether such communities actually continue to exist within ecologically significant time-frames (perhaps hundreds of years), or not, is a moot point. Most literature on the subject is biased towards accounts of instability; even so, as we have little direct evidence of community stability beyond the time-frame of human memory, the matter is, and must long remain, a theoretical one. The points at issue are left to the next and last proposition for the maintenance of species diversity, with the note that equilibrium concepts of high species diversity argue neither for nor against local control.

Proposition 3 *Coral species diversity is the outcome of disturbance.*

> Clements (1916) proposed a theory of the causes of succession so satisfying to most ecologists that it has dominated the field ever since (see Odum 1969). (Connell and Slatyer 1977)

Odum's (1969) concept of orderly succession towards a stable endpoint climax community has, for corals, moved slowly to Connell and Slatyer's hypothesis. This hypothesis, which remains widely accepted today, is that community development is largely a consequence of disturbance and subsequent succession. Their hypothesis concludes that only when cycles of disturbances are broken do essentially stable communities develop, and that this succession is dependent on both the nature of the disturbance and the individual characteristics of the species involved. Hence we have 'opportunist' species with broad dispersal powers and rapid growth rates, then various models for subsequent succession, depending again on species characteristics. Both species–environment and species–species interactions are envisaged acting, more or less, at successive points in time.

This proposition was written with forests in mind; when applied specifically to coral reefs, it became Connell's (1978) *intermediate disturbance hypothesis*, effectively the basis of the nonequilibrium theory of coral community ecology. Connell argued that species diversity at a single location is best explained as a balance between competitive exclusion, and disturbances that prevent exclusion (storms, predation, disease etc). Species diversity is low at low levels of disturbance because the best competitor can become very dominant; it is also low at high levels of disturbance because mortality is too severe for most species to survive. Intermediate levels of disturbance result in maximum species diversity.

Connell's hypothesis was derived from a study of series of 1 m² quadrats on reef crests at Heron Island on the Great Barrier Reef. This study, started in 1962 is still continuing. The hypothesis is appealing because it is clear, tidy and quantitative, but its predictions have been extrapolated, perhaps unjustifiably, into a general theory of population dynamics by other authors. Two rather obvious questions arise: first, can any theory derived from an extreme habitat (uniquely exposed and disturbed and without significant physical-environmental gradients) reasonably be extended to the bulk of coral communities; and second, can any specific hypothesis cope with increasing scales of space and time?

Connell and others have made many additions and refinements to the original hypothesis which, despite considerable interest, has proved durable. Broad characteristics of coral communities appear supportive of the hypothesis: both exposed upper reef slopes and very protected habitats (where monospecific stands are common) have a high species dominance and low diversity, while highest diversities are almost invariably found in non-extreme depths and exposures. These, however, are only isolated observations in a narrow time band, and may only be short-term outcomes of more stochastic, palaeoclimatically-driven cycles.

Hierarchical and circular interactions (figure 18). Circular interactions (eg Jackson and Buss 1975), dominated by competitive stand-offs and reversals, would tend to promote species diversity under stable conditions when hierarchical rankings would reduce it. Connell and Keough (1985) logically propose that hierarchical rankings of competitiveness are transitory, that is, under conditions of continuing equilibrium, inferior competitors are gradually displaced (species diversity is reduced); only under conditions of disturbance can they co-occur (and species diversity be maintained). In concept the predominance of circular interactions over hierarchical ones is almost always the outcome of considerations

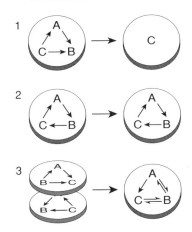

Figure 18 Hierarchical and circular species interactions. (1) Hierarchical interactions among three species leads to the competitive displacement of species A and B. (2) Circular (or network) interactions among three species leads to no displacement. (3) Changes in the competitive dominance of a species (with, for example, depth, time or an environmental variable), breaks down hierarchies and thereby enhances diversity.

such as is illustrated in figure 18.

Guild versus species interactions If coral species form functional guilds, competition within guilds is likely to be stronger than it is between guilds because within-guild species use the same resources. Connell (1987) lists the mechanisms by which that competition might be reduced, although no essential within-guild and between-guild differences are offered. Guild interactions potentially involve all species whether coral or not: for example, competition between corals and algae can help maintain species diversity if common species of coral are reduced disproportionately (eg Hughes 1989).

Process-from-pattern deductions This is a common conceptual pitfall to which disturbance hypotheses are very prone. Small-scale changes in pattern may be externally driven, not intrinsic, or be part of ever-larger continua which are stochastic or chaotic. If so, any process deduced from observed patterns may be so open to qualification that it becomes meaningless. Chornesky (1989) has demonstrated an important point, that observations at single points in time are likely to give a false impression of hierarchical dominance or permanent stand-offs; observation of coral species interactions over continuing time show an unexpectedly high frequency of repeated reversals.

Essential points about coral species diversity at the community scale are: first, that ecomorphs, rather than species, change along physical-environmental gradients; second, that environmental diversity limits total species diversity; and third, that species diversity is inversely related to ecomorph diversity.

THE LOCAL SCALE

In moving to the larger scale (the area of a single location, but still at the bottom of figure 17) the focus moves from downward pressure on species diversity, through mechanisms of competitive exclusion, to upward pressure on species diversity, through source/sink relationships and genotype saturation[4] (p 234).

SOURCE/SINK RELATIONSHIPS AND SPECIES SATURATION

As different communities combine to form biotopes or zones, the total number of coral species involved increases geometrically. It is not difficult to identify, in the central Great Barrier Reef, ten definable communities along the length of a single biotope (or zone), or ten distinct biotopes (or zones) down a single reef face. Nor is it difficult to identify three such sets of biotopes around a single reef complex, or three such reef complexes across the continental shelf, or the same multiple repeated in association with an island archipelago. Thus, even at the scale of a reef complex, there are perhaps 300 recognisably distinct communities. Communities that are contiguous (ie form adjacent zones) may interact competitively, but most communities do not: what they do is supply each other with propagules and thus recruit from each other and exchange genetic information. The net result at the local scale is species diversity saturation, that is, small areas are saturated with species. Thus local and community spatial scales have source/sink relationships.

Evidence for this seems overwhelming, although it is apparently undocumented. Within the central Great Barrier Reef, for example, the species complement of one reef complex is almost the same as the next, that is, each reef complex is saturated with species having most of the species of the whole region (rare species and those limited to inshore or specialised habitats are usually the only exceptions) (p 183).

LIMITING BOUNDARIES

It has long been known that latitudinal range is primarily limited by temperature or correlates of temperature (p 92), and that reef slopes are predictably zoned according to exposure to wave action at one extreme and light availability at the other (p 18). These are aspects of the physical environment that control coral species diversity at both geographic (latitudinal) and local (depth) scales by creating limiting boundaries. At the local scale, there may be many other limiting environmental gradations within boundaries, the most important of which (again using the example of the central Great Barrier Reef) is the cross-shelf gradation in the sedimentary environment (p 183).

THE REGIONAL SCALE

When moving from the local to the regional scale (eg from a group of reefs to the whole Great Barrier Reef), the biogeographic focus of interest

THE EQUILIBRIUM THEORY OF ISLAND BIOGEOGRAPHY

MacArthur and Wilson's (1963, 1967) historic 'dynamic equilibrium' theory brings biogeographic and ecological models together in a way that makes intuitive sense on a regional scale. The basic proposition that attracted substantial debate in the 1970s is that species numbers reach an equilibrium point (or equilibrium density) on islands, but that species identities can change. In the latter context, this is a nonequilibrium theory (Chesson and Case 1986), which essentially refers to source/sink relationships between disjunct regions. When applied to corals, it is generally compatible with region-wide concepts that emphasise environment control of species diversity but, as is inevitable with all theories at this spatial scale, it is prone to many qualifications according to region-specific circumstances of physical environments, faunas, geography, etc.

changes abruptly. Predictability of distribution, abundance and biodiversity (as Jackson 1991 points out) all increase with increasing scale. We leave the subject of the maintenance of species diversity behind, as it loses relevance, and turn to that of the outward transport (dispersion) of species diversity. For the first time we see patterns that are not locally repeated and have, for the first time, a significant interface with evolutionary mechanisms. We also move away from small-scale (genetic or ecological) processes that are universal (frequently replicated in different locations) towards large-scale outcomes, which tend to be unique or specific to particular regions. The ultimate source of species diversity in sink/source cascades is, for the Indo-Pacific, a 9500 km wide equatorial band of high diversity (figure 75, p 235). It is hypothesised in chapter 13 that limits to diversity within this band are dependent on habitat diversity and genetic connectivity which, in turn, limit rates of evolution.

A BIOGEOGRAPHER'S PARADOX

The fundamental strength of Darwinian evolution, the most universally accepted and fundamental concept of Nature the world has ever seen, is its all-encompassing internal consistency. Clearly, Darwinian concepts apply to all the subjects covered in this chapter, from concepts of species to mechanisms controlling species diversity. Yet the geographic basis of Darwin's evolution, his centres of origin, have been roundly rejected by most biogeographers (p 36). In simplest reality, Darwin-supporting ecology is seemingly at odds with Darwin-rejecting biogeography, yet the former is the small-scale basis of the latter's large-scale outcome. The question, addressed in chapter 12, becomes: what aspects of evolution are biologically controlled and what are controlled by physical environment?

5

REPRODUCTION AND POPULATION DYNAMICS

Reproduction in corals is central to a wide range of other subjects, including population genetics, population dynamics, concepts of species and views about speciation. Hence reproduction is important to explanatory biogeography and evolutionary theory, and recent studies have contributed greatly to our understanding of all these fields. These fields would, however, be much better understood if studies of reproduction were balanced by studies of population dynamics, but here there is a discrepancy. We know a great deal about the mechanisms, strategies, timing and geographic variation in reproduction, and how all these vary among different taxa, but we do not have a clear understanding of the dynamics of a single population. The first subject of this chapter (reproduction), is therefore mostly factual; the second subject (population dynamics) is mostly conjectural.

THE MAIN TOPICS

♦ The plant-like array of coral reproductive strategies.
♦ Reproductive modes as 'trade-offs' between local abundance and dispersion and between outcrossing and hybridisation.
♦ Synchrony in spawning and reproductive isolating mechanisms.
♦ Geographic patterns and genetic control.
♦ Changes in species' distributions.
♦ Local abundance and effective population size.
♦ Larval dispersal and 'rafting'.

REPRODUCTION

There is no better testimonial to the 'inventiveness' of corals than in the variety of mechanisms they use to reproduce. Given the geological longevity of coral species (p 146), the central role of sex in speciation, and the dependence on reproduction of most dispersion and genetic communication beyond the boundaries of a single reef, it would be surprising if this were not so. Nevertheless, the knowledge gained over the last decade has taken corals from the comfortable sphere of understanding traditionally associated with the vertebrates and plunged it into the sort of chaos usually reserved for plants. As in botany, complexity in coral reproduction breeds complexity in so much else, starting with genetics, then flowing on to systematics, taxonomy and biogeographic distribution.

The subject of coral reproduction has been exceptionally well reviewed by Fadlallah (1983), Harrison and Wallace (1990) and Richmond and Hunter (1990). New taxon-based or region-based studies are now published every few months, strengthening or qualifying established findings about the nature of sex in corals, providing variations on central themes and presenting new information about how reproductive output is coordinated. Most of this information is relevant to the biogeography of corals, or their speciation, or both.

The first casualties of our newfound understanding of coral reproduction were global statements like 'corals are viviparous', 'corals are hermaphroditic' and summary statements about what corals do what, and why. The second was cosmopolitan reproductive attributes of coral species. Interest has changed from global generalities to specific detail of taxa and region and, as with all subjects to do with evolutionary questions, a speculative level has separated from the operational one and will probably continue to exist until coral biologists, like botanists before them, have gained a better understanding of the issues and mechanisms involved.

Corals reproduce both asexually and sexually. Asexual reproduction can occur through fragmentation, polyp detachment from skeletons (Sammarco 1986) and asexual production of larvae (Stoddart 1983; Ayre and Resing 1986). The product of all except the last (which probably occurs in a few brooding species only) is geographically limited to the reef of origin and, along with colony formation and growth, their effect is to amplify sexual reproductive success, rather than facilitate dispersion over broad geographic scales. In sexual reproduction, gametogenesis takes place in temporary gonads embedded in mesenteries. It usually occurs annually, but may be seasonal, monthly or erratic. The product of the reproductive output is either the spawning of gametes for external fertilisation and development of planula larvae, or the brooding of planula larvae for release after internal fertilisation. Corals may be gonochoric or hermaphroditic. Planulae may be teleplanic or philopatric.

Corals, like plants, have all manner of variation around these central reproductive mechanisms, including variation within and between families, genera and species. Within species, reproductive mechanisms may vary geographically, ecologically, demographically and anatomically. They may also vary between zooxanthellate and non-zooxanthellate species and with the size of polyps and/or colonies. Such variation obviously plays a large role in the genetic composition and distribution of species and, at a conceptual level, contributes to our understanding of the nature of coral species and what mechanisms might be involved in their organisation.

About three-quarters of all zooxanthellate species spawn eggs and sperm rather than brood larvae (figure 19). Spawning is associated with higher fecundity, while brooding results in fewer, larger and better-developed larvae. Synchronised multi-species 'mass' spawning is highly conducive to hybridisation between species and the mixing of races within species (p 229).

About three-quarters of all zooxanthellate species are hermaphrodites and these include both spawners and brooders (figure 19). Hermaphroditic corals may have simultaneous or sequential gonad development and/or gamete release, giving varying potentials for self-fertilisation.

There is evidence that polyploidy[1], a genetic isolating mechanism common in plants and insects, can occur in corals (Kenyon 1994).

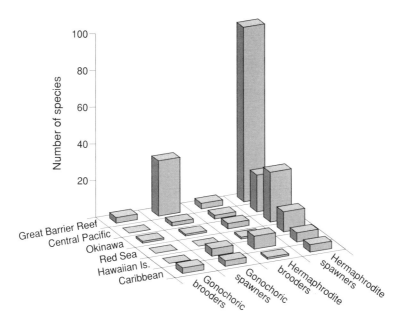

Figure 19 Geographic variation in sexual modes of reproduction (after R. Kinsey pers comm). Data are from Richmond and Hunter (1990) and Harrison and Wallace (1990).

SEXUAL MODES

Sexuality (hermaphrodite versus gonochoric) tends to be generally fixed within coral species and higher taxa, although there are geographic and up to family-level exceptions. The pattern of gonad arrangement is also generally consistent within families. There is no all-encompassing systematic pattern in mode of reproduction (gamete release versus brooding). Thus sex and gonad arrangement are generally stable and predictable, while mode of reproduction is a relatively plastic life history trait (Harrison 1985).

Differences between the two principal modes of reproduction are not just simple alternatives for getting sperm to ova. Fertilisation in gamete-spawning species generally occurs at the ocean surface, where it is greatly facilitated by a two-dimensional (depthless) physical environment. In most reef situations, this leaves little time for the development and settlement of planulae before they are transported away from their reef of origin by surface water movement. This 'broadcasting' mode of reproduction is thus conducive to inter-reef dispersion and inter-reef genetic connectivity, is relatively effective at low abundances and is also probably conducive to hybridisation. Brooding of larvae, if sexual, requires self-fertilisation (available to simultaneous hermaphrodites), or the movement of sperm from one colony to another in an unstable three-dimensional environment. For this mode, it is reasonable to assume that fertilisation occurs only between nearby colonies, and that planulae, once released, have the option of prompt local settlement. The brooding mode of reproduction is thus, in concept (but not necessarily in practice, eg *Pocillopora damicornis*, p 82), conducive to local dominance, philopatric dispersion, self-fertilisation and limited inter-reef genetic connectivity.

Reproductive mode may therefore involve 'trade-offs', among which are philopatric dispersion and local abundance versus long-distance dispersion, self-fertilisation versus outcrossing and within-species fertilisation versus hybridisation. When transferred to taxonomy and biogeography (all else being equal), the trade-off may become one of clinal versus cosmopolitan species, and short-term local abundance versus long-term evolutionary survival. These issues are complex, especially as they are species-specific, are always qualified by individual geographic and ecological situations and   involve different energetic costs (Richmond 1987b). Harriott (1992) has proposed that brooding of larvae would be advantageous in isolated communities because they have a greater planktonic endurance and/or because they can settle more rapidly. Where dispersion is not a survival priority in evolutionary terms, planulae with a short planktonic existence may be the most efficient means of maintaining local species diversity; where it is a priority, there is great potential advantage in multiple reproductive options.

Multiple sexual reproductive mode options may be found within the same coral species in different geographic regions (p 82), in species that combine brooding with dispersion by rafting (p 87), where planulae have both rapid-settling and long-term dispersal modes (p 87), and in species that combine both modes. Harrison and Wallace (1990) review the last; the best documented example is *Pocillopora damicornis*, where individual colonies are known to brood larvae and spawn gametes (Ward 1992).

SYNCHRONY

Sexual reproduction in attached benthic organisms raises problems of fertilisation, especially timing of gamete release. Since the discovery of multi-specific, synchronous 'mass' spawning on the Great Barrier Reef (Harrison et al 1984), there have been wide-ranging studies of reproductive synchrony. In principle, annual mass spawning has separate distant controls, proximate controls, and final cues. The distant (annual) control may be temperature, day length and/or rate of temperature change, either increasing or decreasing. The proximate control is usually lunar. The final cue is usually the time of sunset (although some species in some regions spawn during the day).

Oliver et al (1988) and Harrison and Wallace (1990) provide a compilation of the array of variations and contradictions corals have on this general theme of synchrony. Synchrony may be closely linked to a specific controlling mechanism or may ultimately be a function of whatever local physical-environmental controls and cues work best. Cues may be biological as well as physical and synchrony by chemical messengers may not only involve corals, but a host of other invertebrates as well. So far nobody has transplanted corals from one geographic region to another (eg between the west and east coast of Australia) in order to explore the role of genetic legacy in synchronised spawning.

REPRODUCTIVE ISOLATING MECHANISMS

Brooding species can store unfertilised ova for perhaps weeks and thus require little synchrony for fertilisation; spawning species require synchrony within a time-frame of hours. This need of synchrony contrasts sharply with geographic variation in the time of mass spawning: for example, on the Great Barrier Reef, mass spawning occurs in October/November (Harrison et al 1984; Willis et al 1985; Babcock et al 1986) in Western Australia in March (Simpson, 1985, 1991) and in Japan in June/July (Heyward et al 1987; Hayashibara et al 1993). The discrepancy between the need for synchrony and the amount of geographic variation in the annual timing of mass spawning suggests the presence of short-term barriers to successful long-distance founder dispersal. There is no suggestion that the needs of synchrony create long-term genetic barriers; however, over time-frames of (perhaps) hundreds of years, needs for synchrony could help maintain geographic barriers within species and could be a common origin of reproductive barriers in co-occurring races. Perhaps, as Babcock et al (in press) suggest, close synchronisation of mass spawning at the high latitude Houtman Abrolhos Islands, western Australia (figure 56, p 174), is an inheritance from tropical ancestors. If so, barriers of synchrony are likely to be most common in equatorial regions where there is less synchrony of spawning periods (Oliver et al 1988; Richmond and Hunter 1990) and greater opportunity for asynchronous, genetically isolated populations to co-exist. The asynchronous planulation of populations of *Pocillopora damicornis* in Hawaii (Richmond and Jokiel 1984) is the only example of such reproductive isolation to date, probably reflecting the amount of study this particular species has so far received.

GEOGRAPHIC PATTERNS AND GENETIC CONTROL

The attenuation of coral species diversity with increasing latitude was long considered to be temperature-correlated through failure of reproduction, and high-latitude populations were believed to exist as non-reproductive, non-self-generating 'pseudo-populations' (eg Yamaguchi 1987). This is seldom, if ever, the case: at least 60 per cent of species studied at the Houtman Abrolhos Islands (the highest-latitude reef region of the Indian Ocean) are known to engage in mass spawning (Babcock et al in press). At Lord Howe Island, the southernmost coral reef in the world, Harriott (1992) and Harriott et al (in prep) have found that the density of recruitment is comparable with that of the Great Barrier Reef. Multi-species sexual reproduction also has been reported in Hawaii (Kenyon 1992) and in non-reef high latitudes in Japan (Yeemin et al 1990; Woesik in prep). In fact, no studies to date indicate the existence of any communities, or even populations, that are not sexually reproductive in the long term, whether in high latitudes, or otherwise.

Some regional differentiation has been found in reproductive mode: planulation is relatively common in high-latitude regions, for example Lord Howe Island, and appears to be a common mode for *Pocillopora damicornis* at Rottnest Island, southwest Australia (Stoddart 1983, 1984), and for *Stylophora pistillata* in the northern Red Sea (Rinkevich and Loya 1979). In these cases, the regionally favoured mode may be that which increases local abundance at the expense of dispersal success. However, spawning, not brooding, is the mode of *P. damicornis* and *P. verrucosa* in the far eastern Pacific (Glynn et al 1991).

Levels and mechanisms of intra-specific genetic control of reproduction are largely undetermined. Independence of genetic control is seen in plasticity of reproductive mode and physical-environmental regulation of spawning. *Pocillopora damicornis*, alone, has been sufficiently well studied to demonstrate major intra-specific geographic variation in reproduction (eg between Rottnest Island, Hawaii and the far eastern Pacific). These differences are accompanied by morphological distinctions at Rottnest Island (Veron 1986), but not at Hawaii or the far eastern Pacific (pers obs). *Porites lobata* is gonochoric over most of its range, but is both hermaphroditic and gonochoric in Costa Rica (Glynn et al 1994). Other cases are more equivocal: *Galaxea fascicularis* (Heyward et al 1987; Harrison 1988) and *Porites astreoides* (Chornesky and Peters 1987) have been found to have both hermaphroditic and female colonies, but in both cases there is an identification issue that questions the assumption that these are, in fact, true intra-specific variations.

SELF-FERTILISATION, OUT-CROSSING AND HYBRIDISATION

Barriers to self-fertilisation of hermaphroditic species may be morphological, chemical or behavioural, each of which may become ineffective with time (days to weeks). This breakdown may occur in some species and not others (Heyward and Babcock 1986; Richmond and Hunter 1990). Clearly, there can be selective advantage in barriers that do progressively break down and allow self-fertilisation to occur if out-crossing fails.

Naturally occurring hybrids and hybrid swarms are well-known in plants at both race and species level (p 226), and these have been exploited widely in horticulture. An increasing number of hybrids have also been induced artificially in corals, especially in *Acropora* (Willis et al 1994), but also in *Montipora* (Stobart et al 1994) and *Platygyra*, and between *Platygyra* and *Leptoria* (Miller 1994). These recent findings have led to the speculation that hybridisation may be common in Scleractinia.

Some hybridisation has been between very similar coral species or within a species complex and may, therefore, occur naturally and remain undetected. Results of artificial hybridisation could also be attributed to taxonomic or identification uncertainties or mistakes. Other crosses, for example, between different sub-generic groups of *Acropora* (Willis et al in prep) or between *Montipora digitata*, *M. spumosa* and *M. stellata* (Stobart et al 1994; Willis et al in prep) cannot be thus explained; these are not part of a species complex, and the crosses have not been observed to occur naturally. They may, however, occur naturally, if they are indistinguishable from one or other of the parent species. Some crosses have been maintained in aquaria for over two years and may be able to be tested for sterility when they become reproductively mature[2].

The study of hybridisation is an interesting field of endeavour that is only just beginning. Results to date indicate that fertilisation between coral species can even have a higher success rate than fertilisation within species (Willis et al in prep). In what is probably the first case of a specific genetic isolating mechanism in corals, Willis et al showed that two co-occurring 'sibling species' of *Montipora* (figure 8, p 32) will not artificially cross-fertilise, and Stobart and Benzie (1994) subsequently demonstrated electrophoretic differences between these sibling species.

Important issues these studies raise concern isolating barriers and, as such, are at the heart of evolutionary mechanisms and the maintenance of genetic and morphological diversity. They are not, however, at the heart of the taxonomy of corals any more than they are of plants, or of the enormous variety of animals of all phyla that can be hybridised in artificial situations.

The conclusion of this book (chapter 12), that evolution in corals is mostly reticulate, can fall on the simple question of whether or not species *can* hybridise. Whether or not they *do* at a particular place at present is another question. Whether they *have* in undefined space and time is another question again.

SPECIES AND POPULATION DYNAMICS

In contrast to our knowledge of coral species distributions, we only have a hazy understanding of *change* in species distributions, and less understanding still of change in populations. At species level, most recorded distribution changes on a biogeographic scale are those determined from the fossil record, as major changes take many decades to centuries. Minor changes have been observed directly, but most of these are a consequence of anthropogenic environmental disturbances.

At population level, changes are not observable and can only be recorded with a great concentration of effort on a particular species at a particular locality. The structure and dynamics of populations have mostly been inferred from deviations from Hardy-Weinberg predictions using electrophoresis (eg Benzie et al in press; Ayre and Duffy in press), findings that can be extrapolated (if so wished) to corals as a whole and to general concepts, but so far such studies have not been undertaken on biogeographic scales.

TEMPORAL CHANGE IN SPECIES DISTRIBUTIONS
THE LAST 6000 YEARS

Changes to coral species distributions since modern sea levels were attained (about 6000 years ago) are assumed to be small compared with those of the last glacial interval. On biogeographic scales, the process of long-term readjustment to major global perturbation is probably still in progress; on regional and local scales, there is the complex of relatively rapid changes described in the next chapter. Both scales of space and time interact continuously, undermining the attainment of long-term permanency in distribution patterns.

For all groups of organisms, there is a lack of information about population changes that take place over intervals of thousands of years compared with those that take place over a few decades. The former can only be deduced from fossil remains, and for corals this usually means finding sites that have emerged tectonically, or have been exposed by human excavation. Fortuitously, the world's northernmost coral community (at Tateyama, Japan, figure 57, p 176) was uplifted 5000–6000 years ago in a single tectonic event, exposing over seventy-two zooxanthellate coral species (p 145). The difference between these corals and those that were present at Tateyama earlier this century[3], indicates a rapid readjustment after a sea temperature rise, and return of the Kuroshio to its present position. Other mid to late Holocene reefs have been uplifted (notably the Huon Peninsula of Papua New Guinea; Chappell 1974) or excavated, but none reveal changes to coral species composition.

THIS CENTURY

The dearth of information gap about changes in species distributions spans the Holocene to the time of direct human recording. This started about a century ago (eg Davis's, 1982 historical reconstruction for the Florida Keys, and Saville-Kent's 1900 original observations on the Great Barrier Reef). Such accounts indicate major regional changes in the abundance of some species (eg of *Acropora palmata* of the Caribbean), but inevitably most are anecdotal and are not species specific.

Observations of 'itinerant' species, though patchy, provide some insight into long-term dynamics, especially the rate at which distribution changes *can* occur. It is always easier to demonstrate that a coral species no longer occurs where it was formerly recorded than to demonstrate that it now occurs where it was formerly believed absent. It is also much more reliable, and more biogeographically meaningful, if such observations are made in regions of low species diversity, especially at remote or high-latitude locations near the edge of distribution ranges. At Miyake-jima (100 km south of Tateyama), the once common *Cycloseris somervillei* (Tribble and Randall 1986) became locally extinct in a single winter owing to an ENSO-correlated chilling (J Moyer pers comm). Utinomi (1970) recorded forty-five species of coral along the Ehime coast of southwest Shikoku, Japan (figure 57, p 176), among which were *Acropora echinata* and three free-living fungiids, none of which is known north of the Ryukyu Islands today. *Acropora valida*, once found in the far eastern Pacific, may have become locally extinct there, also during an ENSO event (Prahl and Mejia 1985).

Evidence of species diversity increase comes almost entirely from Australia. Marsh (1994) has recorded the recent establishment of *Acropora yongei* at Rottnest Island. Successive studies of the species composition of Lord Howe Island suggest an increase of up to six species during the past decade. These observations indicate that latitudinal variations in distribution ranges are dynamic and can be short term, even during periods of general climatic stability.

LOCAL ABUNDANCE

For most considerations of abundance and demography, there can be no useful generalisations between different coral species, or even between different populations and colonies. Potts et al (1985) make the point with *Porites* at a single location on the Great Barrier Reef: the more common species are themselves dominated demographically and genetically by a few persistent genotypes; 5 per cent of colonies contain 52 per cent of the skeletal mass, 9 per cent of colonies contain 50 per cent of the living tissue, and the largest colony is more than 677 years old.

Abundance is an important, but missing, component of distribution data (p 199). Just single colonies of some coral species are sometimes found in very remote regions and appear to be solitary recent immigrants. They may represent only temporary extensions to the species' long-term distribution range, or a founder event in the process of happening.

There are numerous studies of local destruction of coral communities from acute natural and anthropogenic environmental disturbances. Very few of these changes probably have biogeographic significance, the single major exceptions being the impact of the coral-eating starfish *Acanthaster planci* and coral-eating gastropods, *Drupella* spp Computer models of the recovery of *Porites* populations from *Acanthaster* attack (Done 1988) may indicate minimum recovery times of more than fifty years in the absence of subsequent attacks and depending on the initial abundance and damage.

ABUNDANCE OF SPECIES

Abundance *estimates* recorded in this book (appendix) are, of necessity, very general indications only. Except for very rare or very common coral species, abundance is extremely difficult to record. It varies greatly from country to country or from one geographic region to another. It may also change enormously between one habitat and the next, and even then it may only be an apparent abundance because large colonies of conspicuous species will appear to be more abundant than small or cryptic, yet more numerous, colonies of other species. For present purposes, a *rare* species is one seldom encountered, even if searched for; an *uncommon* species is one that is sometimes found in likely biotopes; a *common* species is one usually found in a likely biotope. This crude classification assumes that the species is recognisable *in situ*. Regional abundance estimates are believed to be insufficiently well established to be used in numerical analyses (p 8).

DISPERSION

The subject of long-distance dispersion has always begged the question 'did they evolve there, or did they disperse to there?' The first alternative, implicitly supported by most theoretical and vicariance biogeographers, requires no long-distance dispersal as species (at successive points back in time to the time of their evolutionary divergence) are considered to inherit antecedent distributions. Depending on particular versions of vicariance theory, corals are mostly dispersed in modest steps involving, perhaps, islands appearing where none at present exist. However, island 'stepping-stones' are largely excluded as a basis for dispersion across the flat sea floor of the eastern Pacific, creating the focus of interest in the literature on the origins of the far eastern Pacific fauna (p 58). The second alternative (long distance dispersion) requires either the long-distance dispersal of adult colonies by 'rafting' on floating objects, or the dispersal of teleplanic planulae. Vicariance theorists have emphasised rafting as an important mode of dispersion, presumably because frequent long-distance dispersion as a general property of corals is anathema to classical vicariance speciation.

The distribution patterns of most coral species (chapters 9 to 11) makes the second alternative mandatory: most long-distance dispersal (some remote regions excepted) has occurred in non-evolutionary timeframes. The issue arises as to whether this occurs by larval dispersal or by rafting. Both mechanisms are dependent on surface circulation vectors; the practical difference is time, given the assumption that rafting is effectively not time limited (other than by coral growth causing sinking of its raft).

Long-distance larval dispersion and rafting This subject today is at about the same place that our knowledge of reproduction was a decade ago. Some interesting observations have been made, and these have led to a wide range of speculations.

Pocillopora damicornis is again the only well-studied coral species. Larvae are large (1 mm diameter) and zooxanthellate. They usually settle within two days of release, but can remain planktonic under laboratory conditions for one hundred days and still be competent to settle and metamorphose (Richmond 1987a). They show an extraordinary degree of plasticity in their planktonic life, including the ability to metamorphose into askeletal planktonic polyps capable of tentacular feeding, and also the ability to undergo reverse metamorphosis — a return to the larval state after settlement and skeletal formation has commenced (Richmond 1985).

Several early authors, and more recently Jackson (1986), Rosen (1988b) and Jokiel (1990a), have claimed that rafting is the most common means of long-distance dispersal. To date, examination of very large quantities of pumice, wood and other flotsam has revealed nine genera that raft: *Pocillopora, Stylophora, Seriatopora, Acropora, Montipora, Porites, Goniopora, Favia* and *Cyphastrea* (Jokiel 1990a,b). Except for *Goniopora*, this list is exactly replicated by the author's observations in Australia and Japan.

The case for rafting as the primary long-distance dispersion mechanism put forward by Jokiel (1990a,b) is based on the competency of fish and invertebrate larvae of three to five weeks; this duration being inadequate for transport of more than a few hundred kilometres; an assessment of species distribution data; the failure to predict species distributions from larval lifespans; conclusions from fish biogeography; and observations on the genetic composition of *Pocillopora damicornis* populations.

The case for teleplanic larval dispersion is based on observations on species distributions and diversity (p 160); five of the nine genera listed above being absent in the far eastern Pacific; free-living (and therefore non-raftable) *Diaseris* and *Fungia* being present in the far eastern Pacific; free-living species and others that are highly unlikely to raft (eg those with large polyps) having the same sorts of distributions as species that do raft; our lack of information about coral larval biology (p 78) and rejection of conclusions borrowed from studies of fish and other taxa; and the same observations on larval competency as above.

The author's conclusion is that rafting may be a major component in the dispersion of some coral species, notably of *Pocillopora* and perhaps other Pocilloporidae, and may be a vehicle for transport to very remote places, including the far eastern Pacific. Otherwise, it is a relatively minor method of long-distance dispersion.

Stepping-stones   Stepping-stones appear to be of great importance between adjacent regions, but importance appears to decrease when very great distances are involved. Thus, in the tropical Indo–West Pacific, the composition of coral species of isolated islands is clearly dependent on that of upstream sources of propagules (stepping-stones) (chapter 10). In the far eastern Pacific and remote high latitude locations (eg Hawaii), this may not hold true, because there is no clear sequential 'dropping-out' of species. Species compositions in such locations appear to be more the outcome of rare long-distance dispersion events and species accumulation rather than of frequent propagule export from stepping-stones (figure 51, p 165).

Post-dispersion reproduction   Whether dispersion occurs in the larval phase or by rafting, the post-destination reproductive success of established colonies would be mitigated for spawning species, which require physical-environmental cues for synchronisation (p 81). The chances of annually spawning species being transported, then established in sufficient abundance for environmentally cued spawning to be sufficiently synchronised to operate in a foreign environment seem small, unless the initial immigration was followed by substantial asexual reproduction, or unless it occurred by the rafting of colonies with large numbers of already brooded larvae.

6

THE
PHYSICAL
ENVIRONMENT

The physical environment limits the distribution ranges and diversity of coral species, controls most morphological variation in coral species, and (it will be concluded in chapter 13) is more important than biological mechanisms in driving evolution. As with evolutionary concepts, we are dealing with issues that interact differently in different scales of space and time. This chapter deals with modern physical environments and how these correlate with modern distributions. Chapter 7 deals with palaeoenvironments and how these correlate with the fossil record and evolutionary processes. The separation of the two is not simply one of scale: modern physical environments, like modern distributions, can be studied directly; past physical environments, and their evolutionary correlations, can only be inferred.

Physical-environmental parameters that generate biogeographic patterns, tend to be either *latitude-correlated* (including temperature, light, reef/non-reef habitats and boundary currents), or *non-latitude-correlated* (including non-boundary sea surface circulation, substrate availability, water quality and nutrients, regional ecology and regional dispersion barriers). Smith and Buddemeier (1992) provide a recent review of the affects of environmental change on reef ecosystems.

THE MAIN TOPICS

◆ Latitudinal control of reef distribution.
◆ Low temperature control of coral latitudinal distributions.
◆ Biogeographic consequences of high temperature.
◆ Light controlling depth limitations and its role in symbiosis.

- The biogeographic role of reef physical environments.
- Surface circulation controlling latitudinal dispersion.
- Variations in surface circulation: ENSO events and palaeoclimatic cycles.
- Regional physical-environmental parameters: substrate, sedimentary regimes and tidal regimes.

LATITUDE-CORRELATED ENVIRONMENTAL PARAMETERS

Why do so many coral species coexist in the tropics? A starting point in this complex subject is 'Rapoport's rule', the observation that the mean latitudinal range of major taxa increases with increasing latitude (Stevens 1989). This is another way of stating that increasing latitude is correlated with increasing environmental tolerance. Within the ocean, according to Clarke (1992), this correlation is less clear, although it appears true for shallow-water benthos in general and certainly appears true for corals. Explanations that have been offered for corals are wide-ranging, but all point (rightly or wrongly) to physical-environmental parameters acting directly on coral species' tolerances, or acting indirectly on rates of evolution.

The species diversity of corals *is* correlated with latitude (p 171), so are many physical-environmental parameters, and so is the abundance of coral reefs (p 168). The questions are what aspects of the physical environment are limiting, and are these limitations direct (physiological) or indirect (ecological)?

TEMPERATURE

> There seems no good reason for abandoning the view that it is temperature which controls the horizontal distribution of reef-building corals, merely qualifying this by reference to the probable effects of currents in the distribution of the planulae larvae. (Yonge 1940)

Although this statement has recently been challenged (hypotheses 16 and 17, p 56), it remains one of the most enduring statements in the literature of coral biology: after fifty years, 'horizontal' should now be replaced with 'latitudinal', and 'probable' can be removed, at least for the world's four principal latitudinal sequences (figure 75, p 235).

CORAL AND REEF LATITUDINAL DISTRIBUTIONS

The long-held view, which cannot now be supported (p 50), was that the distribution of reefs is limited by the distribution of corals, and that this was due to physiological processes, in general, and food capturing and reproduction, in particular. The logical alternative is that temperature limits both reefs and corals through interactive ecological processes, where the energy (and thus light- and symbiosis-correlated) demands of reef construction become progressively less competitive against macroalgae-dominated ecosystems.

The time interval of effective temperature minima is very different for corals than it is for reefs. Effective minima for corals are those that induce significant mortality in a natural environment. Given the cyclical and patchy nature of sea surface chilling in shallow onshore habitats, these minima are ones sustained, or repeated, over periods of a few days (as opposed to hours or weeks). Effective minima for the existence of reefs are those below which reef construction cannot be sustained. This will also relate to natural environmental cycles, but in time-frames of hundreds to thousands of years.

MINIMUM TEMPERATURES FOR REEFS

Ever since the pioneering work of Dana (1843), Vaughan (1918, 1919), Davis (1928) and Yonge (1940), and through many subsequent reviews (Vaughan and Wells 1943; Wells 1954a, 1957; Stoddart 1969; Rosen 1971a, 1984), 18°C, sustained over protracted periods of time, has survived as the identified minimum sea surface temperature to which functional reefs are normally exposed. This correlation was recently re-examined in the light of detailed temperature records of Japan by Veron and Minchin (1992) and was again found to hold true. Lower temperatures have often been recorded in reef environments (below), but in most cases these observations have been very short term, have involved only partial mortality (where part of the colony dies) or have involved reefs that are geological relics or consist primarily of unconsolidated rubble. Short-term temperature fluctuations of the 'reefs' of the Gulf (Persian or Arabian) (Shinn 1976; Downing 1985; Coles 1988; Coles and Fadlallah 1991) are the currently known global minima in this context.

LOW TEMPERATURE TOLERANCE AND MORTALITY

All corals will die if exposed to abnormally low temperatures, just as most other tropical organisms will. This is not something that requires any explanation specific to corals.

Very few zooxanthellate corals are known to tolerate temperatures below 11°C under natural conditions. *Oulastrea crispata* at Noto Peninsula, Sea of Japan, can tolerate approximately 0°C (for an unknown period of time) and apparently retains zooxanthellae as it does (Yajima et al 1986; M Nishihira pers comm). *Siderastrea radians* (locality unrecorded) has been recorded to tolerate 4.5°C (Vaughan and Wells 1943).

The most informative studies of low temperature tolerance of corals were carried out at Tanabe Bay and nearby Kushimoto of central southern Japan, where a suite of *Acropora*, *Porites*, *Echinophyllia*, *Hydnophora* and *Leptastrea* species all suffered varying degrees of mortality at temperatures as low as 9.4°C (Fukuda 1984a,b; Misaki 1984, 1985). These studies, when combined with the author's (Veron 1992b) regional distribution data, show that there are regional differences in the minimum temperature tolerance of some coral species, and that species richness within marginal regions is at least partly a function of temperature tolerance.

Other records of low temperature survival are of a similar order: 10.6°C for *Solenastrea hyades* in North Carolina (Macintyre and Pilkey 1969), 11.5°C for *Acropora* sp, *Porites* spp and *Platygyra daedalea* in the Gulf (Coles and Fadlallah 1991), less than 13°C for *Montipora* spp in the Yaeyama Islands (Japan) (Nomura 1986), and similar temperatures for many Florida species (Davis 1982; Porter et al 1982; Walker et al 1982; Roberts et al 1982; Burns 1985) as well as *Plesiastrea versipora* and two *Coscinaraea* species of southern Victoria, Australia. Records of low temperature tolerance are, therefore, broadly spread, both taxonomically and geographically. They usually involve partial mortality and subsequent recovery. Where relevant observations have been made, most authors have concluded that mortality is greatest in shallow water, where windchill is greatest.

The precise relationship between mean sea surface minima and mortality is, again, best seen in the studies at Kushimoto, which has a mean monthly minimum of 15.3°C. *Acropora hyacinthus*, the dominant species of the region, suffered no mortality when, in 1980, the temperature fell to 13.7°C; some bleaching and mortality occurred in the following month at 13.4°C; widespread mortality occurred in 1984 when the temperature reached 13.2°C (Misaki 1984, 1985). In this study, the temperature critical to survival was thus only 2°C below the mean monthly minimum for the location and, at that temperature, critical effects were within a 0.5°C band (perhaps influenced by rate of decrease). Similar mortality was found with other species, and the study also showed that the extent of mortality varied with the duration of exposure.

Mortality is not the only temperature-correlated factor limiting coral distribution. Competition with macroalgae (Johannes et al 1983; Crossland 1984; Coles 1988), various metabolic activities (notably calcification) (Clausen 1971; Clausen and Roth 1975; Smith 1981; Crossland 1981, 1984) as well as aspects of reproduction, are also temperature-correlated and potentially limiting. Of these, competition with macroalgae appears to be of overwhelming importance in all high-latitude locations of the central Indo-Pacific and probably elsewhere.

TEMPERATURE AND LATITUDINAL ATTENUATION

Japan is the only country in the central Indo-Pacific where there is an almost continuous island arc, remote from continental influence, extending from a region of high species diversity to the high-latitude limits of coral distribution (p 177). Several excellent sets of sea surface temperature records are available, covering periods of several years to several decades. These cover three spatial scales: within-reef, within-locality and in the open ocean (on the scales of Marsden Squares, 1° latitude × 1° longitude). Comparisons between these sets of records (Veron and Minchin 1992) is a complex undertaking. Importantly for all temperature studies on reefs, these data conclusively demonstrate that within-reef records are only weakly correlated with within-locality records, and that both of these are much less well correlated with oceanographic data, especially the statistical concentrations of oceanographic data (daily, weekly or monthly means, maxima, minima etc). For example, there is frequently a discrepancy of 5°C between the minimum temperature recorded within a reef and the nearest concurrent Marsden square oceanographic record, the difference being a combination of real difference and sampling artefact. This discrepancy is more than sufficient to falsify the many published correlations between biogeographic distribution and sea surface temperature. This applies to intra-regional correlations, let alone inter-regional and global ones.

Figure 20 illustrates the general relationship between coral species richness and different measures of temperature data over the latitudinal range of Japanese coral communities. No comparable oceanographic data exist for the world's three other sequences of localities where there is major latitudinal attenuation of coral species diversity (p 171). These data demonstrate that, within Japan, coral species richness can be predicted almost solely from sea surface temperature (figure 21). This result, it should be emphasised, cannot be directly applied to other geographic regions: the Kuroshio is the strongest of the world's four major latitudinally dispersing currents; the island and peninsula stepping-stones are closer together, more regular and (at higher latitudes) probably more suitable for coral growth, than is found elsewhere. Of general relevance, it is clear that whole coral communities, not just particular species or species with particular reproductive or ecological characteristics, have an unrestricted capacity for dispersal where the time interval needed to do so is less than thirty days. In the case of central Indo-Pacific latitudinal gradients, this correlation shows that it is temperature and the correlates of temperature, not dispersal capability, that is limiting.

In summary, records from Japan show that 22.5 per cent of the total species diversity can tolerate a minimum of 10.4°C; 27 per cent can tolerate 13.2°C; and 48 per cent can tolerate 14.1°C. These statistics omit consideration of both frequency and duration of exposure; however, they conservatively indicate that approximately half of all coral species are able to tolerate 14°C, that is, half of all species tolerate 4°C below the minimum recorded in reef environments in Japan. This is strong evidence indicating that reef development is not limited by the low temperature tolerance of corals.

EFFECTS ON GROWTH AND MORPHOLOGY

Next to mortality, the most visual effect of low temperature is on morphology. Several species of *Acropora* at their latitudinal extremes in Japan form tiers of irregularly fused plates over-growing lower tiers of dead plates. This distinctive morphology appears to be due to temperature-induced cycles of partial mortality and recovery. Fukuda (1984b) notes that this recovery can be rapid, rather more so than recovery from *Acanthaster* damage. At Tateyama, where the world's northern-most coral outcrop occurs (figure 57, p 176), all species except *Alveopora japonica* have unusual, flattened to encrusting growth forms and some (notably *Blastomussa wellsi*) have distinctive corallite structures. Similarly, colonies from high-latitude localities in Australia may have distinctive morphologies and corallite structures, but do not appear stunted or otherwise environmentally stressed. If studied without reference to morphologies found in contiguous regions of lower latitude, these morphologies would indicate the presence of many high-latitude species and (as similar morphologies are frequently found in both hemispheres) amphitropical distributions, suggesting displacement from an ancestral tropical distribution (p 38). However, what these morphologies *do* show is the presence of geographic subspecies (p 229) albeit ones that are primarily environmentally determined rather than genetically distinct.

Very few data are available on the growth rates of corals growing at low temperature extremes. Mean growth rates of an unknown corymbose species of *Acropora,* measured both at Kushimoto and at subtropical Kuroshima, are recorded as 19 and 90 mm per year (respectively, Fukuda 1981 and Irie 1980). Crossland (1981, 1984) concluded that low temperatures are a primary determinant, and light a secondary determinant, of growth and survival of *Acropora formosa* at the Houtman Abrolhos Islands, western Australia (figure 56, p 174). Significantly, the corals of these islands appear to grow in profusion, and have a high level of carbonate production relative to tropical reefs (Smith 1981).

Correlations between high-latitude physical environments and other coral species attributes are discussed in chapter 11.

BIOGEOGRAPHIC CONSEQUENCES OF HIGH TEMPERATURE

Heat stress, unlike cold stress, is not a common dispersion-limiting phenomenon and does not affect latitudinal boundaries. Its principal effect (as with other forms of stress, but not low temperature stress) is to cause breakdown of coral/zooxanthellae symbiosis, leading to loss of zooxanthellae and 'bleaching'. Although heat stress can occur on high-latitude reefs (eg Hawaii, Jokiel and Coles 1990; Bermuda, Cook et al 1990), it tends to occur in small areas in equatorial regions, where it is generally correlated with reduced tidal flushing or abnormally low tides. On a biogeographic scale, heat stress is usually correlated with inter-annual climatic fluctuations, of which the El Niño-Southern Oscillation (ENSO) is the best known and most important (reviewed by Glynn, 1990; Guzmán and Cortés, 1992).

BLEACHING: AN ADAPTIVE MECHANISM?

Bleaching is generally observed in association with physical-environmental stress and is thus considered detrimental to zooxanthellate organisms. Buddemeier and Fautin (1993) suggest that bleaching might be adaptive rather than pathological, providing an opportunity for recombining hosts with alternative algal types to form symbioses that are better adapted to altered circumstances. The small difference between temperatures that occur at regular intervals with no effect and those that induce bleaching, the existence of bleaching independently of temperature over wide depth ranges (eg Fisk and Done 1977; Jokiel and Coles 1990; Gates 1990) and the discovery that one taxon of zooxanthellae can replace another (Rowan and Knowlton in prep), all support this view. In this context, temperature may only be a convenient physical-environmental cue for bleaching, just as it regulates reproductive cycles.

The main arguments against adaptive bleaching are that it is much less common in some geographic areas than others, that it can cause long-term major disruptions to populations (below), that it can disrupt the reproductive cycle (Szmant and Gassman 1990) and that it retards growth (Fritt et al 1993). If it is an adaptive mechanism, the price is a high one.

In the western Pacific ENSO warming of 2–3°C above normal for several months has probably been responsible for widespread bleaching and mortality observed in Indonesia in 1983 (Brown and Suharsono 1990) and elsewhere in 1986–87 and 1989–90 (Williams and Bunkley-Williams 1990; Glynn et al 1994). However, ENSO-related temperature elevations probably have a significant biogeographic impact only in the far eastern Pacific where they have, directly or indirectly through bleaching, caused mass mortality throughout the region (Lessios et al 1983; Glynn 1984) and may even have caused the extinction of *Acropora* there (p 85). Because the effects of ENSO fluctuations are synergistic with extremely low levels of immigration into the far eastern Pacific, and because these fluctuations have probably impacted the region since the closure of the Central American Isthmus (p 143), they could be as important as the Eastern Pacific Barrier (p 99) in maintaining the province's isolation.

The limiting values of high temperatures of ecological importance are sustained maxima of 30°–34°C (Jokiel and Coles 1977; Glynn 1984; Hoegh-Guldberg and Smith 1989). These maxima vary geographically and with time of exposure, with tolerance being more than 2°C higher in tropical localities than in temperate ones (Coles et al 1976; Coles and Jokiel 1977).

The optimum temperature for growth probably varies greatly with the nutrient environment and has synergistic effects with salinity and light (Coles and Jokiel 1978); an optimum of 26°C has been recorded for Hawaiian corals (Jokiel and Coles 1977).

SEASONALITY

Valentine's (1984a) hypothesis that progressively more seasonal regimes lead to increasing development of strategies for resisting population decline are more applicable to latitudinal gradients where there is latitudinal species replacement, not species attenuation, as is the case with corals. Nevertheless, seasonality becomes an issue with corals at high latitudes, potentially (Crossland 1981, 1984; Dodge and Lang 1983), but not necessarily (Smith 1981), affecting rates of calcification and growth. Much more significantly, seasonality greatly affects competition with macroalgae (p 92). Calcification and primary productivity are both much less seasonal within the tropics (Kinsey 1977; Chalker et al 1984), but competition with macroalgae may remain an important limiting factor to the growth of inshore reefs.

LIGHT

The ability of corals to build reefs using the energy of the sun, is the key to the existence of all modern coral reefs, and perhaps all reefs in all geological time (p 113). Light, not temperature, is by far the most ecologically limiting of all physical-environmental parameters, yet there is apparently no evidence that it causes *horizontal* biogeographic boundaries. Interest in light, from biogeographic and evolutionary points of view, is in the evolution of algal symbiosis and its role in reef-building over evolutionary time, and in the role of light, in synergy with the sedimentary environment, in creating ecological 'sinks' of diversity.

SYMBIOSIS

Symbiosis has long been, and remains, a subject fraught with unknowns (Trench, 1992). The proposed physiological advantages of symbiosis have been many: the removal of metabolic wastes (Yonge 1931; Goreau 1961); the enhancement of calcification (Goreau 1961); direct nutrient contribution (eg Franzisket 1969; Porter 1974; Porter et al 1984); and concentration and recycling of limited nutrients, including nitrogen and phosphate (eg Muscatine and Porter 1977). Symbiosis probably does all of these things (Chalker et al 1988; Barnes and Chalker 1990; Muscatine 1990; Miller and Veron 1990) and probably allows corals to exist phototrophically over protracted periods of time, if not indefinitely, given the addition of minor (otherwise ecologically limiting) nutrients.

The evolutionary cost of symbiosis, however, is great. On ecological scales, light is responsible for the depth limitations of almost all corals on almost all reefs and is the principal environmental parameter that controls intra-specific polymorphisms which, in turn, constrains species diversity (p 229). On a biogeographic scale, dependence on symbiosis allows corals to be readily overgrown by macroalgae, the factor likely to exclude corals, if not reefs, from high-latitude physical environments (p 90). On a geological scale, the failure of symbiosis through lack of light may have played a central role in mass extinctions (p 118).

Apparent disadvantages of light dependence arise because of the addictive nature of symbiosis. As far as is known, only a few coral species are able to exist facultatively (aposymbiotically), and of these only *Astrangia danae* (Jacques et al 1983), and perhaps *Madracis* spp, appears to do so as a regularly available option. Why facultative symbiosis is almost unknown among zooxanthellate species is perhaps one of the most critical questions of coral evolution. On one level, the answer is likely to be a simple link in a physiological chain; on another level (why that link is so critical), the answer is likely to be deeply embedded in the genetic basis of multi-species synchronous evolution, and perhaps paralleling the evolution of mitochondria from protozoa.

ZOOXANTHELLAE IN CORALS

Calcification powered by photosynthesis is probably at least as old as the stromatolites of the Precambrian. It has allowed reef-like structures to be built by plants and has allowed reefs to be built by a great variety of animals that behave as plants, because of symbiosis. Zooxanthellae symbiosis is so effective that algae not only meet most of the nutrient requirements of their coral hosts, but allow their hosts to act as net primary producers. Thus modern reefs have a gross productivity comparable to productive terrestrial ecosystems. Little of the productivity is exported: it is recycled within the ecosystem (Kinsey, 1985).

The subject of symbiosis attracts a wide range of theories and issues. The term *symbiosis* can be defined as 'the living together of differently named organisms'. Endosymbionts (living within the host animal) are generally referred to as *zooxanthellae*, *'zoochlorellae'* and *cyanobacteria*, names that originally had a taxonomic basis, but now describe a complex of taxa that are so wide that they overlap plant/animal boundaries. Zooxanthellae alone are a composite of many different families and perhaps classes (Blank and Trench 1986; Rowan and Powers 1992). The formal and informal links of this term with algae taxonomy, and the lack of applicability of it to fossil corals (that may have had other symbionts), excludes it from being readily acceptable as a *functional* descriptor. Schumacher and Zibrowius (1985) have compiled an inventory of the uses and misuses of the terms 'zooxanthellate', 'hermatypic' and 'symbiotic'. *Zooxanthellate* effectively now means *with endosymbiotic primary producers*.

DEPTH AND LATITUDE

Do latitudinal gradients in light availability create latitudinal boundaries for corals? Probably not, but the evidence is circumstantial and is likely to vary with the species of coral as well as its zooxanthella symbiont. The most important light changes with latitude are seasonal variation in day length, and a range of effects of the angle of incidence of sunlight (reviewed,

Campbell and Aarup 1989), most of which are depth-dependent.

Possible effects of light availability on the depth and latitudinal distribution of corals are likely to be minor compared to those of water clarity, temperature, seasonal growth of macroalgae, specific light requirements of zooxanthellae and mechanisms of photoadaptation. At Izu, Japan (35°N latitude, figure 57, p 176), Lord Howe Island ($31\frac{1}{2}$°S latitude, figure 55, p 174) and the Houtman Abrolhos Islands ($28\frac{1}{2}$° S latitude, figure 56, p 174), corals regularly occur in clear waters at depths of 30 m and sometimes more than 40 m where a suitable horizontal substrate is available (pers obs). Light availability as such does not appear to be limiting at these depths; what is limiting is the capacity of the coral to compete with macroalgae, especially *Sargassum*.

Recent findings of Rowan and Knowlton (in prep.) that 'species' of zooxanthellae change with depth in *Montastrea annularis* may be significant in a biogeographic context: depth and latitude could both be restricted by distribution patterns of zooxanthellae.

REEF PHYSICAL ENVIRONMENTS

The essential biogeographic points about reef environments, as opposed to non-reef environments, are as follows:

♦ Over geological time, the distribution of reefs is poorly correlated with the concurrent distribution of corals (p 169), because of the widespread existence of non-reef coral communities, and because these communities can undergo rapid changes in diversity in geological time (p 85).

♦ Stanley (1981), Cowen (1988) and others have hypothesised that the relatively slow recovery of reef ecosystems from major extinction events may reflect the time necessary to re-establish symbiosis (p 118). Reefs, however, do not cease to exist if their biota are destroyed; unlike most other ecosystems, they have a very long-term geological existence that is independent of relatively short-term intervals of biological existence.

♦ The regional concentration of reefs and the diversity of extant corals are only weakly correlated (figure 22, p 102). There are many historical explanations for this, including explanations of the species diversity patterns of extant corals (p 50).

♦ The most abrupt latitudinal decrease in species diversity of central Indo-Pacific corals is correlated with the latitudinal limits of reefs (p 184). This is not temperature controlled: within high-diversity regions, local habitat diversity and local species diversity have a source/sink relationship (p 75).

♦ There are major ecological differences between reefs that break the force of wave action and those that do not. Wave-resistant reefs typically have complex communities with high niche and species diversities; non-wave-resistant reefs (ie those that are not high or large enough to disperse wave energy) tend to be ecologically homogeneous and non-reefal in an ecological sense (Done, 1983).

Reef environments are important to the maintenance of coral species diversity, yet are not essential to the continuing existence of most coral species. The extensive platform reefs and atolls that we see today are not the norm for corals, but rather are a product of an unusually long period of unchanging sea level (p 123). Reefs are unlikely to have played an essential role either in the distribution of coral species or in their evolution.

SURFACE CIRCULATION

Ocean circulation is linked with atmospheric circulation: differences in heating between equatorial and high latitudes, in conjunction with the rotation of the earth, lead to trade winds, which drive the major anti-cyclonic subtropical gyres in each of the earth's major ocean basins (figure 23). These create the great poleward boundary currents that are the principal vehicles of transport of all tropical marine life to temperate regions. These currents meander and break up into large gyres which reach adjacent regions and may, in turn, create eddies that have a much wider range of penetration (reviewed by Mann and Lazier 1991).

The most important mechanism governing the latitudinal distribution of corals in the central Indo-Pacific is the existence of 'connectivity ratchets' along each of the three continental margins of eastern Australia, western Australia and eastern Asia. In each of these three regions, pelagic marine life is essentially committed to one-way (poleward) journeys. Upstream sources continually supply propagules to downstream locations where they survive, or not, according to ecological and environmental factors, of which temperature is the most important (p 92). Given adequate continuity of 'stepping-stones' along the way, the speed (hence the temperature at each point of latitude) of the dispersing boundary current largely determines the latitudinal extent of species' distribution. Thus, the fast Kuroshio of Japan extends coral distributions to higher latitudes than does either the Eastern Australian Current or the Leeuwin Current of western Australia. The role of boundary currents in controlling latitudinal dispersion is taken up in chapter 10.

NON-LATITUDE-CORRELATED
ENVIRONMENTAL PARAMETERS

The shapes and positions of the continents are the most enduring of all controls of coral distributions. Modern continental configurations have not changed greatly since the Miocene closure of the Tethys Sea, and the Eastern Pacific Barrier has probably been in place (to varying extents,

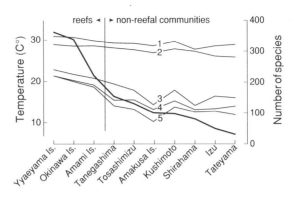

Figure 20 The relationship between different measures of sea surface temperatures of Japan, and how they relate to species diversity. 1 = average maxima of warmest month; 2 = average mean of warmest month; 3 = average mean of coldest month; 4 = average minima of coldest month; 5 = mean monthly minima. Temperature data are from Veron and Minchin (1992), who give primary data sources and several qualifications that apply to them. Localities are as in figure 57 (p 176).

depending on the presence of displaced terrains, p 112) throughout the entire existence of the Scleractinia. On this scale, the basic biogeographic template has always existed; it is the details that have changed. These changes have been brought about through tectono-eustacy and glacio-eustacy and the many manifestations of climatic change.

SURFACE CIRCULATION

As recently pointed out by Jokiel and Martinelli (1992), Pacific surface circulation patterns will inevitably concentrate coral species, irrespective of origin, in the western equatorial region (the Indo-Pacific centre of marine diversity). There is thus a priori support for any proposition that coral species that originate peripherally (in the central and southeastern Pacific) are eventually concentrated in the western Pacific (p 164 and 234). This proposition does, however, require different remote places to have different species complements. As this is not the case (most species being widespread and many occurring in several remote localities, p 189), dispersion in the Pacific must be occurring, or have occurred at some time, in 'outward' (easterly) directions as well as 'inward' (westerly) directions.

Principal circulation patterns (figure 23), and gyres developed from them, are perhaps adequate vehicles of transport in both easterly and westerly directions across most of the Indian Ocean and tropical Pacific, given larval endurances in the order of weeks to months (p 80). However, that duration makes transport to some remote regions (notably the far eastern Pacific, the Hawaiian Archipelago, the Line Islands, and perhaps the Mascarene Islands) improbable, and to others (the southeast Pacific, including the Tuamotu and Austral Islands, Pitcairn Islands and Easter

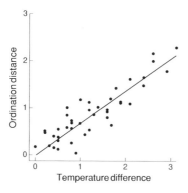

Figure 21 The relationship between pairs of localities (those in figure 57, p 176) in uni-dimensional ordination (an integrated measure of dissimilarity in species composition) plotted against temperature differences (average mean of warmest month). The fitted regression line shows the close relationship between temperature and species dissimilarity (after Veron and Minchin, 1992).

Island) seemingly impossible. To explain the occurrence of corals in these regions, it becomes necessary to postulate that island stepping stones formerly existed where there is now open ocean (Rosen 1988b); or that islands formerly occupied more equatorial positions than they now do (the 'island integration theory' (p 58)); or that there have been major perturbations in surface circulation patterns. The time-frame under consideration, set by moderate to low levels of endemism (p 191), is less than 20 million years (p 146). The first two options, under these circumstances, are either improbable or impossible according to what is known (or might conceivably be imagined) of tectonic movements in each separate region. The third option, which does not demand tectonic movements of islands at strategic places and intervals, begs close scrutiny.

Whether or not intervals of extreme westerly winds could create surface currents adequate for dispersal to the southeast Pacific is unknown. ENSO events might create such conditions, but available satellite and bathythermograph observations of the southern Pacific (eg Roemmich and Cornuelle 1990) indicate that they are unlikely to do so. Even extreme events reported from historical barometric data do not always indicate wider-than-normal ENSO influences (Allan et al 1991). General circulation models have, as yet, limited capacity to simulate ENSO events (McCreary and Anderson 1991), but these appear equally unlikely to indicate suitable vectors. The same conclusion reasonably applies to eddy-resolving supercomputer models (Semter and Chervin 1992).

Alternatively, present coral distributions may have little to do with modern (or Holocene) climatic conditions and surface circulation patterns. Major changes in surface circulation are likely to be correlated with

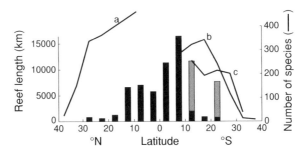

Figure 22 The relationship between length of reef front (the columns) and number of coral species in the Indo-Pacific centre of diversity. Length of reef front was obtained from satellite data (see text). Reefs of the Great Barrier Reef are shown in grey. Numbers of species are as indicated in Figs 57 (p 176) (a: Philippines to Japan), figure 55 (p 174) (b: Papua New Guinea and eastern Australia) and figure 56 (p 174) (c: western Australia).

global palaeoclimatic cycles (p 128). The essential point about these cycles is that it is difficult to imagine how they could not have had very major impacts on virtually all aspects of marine biology through fluctuations in sea levels and surface circulation vectors. Although they have been referred to by a sprinkling of authors (as 'orbital forcing' or 'Milankovitch' cycles) in many areas of biology for nearly a decade (reviewed, Arthur and Garrison 1986), the potential importance of palaeoclimatic cycles to evolutionary and biogeographic changes has only recently been highlighted (eg Bartlein and Prentice 1989; Bennett 1990). The subjects of orbital forcing and glacio-eustatic changes, are taken up in the next chapter, as they are palaeoclimatic events acting in evolutionary time scales.

REGIONAL ENVIRONMENTS

We have seen (p 71) that coral species richness is locally correlated with habitat or niche diversity, and this book concludes that rates of evolution are ultimately controlled by these 'sinks' acting in opposition to broad-scale 'sources' (p 234). On a biogeographic level, niche diversity is correlated with the presence/absence of reefs (p 75), the importance of which has been demonstrated in latitudinal attenuations, but is a largely unknown quantity elsewhere.

Substrate availability is the most biogeographically limiting of all physical-environmental parameters and, indeed, is the probable evolutionary 'reward' of reef-building. Most of the biogeographic pattern of the Indian Ocean is determined by the vast ocean depths in the south, and by the turbid waters of the Asian continental margin in the north (figure 75, p 235). Similarly, the attenuation of coral species diversity eastward across the Pacific is, and perhaps always has been, primarily a function of substrate availability.

Sedimentary regimes, which include various associations between substrate type, sedimentation, turbidity and light availability, affect coral distributions on all scales from local depth restrictions to broad-scale biogeography. Examples at one extreme are microenvironmental gradations in sedimentary regimes which can, in turbid environments, compress the habitable depth ranges of corals to bands of a few metres. On an ecological scale, macroenvironmental regimes are largely responsible for (for example) cross-shelf variation on the Great Barrier Reef, for delineating the southern limit of the Great Barrier Reef and for separating the faunas of the northern Great Barrier Reef and southern Papua New Guinea. On a biogeographic scale, sedimentary regimes are largely responsible for the low coral-species diversity, and perhaps lack of reef development, in the shallow waters of the East China Sea, Gulf of Thailand, South China Sea, Java Sea, the Bay of Bengal, and other regions further afield that have extensive areas of turbid shallow water, such as the Gulf of Mexico.

Tidal regimes impact on a very wide range of habitat types, not only inter-tidal habitats, but also all habitats correlated with current-generated sedimentary regimes and current-generated reef morphologies. Tidal ranges tend to be higher along continental margins than in the open ocean and thus the latter usually have narrow tidal ranges and fewer types of inter-tidal habitats, especially those which are maintained by tidal flushing. High tide ranges, whether on continental margins or not,

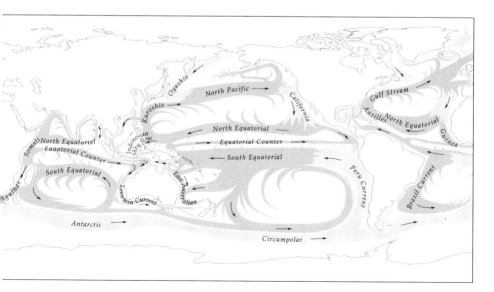

Figure 23 Principal global surface circulations. Neither regional detail nor seasonal variations are indicated: the figure is intended to illustrate the dispersion potential of continental boundary currents from low to high latitudes, and the lack of dispersion potential to the south Pacific and Indian Oceans.

generally result in a wide range of habitat types and also maximise, through flushing of flat terrains, the total area available for coral occupancy. Whether this has any biogeographic consequence remains to be determined.

Nutrients and biotic factors have widely varying regional impacts all acting in largely unknown time-scales. Consequences of anthropogenic changes in nutrient loading on a biogeographic scale (eg on the Great Barrier Reef, Yellowlees 1991) are as yet poorly understood but may be very significant, especially in synergy with global climatic change (Wilkinson and Buddemeier 1994). Biotic changes to coral communities due to the impact of *Acanthaster planci* and other coral-eating organisms have well-known regional impacts which may last a decade or more (eg Guzmán and Cortéz 1989; Done 1992), with possible consequences for distribution patterns and community composition.

PART B

FOSSILS
AND
PALAEOCLIMATES

The main topics of chapters 7 and 8 are listed at the beginning of these chapters. The following is a summary of these topics.

THE MAIN POINTS

Some of the following points (identified by cross references) include conclusions made in other parts of this book.

♦ Early evolutionary history, as seen in the Family Tree of Scleractinia, must be reconstructed from the essentially separate fields of Mesozoic coral palaeontology, Cenozoic coral palaeontology, extant zooxanthellate coral taxonomy and extant azooxanthellate coral taxonomy. This has been a barrier to the creation of a modern synthesis of genera and generic affinities.

♦ The families of extant corals are only distantly related. Only those of the Faviina show divergence in the Cenozoic; all other families have Mesozoic origins.

♦ The Order Scleractinia is not monophyletic: groups of families, forming suborders, appear to have different origins. All extant families appear to have a common origin.

♦ The first scleractinians to form fossils were probably azooxanthellate; they did not build reefs.

♦ Mass extinction events have played a dominant role in Mesozoic coral evolution. According to the fossil record, one-third of all families and over 70 per cent of all genera became extinct at the Cretaceous/ Tertiary boundary.

♦ Reef-building was slow to re-establish after the Cretaceous/Tertiary extinctions although climatic conditions were generally favourable. This may have been the time required to establish, or re-establish, algal symbiosis in different families.

♦ Coral distributions are poor indicators of palaeoclimatic patterns; reef distributions are probably reliable long-term indicators of the width of the tropical ocean band.

♦ A cosmopolitan fauna developed in the Palaeogene. This was partitioned in the Miocene by closure of the Tethys Sea and in the Pliocene by closure of the Central American Seaway (formation of the Isthmus of Panama). Many genera now restricted to the Indo-Pacific formerly had much wider (Tethyan and Caribbean) distributions.

♦ The fossil record indicates that the total number of genera progressively increased through the Cenozoic. Most genera have indefinite geological longevities; those that have become extinct have mostly done so during two specific extinction events.

♦ The average age of genera of the Caribbean (60 million years) is twice that of the Indo-Pacific centre of diversity (30 million years). This is partly due to gaps in the fossil record, and may also reflect the latter's Tethyan origin. Age distributions do not support Darwinian centres of origin theory (p 36).

♦ There are reasons why the fossil record may distort the evolutionary picture; molecular techniques have the potential to correct these distortions (p 29).

♦ Except for fossil records of extant species and their immediate ancestors, most 'species' distinctions in coral palaeontology have little biological meaning and are not useful for palaeobiogeographic reconstructions.

♦ The fossil record is mostly a poor indicator of distribution and climatic change. However, it has shown that minor (no more than 2°C) change in sea surface temperature can have major effects on high-latitude distributions (p 184).

♦ Rates of speciation and extinction tend to be similar and range geographically and taxonomically from 4 to more than 10 per cent of coral species per million years. The Neogene Caribbean tends towards the higher rate, the Neogene Indo-Pacific tends towards the lower rate.

♦ The average age of coral species in the fossil record varies among families and also varies between the Indo-Pacific and Caribbean provinces.

♦ Rates of macroevolution are much slower than rates of distribution change (correlated with microevolutionary processes). Evolutionary concepts, including classical vicariance, which are based on similar time-frames for evolution and distribution change, must be excluded for corals. Evolutionary change is slow in corals compared with most other fauna.

♦ Evolutionary concepts, both biogeographic and genetic, that are 'imported' from studies of terrestrial organisms are seldom applicable to marine fauna. Most currently popular concepts, as well as areas of debate, are not relevant to corals (p 51).

7

GEOLOGICAL BACKGROUND

'And if in some places the sea recedes while in others it encroaches,
then evidently in the same parts of the earth as a whole are not always sea,
nor always mainland, but in process of time all change.'
(Aristotle Meterologica c 335 BC)

This chapter is a summary of what little is known about the succession of ancient physical environments which, over 230 million years, has shaped the evolution of the Scleractinia. Most of the subject matter is remote from the rest of this book, having little to do with coral species. The focus in this chapter is on palaeoenvironments and Mesozoic corals; the ancestors of modern corals are discussed in chapter 8. The database of fossil coral genera on which both chapters are based is very extensive[1]; much of it is in need of specialist revision.

The study of links between plate tectonics and palaeoclimatic change, and how the latter is deduced from rocks and fossils, is a very multi-disciplinary subject. Frakes (1979), Kennett (1982), Bradley (1985) and Frakes et al (1992) give informative overviews.

THE MAIN TOPICS

♦ The Family Tree of Scleractinia.
♦ Origins of the Scleractinia.
♦ Triassic, Jurassic and Cretaceous corals.
♦ Changing positions of continents, changing palaeoclimates, and the distribution of reefs of the Mesozoic and Cenozoic eras.
♦ Role of extinction events in the evolution of the Scleractinia. The Cretaceous/Tertiary extinctions.

- Development of a circum-global circulation in the southern hemisphere and the progressive destruction of the circum-global circulation of the tropics.
- Palaeoenvironments of the immediate ancestors of modern corals.
- Palaeoclimatic cycles.

THE 'FAMILY TREE' OF SCLERACTINIA

Wells's (1956) treatise, perhaps the most important combined taxonomic and palaeontological work on corals ever published, has long ceased to represent modern knowledge. As a compendium, the equal treatment given to extant genera and to fossil genera of different geological intervals masks the fact that these genera are not even vaguely equal. Our knowledge of *Acropora,* at one extreme, covers several volumes; at the other extreme, hundreds of Mesozoic genera are known only from eroded fragments that have very little information value.

The revision of Wells's (1956) family tree presented in this book has required the blending of four essentially separate taxonomies: those of (a) Mesozoic fossils, (b) Tertiary fossils, (c) extant zooxanthellates and (d) extant azooxanthellates. The author's attempt to do this (figure 25) is dominated (from the top down) by the systematics of extant families, which are well-known, and (from the bottom up) by the systematics of extinct families, most of which are anything but well-known. It is this latter aspect that is primarily responsible for this tree being so unlike that of Wells. Emphasis has changed among Wells's blend of skeletal microstructure and gross morphology, the microstructure-oriented French school of Alloiteau, Cuif and Beauvais and the microstructure specifically of aragonitic elements (Roniewicz 1984, 1989; Roniewicz and Morycowa 1989).

Many points of uncertainty are at suborder level. Several of these involve families with extant relatives, and thus the validity of at least some aspects of microstructural methodology is testable against the systematics of extant families. Otherwise, the base of the tree seems forever rooted in vague evidence from competing hypotheses.

The present tree consists of thirteen suborders, of which six are extant, and sixty-one families, of which twenty-four are extant. The number of nominal genera is approximately 1800 (Wells 1986, with subsequent additions). Approximately 60 per cent (n=1216) of these were used in the compilation of figures 25 and 36 (p 134), the remainder are considered invalid, or have been so-described by specialists.

THE MESOZOIC

The importance of the Mesozoic Era to the history of extant Scleractinia is readily seen in figure 25. Scleractinia are first recognised as a distinct group in the Triassic; they probably gained the greatest diversity the

REEFS: WHERE BIOLOGY AND GEOLOGY MEET

Nowhere are the separate cultural environments of biological and geological sciences more apparent than in the study of coral reefs. Although many current authors have a foot in both camps, the biological literature as a whole is characteristically marine-oriented, recent and experimental, while the geological literature (Flügel and Flügel-Kahler 1992 provide a recent compilation) is characteristically terrestrially oriented (stratigraphy and palaeontology) and has been relatively long established. The historical reason for this separate development, no doubt, is the level of association between biology and geology that is unique to coral reefs. The result, however, has been a barrier to understanding and awareness.

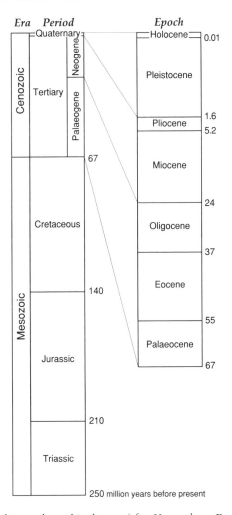

Figure 24 Geological intervals used in the text (after Haq and van Eysinga 1987).

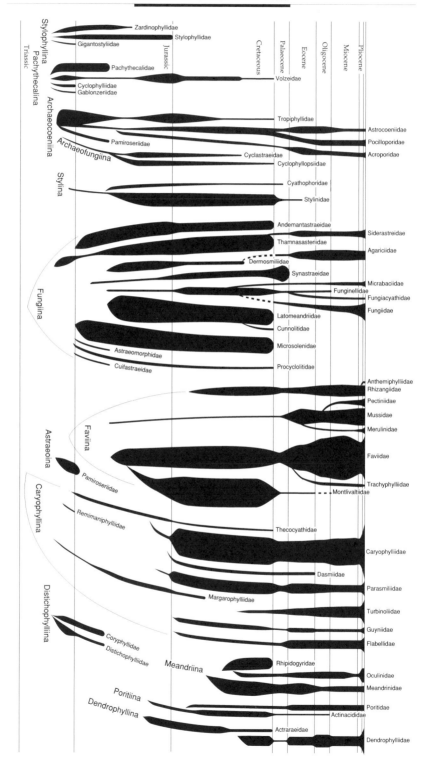

Stylophyllina
Pachythecaliina
Triassic

Zardinophyllidae
Gigantostyliidae
Stylophyllidae
Jurassic
Pachythecalidae
Cyclophylliidae
Gablonzeriidae
Volzeidae

Archaeocoeniina
Archaeofungiina
Pamiroseriidae
Cyclastraeidae
Cyclophyllopsiidae

Stylina

Cretaceous
Palaeocene
Eocene
Oligocene
Miocene
Pliocene

Tropiphyllidae
Astrocoeniidae
Pocilloporidae
Acroporidae

Cyathophoridae
Stylinidae

Fungiina

Andemantastraeidae
Siderastreidae
Thamnasasteriidae
Agariciidae
Dermosmiliidae
Synastraeidae
Micrabaciidae
Funginellidae
Fungiacyathidae
Fungiidae
Latomeandriidae
Cunnolitidae
Microsolenidae
Procyclolitidae
Astraeomorphidae
Cuifastraeidae

Astraeoina
Favina
Pamiroseriidae

Anthemiphylliidae
Rhizangiidae
Pectiniidae
Mussidae
Merulinidae
Faviidae
Trachyphylliidae
Montlivaltiidae

Caryophyllina
Remimaniphylliidae

Thecocyathidae
Caryophylliidae
Dasmiidae
Parasmiliidae
Margarophylliidae
Turbinoliidae

Distichophyllina
Coryphyllidae
Distichophylliidae

Guyniidae
Flabellidae

Meandriina
Rhipidogyridae
Oculinidae
Meandrinidae

Poritiina
Dendrophylliina
Actinacidae
Poritidae

Actraraeidae
Dendrophylliidae

world has ever seen in the Jurassic; and they came to near-extinction at the close of the Cretaceous. These great events, occurring over a time-span of nearly 200 million years, can only be glimpsed today through occasional fossil remains that are disconnected in space as well as vastly disconnected in time. Mesozoic continental positions and global climates can also only tentatively be reconstructed in roughest outline; the level of detail needed for palaeobiogeographic and evolutionary explanation is almost entirely absent.

THE TRIASSIC

The essential conclusion from plate tectonics, which overturned all biogeography and so much else in the 1960s, was that the Early Mesozoic world consisted of one super-continent, Pangaea, and one enormous ocean, the Panthalassa, nearly twice as wide as the equatorial Pacific is today (figure 26). Global geography, from that time to the present, is based on the breaking-up of Pangaea, and the consequent

Figure 25 (left) The family tree of Scleractinia according to extant coral taxonomy and the fossil record. Branch widths indicate numbers of genera per family for each geo-logical interval. Principal systematic references for Mesozoic corals are Wells (1956), Beauvais (1980, 1984, 1986) Chevalier and Beauvais (1987) and Roniewicz and Morycowa (1989). Systematics of now-living genera follows the author (zooxanthel-late genera) and S. Cairns, pers. comm., (azooxanthellate genera). Records of genera are from relevant publications; principal references are cited in the text.

Figure 26 The Late Triassic world showing distribution maximum of coral reefs and/or reef corals (after Stanley 1981; Beauvais 1982; Flügel 1982; and others). The base map is modified after Smith et al (1981). The reefs that now occur along the eastern Panthalassa rim are likely to be displaced terrains (see text). Broken lines enclose prin-cipal coral provinces.

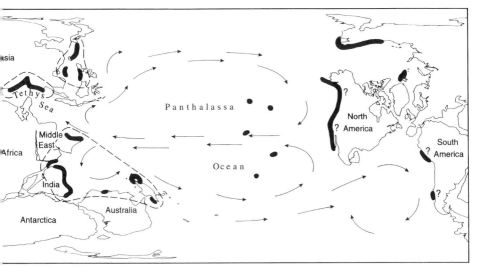

formation of today's three great ocean basins.

There is a paucity of fossils after the end-Palaeozoic mass extinctions which creates not only a gap in our knowledge of evolutionary continuity, but also a gap in our ability to reconstruct Early Triassic climates. Frakes's (1979) conclusion that climates were warm over a wide latitudinal range in the mid to late Triassic needs to be reconciled with subsequent views of displaced reef terrains (below); nevertheless, Tethyan/Boreal provincial boundaries were probably wider in the Triassic than Jurassic (Hallam 1986) and certainly a wide latitudinal component in surface circulation is suggested by the unbroken expanse of the Panthalassa.

Coral Reefs

Reconstructions of Panthalassa reef distributions is made complicated by the presence of 'displaced terrains' (p 269), some of which include extensive coral reefs. This displacement of reefs by thousands of kilometres from the central Panthalassa to the western coast of North America and elsewhere around today's Pacific rim (Stanley 1987) creates similar problems for Mesozoic palaeobiogeography that continental drift did for Darwin's centres of origin (p 36) (eg Belasky and Runnegar 1993). There are differing views about the geographic extent of displacement (Tozer 1982; Hallam 1986; Debiche 1987), but most if not all eastern Pacific Mesozoic fossil deposits today are nowhere near their original place of formation (Grigg and Hey 1992). They may also have become intermixed with subsequent Mesozoic faunas, as the Triassic terrains did not 'dock' till the Late Jurassic or Early Cretaceous.

The first Scleractinia

Drawing the 'family tree' of Scleractinia in side view, as done here, necessitates a gross distortion of information. In truth, the 'tree' should be viewed from above, where the extant foliage is clearly seen, where the configuration of most upper branches can just be discerned, but where the main trunk(s) can only be guessed. What the fossil record shows is that the Scleractinia were, almost certainly, polyphyletic at the time of the evolution of skeletogenesis. Whether extant families are polyphyletic, or not, is a separate issue (p 114).

The fossil record, combined with our knowledge of skeletal microstructure of extant corals, shows that at least seven, and possibly nine, suborders of Scleractinia existed in the mid-Triassic and that these have no known common ancestor. In reality there may have been many such suborders. Some may well have had a common origin; if so, it was either an askeletal (anemone-like) scleractinian[2], or a rugose coral. The latter seems unlikely; the Rugosa had a system of septal arrangement (serial, rather than cyclical) that is unlike that of Scleractinia and, more importantly, they had calcitic rather than aragonitic skeletons. The latest known Rugosa were Late Permian (250 million years ago), whereas the earliest known Scleractinia are mid-Triassic, a separation of 5 to 8 million years. Oliver (1980) presents these issues; Cuif (1980), Beauvais (1980) and others give opposing arguments[3].

The most likely alternative to a rugosan origin of the Scleractinia is soft-bodied sea-anemone-like or Corallimorpharia-like organisms. Given the spectrum of skeletal and askeletal morphologies available, and the capacity of extant corals to exist askeletally (p 87), it seems possible that skeletogenesis is closely linked with symbiosis and reef formation, and that it independently evolved back and forth many times (and not only in the Scleractinia) in response to a unique opportunity to exploit unlimited resources (calcium carbonate and sunlight) for the common 'evolutionary reward' of environmental stability.

The earliest 'scleractinians', whatever their origin, did not occupy reef environments. Their first appearance in the Middle Triassic was as a widely dispersed array of small, solitary or phaceloid organisms of the shallow-water Tethys of the Bavarian, Austrian and Italian Alps, and a second largely distinct fauna of Indo-China (Beauvais 1982). Coloniality is a correlate of reef-building: these corals had no reef-building capacity, nor did they have the morphological characteristics (Coates and Jackson 1987) of reef corals.

During the Middle and Late Triassic, corals became widespread throughout the Tethys region and their fossils are now found around most of the equatorial Panthalassa Ocean rim (figure 26). Comparisons between these widely dispersed outcrops have led to palaeobiogeographic speculations about pantropical distributions (Newton 1988), barriers, dispersion and provincialisms, almost on a parallel with those of extant biogeography, and with even more unsupported assumptions.

The end-Triassic extinction event was not the equal of the Permian/Triassic event 45 million years earlier, which completely exterminated most reef life, but it may have rivalled the Cretaceous/Tertiary extinctions in its severity for marine life (G D Stanley 1988). For corals, approximately eighteen out of sixty-seven genera survived. The hiatus in reef development lasted 4 to 10 million years. The end result was a substantially new fauna.

Were early scleractinians zooxanthellate or azooxanthellate? Algal symbiosis in coelenterates is largely, but far from exclusively, associated with taxa that have skeletons. It may thus be supposed that symbiosis originally evolved in association with skeletogenesis — a traditional view since Wells (1956). There was, however, a time interval of 20 to 25 million years between the first Scleractinia and the proliferation of scleractinian reefs (Stanley 1981), and it can well be argued that zooxanthella symbiosis[4] is an ecological and physiological correlate of reef-building rather than of the evolutionary history of corals themselves. The question is not as remote as might be supposed: microgeochemical analyses of aragonitic Triassic skeletons, which have the potential to discriminate between zooxanthellates and azooxanthellates (Stanley and Swart 1984; G Stanley pers comm), may provide the answer.

Whatever the events of the Triassic, zooxanthella symbiosis appears to have evolved independently many times during the Tertiary: the zooxanthellate *Duncanopsammia axifuga* is morphologically closer to its azooxanthellate relatives than to its zooxanthellate ones; *Astrangia*, *Madracis*, *Cladocora* and *Oculina*, all in different families, have both zooxanthellate and azooxanthellate species.

Are extant Scleractinia monophyletic? This question is beyond the fossil record to resolve: despite many assertions to the contrary, most branches of the family tree (eg of figure 25) cannot be joined by common ancestry. However, the question can be, and is being, addressed using molecular methods. Fautin and Lowenstein (1994) conclude from radioimmunoassay (p 28) that *Fungiacyathus* (an anemone-like azooxanthellate coral with minimal skeletal development) is systematically closer to two groups of Corallimorpharia than to the Poritidae. DNA sequencing of ribosomal DNA (p 29) (Chen et al in prep) indicates, however, that extant Scleractinia are a monophyletic group, but that the order has uncertain affinities with other coelenterate orders.

As far as morphological criteria and the fossil record go, modern taxa appear to be broadly divisible into two groups (as proposed by Wells 1956): the Archaeocoeniina with its three surviving families, and all other suborders. Extant members of the latter are grouped into three main suborders, the Fungiina, Faviina and Caryophyllina, all with Triassic origins, and two minor groups, here called the Poritiina and Dendrophyllina, with unclear Jurassic origins (figure 25).

THE JURASSIC

Many theories have been offered to explain the great Jurassic proliferation of reefs. Single causal factors seldom survive close scrutiny in geological time-frames, but the opening of the Protoatlantic Ocean (the 'Hispanic Corridor'), providing perhaps the first circum-global circulation of the Mesozoic (figure 27, Stanley 1991) probably had much to do with it.

THE CORALS

The inheritance of the Jurassic was a very depauperate, but systematically diverse, suite of genera (figure 25).

Early Jurassic reefs are rare everywhere in the world (Beauvais 1984, 1986, 1989)[5]. According to Beauvais, all Triassic genera had become extinct in, or by, the Early Jurassic. By the Middle Jurassic, reef development proliferated in present-day Europe and the Mediterranean, but remained poorly developed in the Panthalassa, and may have remained

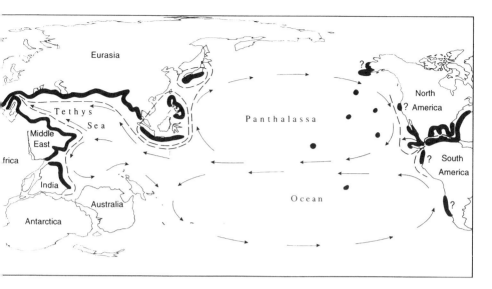

Figure 27 The Late Jurassic world showing distribution maximum of coral reefs and/or reef corals. The base map is after Smith et al (1981) and Haq and van Eysinga (1987). Ocean currents indicated are primarily after Haq and van Eysinga (1987). Broken lines enclose principal coral provinces (see text).

thus throughout the whole Jurassic.

It was in the Late Jurassic that the probable all-time global maximum of coral diversity occurred, with 150 genera recorded in the European Tethys and 51 in the Panthalassa (Beauvais 1989). Beauvais (1989) traces the changing provincialisms of 'species' (p 141) throughout the Jurassic, which primarily reflect plate movements, especially the formation of the Protoatlantic Ocean. By the Late Jurassic there were well-established provinces in the coral world. An Asiatic province extended along the northern margin of the Tethys, with subprovinces along the east Asian coast, a southern Tethys province with an extension to Panthalassa South America, and a northern Tethys province with extension to Panthalassa North America. There is an unexplained partial barrier between the northern and southern Tethys (Beauvais 1989).

By the Late Jurassic the palaeobiogeographic pattern (of now extinct taxa) that had developed was the precursor to Tethys/Atlantic/eastern Pacific pattern of the Palaeogene. The vast open expanse of the eastern Panthalassa was probably a dispersion barrier, similar to today's far eastern Pacific (p 99).

THE CRETACEOUS

Interesting though the Jurassic was from the point of view of its diversity, it is the Cretaceous, when marine faunas came under intervals of

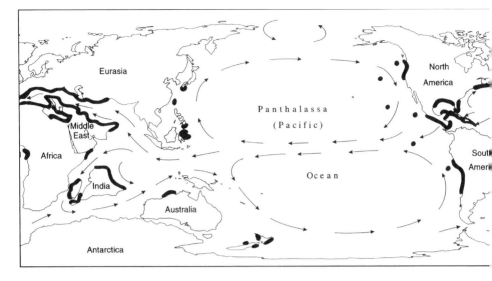

Figure 28 The Late Cretaceous world showing distribution maximum of coral reefs and/or reef corals. The base map and ocean currents are from sources indicated in figure 27 and Funnell (1990). At the Cretaceous maximum sea level, the area of land was much less than indicated here (see text). The direction of circulation through the Tethys has recently been challenged by a simulation model (Barron and Peterson 1989).

acute physical-environmental pressure, that is most interesting from the point of view of the causes and outcome of evolutionary changes. The Cretaceous contains the last great proliferation of corals destined for extinction.

The aspects of the geological history of the Cretaceous (figure 28 and the working review of Ginsberg and Beaudoin 1990) that appear most important to corals are less those of continental positions, and more those of traumatic change in water quality and climate.

The Late Cretaceous was a time of major sea-level changes (Schlanger 1986; Haq et al 1987) flooding nearly 40 per cent of the continents (Howarth 1981), leaving only 18 per cent of the earth's surface as land and, significantly for coral biogeography, creating a 'Super-Tethys' ocean. Consequences for reefs are unknown, because the rate of sea level change is unknown, but the continually decreasing sea levels of the Late Cretaceous may have had a greater long-term impact on survival and preservation than the fluctuating sea levels of the Pleistocene.

Perhaps more significantly for reefs, there were very high Middle Cretaceous levels of carbon dioxide from vulcanism associated with lithospheric plate-spreading (Arthur et al 1985; Lasaga et al 1985). This led to increased surface water acidity and raising of the carbonate compensation depth. Aragonite would have been more soluble and have required more energy to produce, a possible reason for the relative success of the rudist bivalves that largely displaced corals from reefs. The net result, whatever the cause, was decreased carbonate production. Buddemeier and Fautin (in press) conclude that this may have led to the divergence of askeletal corals (ie groups of actinians) from scleractinians.

Rates of accumulation of organic matter of both marine and terrestrial origin were as high, or higher, than at any other interval in the Mesozoic or Cenozoic (Arthur et al 1985). This would have contributed to the severity of anoxic events and shallowing of the carbonate compensation depth.

Ocean and atmospheric temperatures were generally higher than now over the full range of latitude from the equator to the poles (Kennett 1982; Kauffman and Johnson 1988), but this would have varied greatly. The pattern for the warmest time is a wide zone comparable to modern tropical and subtropical conditions extending to at least latitude 45°N and possibly latitude 70°S. Beyond this, climates were warm to cool temperate; the poles were devoid of permanent ice. Mean annual temperatures have been estimated at between 10 and 15°C warmer than present, and the temperature gradient at about half that of the present. These temperatures would have created oxygen stress and were probably the ultimate cause of deep-water anoxic events recorded in Early and Middle Cretaceous oceanic sediments (Crowley 1991).

The poleward boundaries of subtropical high-pressure atmospheric belts, and the major ocean gyres with which they are correlated, would have been at higher latitudes than they are today, and major surface currents would have been generally weaker.

By the close of the Mesozoic, the epicontinental seas withdrew and warm dry global climates that predominated during the Cretaceous had begun their long and irregular decline towards a glacial mode (Hays and Pittman 1973).

THE CORALS

The beginning of the Cretaceous was not marked by any mass extinction event, but there was, nevertheless, a drastic change in coral communities. Rudist bivalves, a previously obscure group, displaced corals as the dominant reef biota in the Early Creataceous and remained dominant for thirty million years. During this period, zooxanthellate corals coexisted with rudists, but largely in separate habitats at greater depths (Scott 1988). As the process of reef-building was undertaken by rudists rather than corals, the reefs must have had a very unfamiliar appearance. Rudists were much better sediment-trappers than corals: rudist reefs probably resembled today's inshore fringing reefs, characterised by a high sediment component and lack of cementation. Rudists were probably zooxanthellate. They had 30 to 60 per cent aragonite in their shells (Kauffman and Johnson 1988) — possibly a selective advantage over totally aragonitic (and therefore more soluble) corals in Early and Middle Cretaceous acidic environments.

Early Cretaceous corals are mostly continuous with those of the Jurassic; Beauvais (1992) records only five new genera. Unfortunately the quality of taxonomic coral palaeontology reaches an all-time low in the Cretaceous. Some 1350 'species' have been described, but the vast majority of these are little more than vague descriptions of specimens (p 141) and make palaeobiogeographic reconstructions rather meaningless.

Corals returned to positions of dominance during the Late Cretaceous, following total extinction of the rudists. By the close of the Cretaceous, extensive reefs, built again by corals, occurred world-wide.

In general appearance and diversity, the Late Mesozoic Scleractinia were much like those of today; perhaps only *Acropora*, with its elaborate skeletal architecture (p 243), would look out of place on a Late Cretaceous reef. All major families extant today were established. Most genera were cosmopolitan and, because of the endurance of the Tethyan seaway, those that survived the Cretaceous/Tertiary extinctions remained so until the time of the origins of modern genera.

EXTINCTION EVENTS

Recurring mass extinction events and the evolution of algal symbiosis (p 96) are two great evolutionary dimensions that have shaped the evolution of modern zooxanthellate corals. The capacity of animal life to build reefs (reviewed by Fagerstrom 1987; Kauffman and Fagerstrom 1993) was brought to full realisation in the ancient seas of the Devonian (410 to 360 million years ago): as far as reefs are concerned, it was the greatest extinction process of all that brought this Devonian fauna to its end (Newell 1971; Wilson 1975; S M Stanley 1981, 1988). Reef-building capacity through symbiosis evolved afresh with the Scleractinia and, although no extinction event has given them the fate of their ecological antecedents, many events, both regional and global, have substantially affected the order's family-level and generic-level composition. Two extinctions at least (Triassic and Cretaceous) were so drastic that the very existence of the Scleractinia appears, from the fossil record, to have hung on the survival of only a tiny fraction of the diversity we have today and, more to the point, did so for millions of years.

Did the survivors just 'tough it out'? Perhaps, but some may have had the evolutionary flexibility, and the time, to reduce or discard their dependence on symbiosis and survive in the relative security of the greater ocean depths of the azooxanthellates. They had, in theory at least, a relatively simple ecophysiological escape route not available to most terrestrial life. The fossil record as depicted in figure 25 would give a misleading impression of mass extinctions if this was so, as change to an azooxanthellate existence would greatly reduce the chances of fossilisation and recovery. The central question concerns the nature of zooxanthellate symbiosis: it is obligatory for most extant zooxanthellate species, but would it necessarily remain so under conditions of gradual climatic change?

There is a substantial and somewhat speculative literature concerning marine mass extinctions in general and the end-Cretaceous extinctions in particular (Kauffman 1984). Most address the cause or the timing of the extinctions, or their consequences for evolution. Most references to corals are largely recycled information based on a weak taxonomy. The view of G D Stanley (1988) and Cowen (1988), that the slow recovery of reef ecosystems from mass extinction may reflect the time necessary to re-establish algal symbiosis, is intuitive, but this hypothesis would be clearer with a better understanding of why coral/algal symbiosis is, in most cases, such a close partnership (p 97).

The Cretaceous/Tertiary boundary is marked by the extinction of most marine reptiles, both orders of dinosaurs, all ammonites, a high percentage of bivalves, gastropods and echinoids, and a major proportion of planktonic foraminifera and radiolaria. Many of these groups became extinct within an apparently brief period, but many other animal groups, especially terrestrial and deep-sea animals, appear to have been little affected.

Many hypotheses have been proposed to explain the Cretaceous/Tertiary extinctions (Kauffman 1984; Officer et al 1987), most relying on nutrient depletion or a cosmic event (Alvarez et al 1984; Hsu 1986). As shallow marine communities (with a high proportion of phototrophic organisms) seem most affected, it seems reasonable to conclude that a drastic reduction in light was at least partly responsible. The meteorite impact theory (creating a long-term upper atmospheric dust layer) has received support from multiple sources. Whatever the cause, it left no long-term global climatic effects: the boundary is a biological one; it is not geological nor climatological.

The statistics from the coral fossil record are impressive: they indicate that one-third of all families, and over 70 per cent of all genera, became completely extinct (figure 36, p 134). The Faviidae is the only family that dominated Jurassic and Cretaceous reefs to have survived to dominate Cenozoic reefs. Approximately six out of sixteen faviid genera survived; all other zooxanthellate families survived with one or two genera. Contrary to general beliefs, azooxanthellate genera fared not markedly better: the Caryophylliidae survived by thirteen genera (out of approximately twenty-seven), the Rhizangiidae by three, the remainder by one or two. There is no adequate basis for evaluating the fate of 'species'; when the fossil record is restored in the Eocene, there is no recognisable species-level taxonomic continuity with the Mesozoic.

The geographic and taxonomic extent of the Cretaceous/Tertiary extinctions are poorly known for marine invertebrates in general, partly owing to a lack of detailed dating of end-Cretaceous records (Surlyk 1990). Palaeocene corals are scarce, and, as Rosen and Turnšek (1989) found, the taxonomic framework is very inadequate.

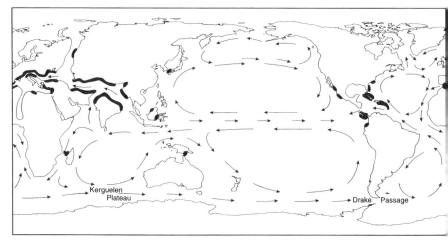

Figure 29 The Eocene world showing distribution maximum of coral reefs and/or reef corals. The base map and ocean currents are from sources indicated in figure 27 and Adams (1981). The development of a circum-Antarctic circulation (opening of Australian–Antarctic seaway, development of Kerguelen Plateau and the opening of Drake Passage) is the key to Palaeogene climates (see text). The most important feature of the tropical world remains the tropical circum-global ocean circulation through the Tethys Sea and Central American Seaway. The slow blockage of this circulation underpins all Cenozoic tropical palaeobiogeography.

THE CENOZOIC

The evolutionary history of modern corals is divisible into three geological intervals: the Palaeogene (67 to 24 million years ago), when the few survivors of the Cretaceous/Tertiary extinctions proliferated into a diverse cosmopolitan fauna; the Miocene (24 to 5.2 million years ago) when this fauna became subdivided into the broad biogeographic provinces we have today and most extant 'species' evolved; and the Plio-Pleistocene to present, when the world went into full glacial mode and modern distribution patterns emerged. This succession of events can be viewed through palaeoclimatic reconstructions and also through hypotheses about global palaeoclimatic cycles. Frakes et al (1992) gives a recent review.

PALAEOCLIMATES

THE PALAEOGENE

The development of a circumpolar circulation in the southern hemisphere (Veevers and Ettriem 1988) (figure 29) was crucial for Palaeogene climates as it changed the global thermal pattern from a predominantly latitudinal, warm-water circulation at all depths, to a predominantly meridional, thermohaline, cold-water one. Throughout the Palaeogene, long episodes of relatively slight atmospheric warming are punctuated by abrupt drops in temperature, leading to successively cooler regimes. High latitude regions appear to have been more affected by climatic change (Frakes et al 1992). Sea level fell in saw-tooth fashion, primarily correlated with polar ice accumulation.

The existence of tropical floral assemblages in the Early Palaeocene Arctic has long been known (eg Frakes 1979). These, together with stable-isotope studies of palaeotemperature, studies of fluctuations in the carbonate compensation depth, analyses of sea levels, biogeographic studies of planktonic foraminifera and many other sources, combine to produce a picture of fluctuations in the width of the tropical belt, super-imposed on gradual environmental cooling. The Epoch opened with low sea levels and emergent continents. Temperatures were mostly warmer than today, but not as warm as the Late Cretaceous maximum.

As far as reefs were concerned, the inheritance of the Cretaceous/Tertiary extinctions was a globally diminished rate of all calcium carbon-ate production lasting at least 1 million years (Zachos and Arthur 1986) but probably much longer. Coral reefs are unknown in this interval; calcite-secreting foraminifera (Plaziat and Perrin 1992) and coralline algae (Bryan 1991) both appear to have recovered from the Cretaceous/Tertiary extinctions more rapidly than aragonitic corals.

High-latitude cooling began in the Early to Middle Eocene when the dynamics of ocean circulation changed and permanent ice appeared at the poles, but reconstruction of palaeoceanographic and climatic events is made difficult by conflicting evidence. The most significant time for corals was that of the end-Eocene mass extinctions (Raup and Sepkoski 1984), about 37 million years ago, which correlate with an abrupt decrease in abyssal ocean temperatures, accumulation of sea ice around eastern Antarctica (Matthews and Poore 1980), and development of a thermohaline circulation similar to that of today.

Tropical biotas, generally widespread during the Early Oligocene, declined throughout the Epoch. A progression of tectonic blockages occurred in the Tethys of the present Middle East, choking off the ancient westward-flowing surface circulation (Rögl and Steininger 1984). Surface currents in the Pacific were probably weak, owing to relative lack of partitioning of the Indo-Pacific.

CORALS AND CORAL REEFS AS INDICATORS OF PALAEOENVIRONMENTS?
Oxygen isotopes and faunal/floral distributions give different indications of Eocene/Oligocene ocean temperatures (Barron 1987; Adams et al 1990). Coral palaeodistributions unfortunately contribute little to this subject because they are generated by events that may be very short term (p 84); the distribu-tion of corals may have little to do with the distribution of reefs (p 168); and high-latitude fossil deposits are at best 'chance' discoveries (p 145).

Major accumulations of wave-resistant reefs, on the other hand, are reli-able indicators of prolonged sea surface temperatures above 18°C (p 91). This has a chemico-physiological basis that is likely to be relatively stable with changing faunal compositions.

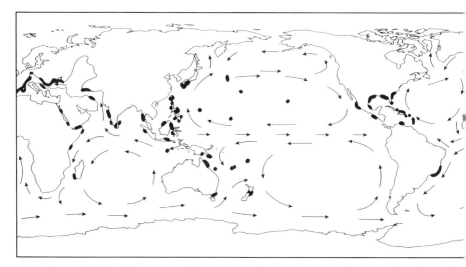

Figure 30 The Miocene world showing distribution maximum of coral reefs and/or reef corals. The base map and ocean currents are from sources indicated in figure 27. The continents are close to their present positions. The Tethys Sea is reduced to a narrow band connecting the Indian Ocean with the proto-Mediterranean. Reef development globally is at a maximum for the Cenozoic.

THE MIOCENE

The Miocene, spanning nearly 20 million years, is a 'catch-all' term for most of the immense interval of time from the ancient world of extinct species to the time of extant coral species or their immediate ancestors.

At the dawning of the Miocene, about 24 million years ago, much of the earth experienced significantly warmer and wetter climates than at present. There was a Middle Miocene warm interval, reflected in expansion of tropical vegetation bands, followed by a marked global temperature decline. Increase in atmospheric and ocean temperature gradients (Romine and Lombari 1985) resulted in the deforestation of Antarctica (Wolfe 1985), and by the Late Miocene, there was a permanent extensive eastern Antarctic cap (Shackleton and Kennett 1975).

There was extensive reef and coral community development from the Late Oligocene to the Middle Miocene in the southeast Asian region, north to Japan and south to New Zealand, as a result of decreased latitudinal temperature gradient (Fulthorpe and Schlanger 1989). In the Late Miocene (figure 30), glacio-eustatic lowering of sea level, reducing shelf area, may have resulted in a global reduction in reef-building (Adams et al 1977), and the Mediterranean became a closed, evaporative basin, exterminating all marine life (Hodell et al 1986).

THE PLIO-PLEISTOCENE

Since the Middle Miocene, the earth's climatic regime has been in glacial mode, primarily characterised by major climatic fluctuations. Ocean circulation has fluctuated in accordance with latitudinal temperature gradients, and there have been periodic pulses of major upwelling. The end-Miocene event was catastrophic for shallow-water marine life as it was one of the greatest sea level regressions of Cenozoic history (figure 31). In the Early Pliocene, there was a warming and a major sea level rise, followed by climatic deterioration (Haq et al 1987; Mercer 1987). Oxygen isotopes of foraminifera indicate the onset of northern hemisphere glaciation 2.4 million years ago, with Milankovitch-cycle-correlated periodicities (p 102) of about 40 000 and 100 000 years (Sikes et al 1991) (see below) (p 128). The Pliocene event of greatest biogeographic significance to corals was the closing of the Central American Seaway 3.5 to 3.1 million years ago (Coates et al 1992) which led to the separate development of Atlantic and Indo-Pacific faunas (p 143).

At the dawn of the Pleistocene epoch, glaciation became so widespread that it affected the whole earth. Indirect evidence of cooling is to be found even in low latitudes in oxygen isotope records, fluctuations of the carbonate compensation depth, sea level changes, biogeographic changes, and changes in snow-line and desert latitudes. Interest in the Pleistocene becomes progressively more regional than global. It also focuses on detail, especially of sea level changes.

SEA LEVEL CHANGES

Sea level changes (figure 31) have had an overwhelming impact on coral reefs throughout the period of evolution of modern coral species. Sea level has undergone at least seventeen cycles of transgression and regression during the Pleistocene (Grigg and Epp 1989), fluctuations that have been a primary determinant of the morphology of all reefs (eg Purdy 1974). The ecological disruption through changes in sedimentary regimes that would have accompanied major glacio-eustatic changes (figure 32) must have been catastrophic.

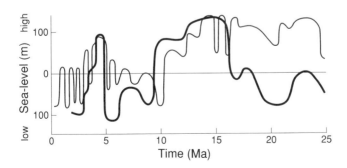

Figure 31 Comparison between the sea-level curves of Vail and Hardenbol (1979) (thick line) and Haq et al (1988) (thin line). Wheeler and Ahron (1991) give evidence from mid-oceanic carbonate platforms in support of the 1979 curve.

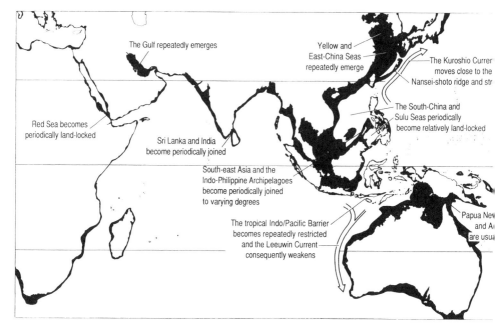

The Gulf repeatedly emerges

Yellow and
East-China Seas
repeatedly emerge

The Kuroshio Curren
moves close to the
Nansei-shoto ridge and str

Red Sea becomes
periodically land-locked

The South-China and
Sulu Seas periodically
become relatively land-locked

Sri Lanka and India
become periodically joined

South-east Asia and the
Indo-Philippine Archipelagoes
become periodically joined
to varying degrees

The tropical Indo/Pacific Barrier
becomes repeatedly restricted
and the Leeuwin Current
consequently weakens

Papua New
and A
are usua

Figure 32 Area of changing coastlines during the past 18 000 years. Other regions of the world, including the Caribbean, are less affected at this scale because continental shelves are narrow.

Sea level changes have several causes, which operate at different scales and frequencies. Major regional tectonic upheavals, sea-floor spreading, and associated changes in the buoyancy of continents are responsible for the most major and long-term changes. Retention of water in the polar ice caps and the relatively minor influences of the thermal expansion of the ocean and retention of water in the atmosphere have very different frequencies and climatic impacts. Haq et al (1987) divides the frequency of cycles into three orders: those that impact most on reefs are likely to be those of highest frequency, corresponding, perhaps, with Milankovitch cycles (below) operating (at least for the Pleistocene; Chappell and Shackleton 1986) at a 100 000 year frequency.

Sea level changes (Hallam 1992) have been occurring throughout the earth's history. Of these, glacio-eustatic changes are the best known because of the dramatic, recent impact of continental ice, but sea level changes of palaeoclimatic origin go back long before the Ice Ages. Their frequency and magnitude has been estimated by many different methods (eg ice volume modelling based on oxygen isotope studies of deep-water foraminifera; seismic stratigraphic analysis; depth measurements of sub-surface solution unconformities) at different times and places. However, high-frequency changes in the Miocene (eg Pomar 1991) and Miocene

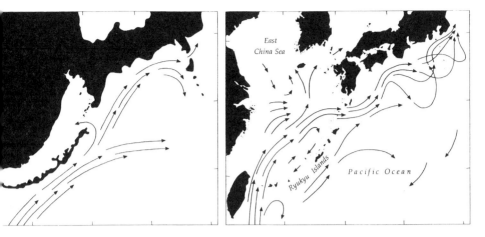

Figure 33 Glacio-eustatic changes and consequent changes in the Kuroshio Current in the East China Sea and southern Japan. Coastlines are 18 000 years ago (left) and Recent (right) (after Kimura, 1991).

low stands of 75 to 125 m below present, are independently supported by different methods (eg Lincoln and Schlanger 1987). Dating using thorium 230, with estimates of growth depths of corals, indicates continuing changes to the Late Pliocene (Moore 1982).

Modern platform reefs are essentially the product of their antecedent foundations, together with two major eustatic events, the last inter-glacial episode (130 to 75 thousand years ago), and the initiation of modern reef growth (9000 to 6000 years ago). During the last inter-glacial, maximum sea level was about 6 m higher than today. The last glacial episode lasted more than 80 000 years, and sea level oscillated by up to 135 m. The last (Wisconsin) glacial maximum low sea level has special significance, because it can be studied with a high degree of resolution, but more particularly because the time-frame is an ecological one (p 269) and has direct bearing on modern coral distributions.

Eighteen thousand years ago, sea level was approximately 120 to 135 m below present (Grigg and Epp 1989), varying geographically according to hydro-isostatic deformation of continental margins. Western Indonesia and Japan were both part of the Asian continent and Australia and Papua New Guinea were part of the same continent (figure 32). The north polar ice cap extended south to northern USA and sea ice extended to the south coast of France. The Antarctic sea ice boundary lay about 7° of latitude closer to the equator than at present (COHMAP Members 1988). The Red Sea was landlocked; the South China and Sulu Seas were partly landlocked (figure 32). All boundary currents would have been displaced away from present coastlines to beyond the continental shelf edges (eg figure 33), and wind-driven

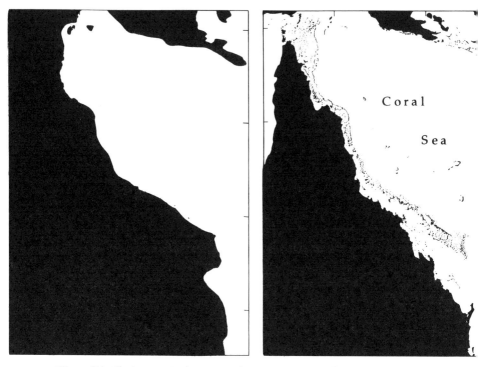

Figure 34 Glacio-eustatic changes on the Great Barrier Reef. Coastlines are 18 000 years ago (left) and Recent (right).

surface currents would have been stronger.

In general the Indo-Pacific centre of coral diversity was much more affected by glacio-eustatic changes than other parts of the globe. The western side of the entire province has been repeatedly exposed. Only the far northern Philippines and eastern Indonesia/northern Papua New Guinea kept coastlines approximating those of today. Even here, all except the lowest of present-day reef slopes would have been exposed, and those lower slopes would have undergone major environmental change and ecological upheaval. All communities that survived the 18 000 year glacial maximum were refuge communities in a biogeographic sense. The western, northern and southern margins of the Indo-Pacific centre of diversity would have been repopulated from its eastern tropical margin. Some fauna of the Great Barrier Reef (figure 34) may have been repopulated from the same source, although faunal differences indicate that intra-regional repopulation would have been more prevalent. Repopulation from eastern Indonesian refuges is probably partly why this region has the world's highest species diversity. As this happened recently in evolutionary terms, it also explains why there are low levels of endemism in adjacent (repopulated) regions.

This brief account of the effects of the last great sea level change is composed from generalisations that may not hold true for particular regions. The greatest environmental perturbation on the reefs of the Gulf of Mexico and (perhaps) Caribbean may, for example, have been one of low salinity from greatly increased inflow of post-glacial meltwater from the Mississippi River. The Red Sea may have had a salinity crisis in the opposite direction before it became open to the Indian Ocean and rainfall increased.

Following the last glacial maximum, continental ice sheets have been withdrawing. Climatic variations during this time include three or four periods of minor glacial expansion, the so-called 'Little Ice Ages'. These were interspersed with warmer conditions, the last occurring 5 to 6 thousand years ago. Brief though this interval was, it still had an effect on coral distribution in Japan comparable in magnitude to any that have been recorded for all extant corals anywhere in the world (p 145).

There have been many hypotheses about mechanisms by which sea level changes drive speciation and/or extinctions (p 54). Many authors have observed that major extinction events are correlated with sea regressions, and almost all accounts of coral evolution have stressed the importance of sea level changes (p 123). Obviously sea level changes *were* catastrophic (Hallam, 1989), but in an *evolutionary* sense, the effects on corals were probably not great:

- Communities are geographically very mobile in time-frames of thousands of years, and sea level change (mostly less than 12 m per thousand years, COHMAP Members 1988) was slow enough to allow transitional communities to establish on continental and island slopes (hypothesis 9, p 54).
- Wave-resistant reefs are not as essential to corals as many have supposed; most coral species actually thrive in non-reef environments (p 98).
- The reefs of today are not the norm for most coral communities in most of geological time.

This suggests that sea level changes as such are nowhere near as important to genetic continuity as accompanying surface circulation changes would have been, and that the evolutionary consequences of sea level changes have generally been overstated. Paulay (1990) comes to a similar conclusion; Stanley (1984a,b) and Valentine and Jablonski (1991) put this case for marine ecosystems in general.

SEA SURFACE TEMPERATURES

Sea surface cooling below the threshold limit for reef growth was the key to Daly's glacial theory of reef growth (Daly 1915), the principal challenge to Darwin's subsidence theory (p 37). Fine details of palaeotemperatures relevant to reefs (p 92) at lowered sea levels are required to examine Daly's theory, and such data have been recovered at the Huon Peninsula of Papua New Guinea, where continuing episodes of tectonic uplift have resulted in a series of exposed reef terraces, each terrace being the outcome of interactions between rate of uplift, sea level at the time of formation, and erosion from previous sea levels. Study of these terraces (Ahron and Chappell 1986) shows that, unlike the deep ocean where most palaeotemperatures are obtained, tropical surface temperatures were similar to present during the early part of the glacial cycle and cooler by about 3°C during the late part. According to Prell and Kutzbach (1987), estimates of tropical sea surface temperatures vary geographically above and below present-day values, reflecting, primarily, altered circulation vectors.

The significance of a 3°C cooling, interpreted in the context of present-day relationships between sea surface temperature and reefs, would mean a cessation of accretion of some high-latitude reefs (eg Elizabeth and Middleton reefs and Lord Howe Island of eastern Australia and the Houtman Abrolhos Islands of western Australia), but no effect on coral survival at those reefs, and substantial retraction in the latitudinal distribution of corals in non-reef environments. In Japan, there would be fluctuations between extinction of almost all corals *from* the mainland (non-reef) islands during cold intervals, and the spread of most tropical species *to* the mainland islands during warm ones. The only effect on reefs would be to increase the effects of local windchill. Temperature change would be much less important than sea level change, but would certainly exacerbate the latter's impact.

PALAEOCLIMATIC CYCLES

Most macroevolutionary explanations of major global biological change have, with good reason, been associated with rare catastrophic extinctions, and the slow tectonic events that have altered the configuration of the continents and the earth's climate. These events are much too slow to have a direct impact on the microevolutionary processes of speciation, and it is perhaps for this reason that microevolution is almost always assumed to be, somehow, biologically controlled. There are now good grounds for challenging this view. Major global geological events, as demonstrated by glacial cycles, can occur at microevolutionary frequencies for corals, and there is ever-growing evidence that these changes are not just limited to the Pleistocene, but have occurred as far back in time as climatic records can reach.

Palaeoclimate models, including those derived from isotope records of marine microfossils, demonstrate that ice-albedo feedback and temperature-precipitation feedback are intrinsic cyclical properties of the earth which create the glacial cycles (Nicolis and Keppenne 1989). These cycles are further influenced by orbital forcing (Milankovitch cycles)[6], which have periodicities of 19 thousand years, 23 thousand years, 41 thousand years and 100 thousand years.

Palaeoclimatic cycles would not have changed the paths or direction of the main equatorial currents or the eastern boundary currents, for these are dependent on the direction of rotation of the earth. They did, however, greatly alter the path of the Gulf Stream in the north Atlantic (CLIMAP 1976, 1982), and similar impacts on the Indian and Pacifc Oceans are predictable (eg Pestiaux et al 1988; Howard and Prell 1992). Milankovitch cycles acting at higher frequencies than glacial intervals are also likely to have modified currents through wind-driven surface forcing, turbulence and upwelling.

Links between palaeoclimatic cycles, surface circulation patterns, atmospheric carbon dioxide[7], deep water circulation (Howard and Prell, 1992) and global precipitation are far from established. For present purposes, however, a causal mechanism for variations in surface circulation patterns operating at both microevolutionary and macroevolutionary time scales clearly exists. It is changes to these patterns (eg de Menocal et al 1993) that create changes to the patterns of genetic connectivity in all marine life that is passively dispersed by ocean currents.

8

THE CENOZOIC
FOSSIL
RECORD

'I conclude that instances of fossils overturning theories of relationship based on Recent organisms are very rare, and may be nonexistent. It follows that the widespread belief that fossils are the only, or best, means of determining evolutionary relationships is a myth.'

(Patterson 1981)

However much evolution as a subject is central to biology, our understanding of evolutionary change primarily comes from the decidedly non-biological study of fossils. Without fossils, our conception of the past could only be imaginary, systematics would exist in a temporal vacuum, and the geological eras of the earth would have little relevance to it. Yet palaeontology has never stood alone: observations of the present have always been blended with it, and, in recent years, molecular techniques have provided phylogenetic reconstructions that are certain, one day, to share centre stage with the fossil record. Palaeontology, neontology and molecular biology are generally ill-suited bedfellows, yet a blend of the three is an essential component of all systematics and all biogeography.

The previous chapter concerned Mesozoic corals and the great physical-environmental events and cycles that shaped evolution from the Mesozoic to the Present. The subject of modern physical environments was taken up in chapter 6; that of modern distributions and evolutionary mechanisms respectively in parts C and D of this book. This chapter is about Cenozoic corals, from the time of the survivors of the Cretaceous/Tertiary extinctions to those now living. Emphasis changes from that of mostly extinct families (the subject of chapter 7) to modern genera and thence to extant species; issues raised along the way become increasingly biological and less geological.

THE MAIN TOPICS

♦ Evolution of modern genera through the Cenozoic: the weak inheritance from the Mesozoic, the building of generic diversity in the Eocene, the proliferation of Tethyan and Caribbean corals in the Oligocene and Miocene.
♦ Statistics of genera in the Cenozoic fossil record: changes in total generic diversity; inheritance of generic diversity; zooxanthellate versus azooxanthellate genera; origins of extant zooxanthellate genera.
♦ Information content of the fossil record.
♦ Global patterns of generic age.
♦ 'Species' in the fossil record.
♦ The geological longevity of coral species.
♦ The immediate ancestry of extant corals.
♦ The geological longevity of coral genera and species.
♦ Evolutionary history of extant coral species.
♦ Concepts of macroevolutionary change.

GENERA

THE PALAEOGENE

The Palaeocene opened with the few survivors of the Cretaceous/Tertiary extinctions (p 119), existing in ecological conditions very dissimilar from those of today. For the first 12 million years (longer than the interval between the last rugosan and first scleractinian, p 113) only thirteen new genera have been recorded. Probably only three of these were zooxanthellate; *Stylophora* is the only one now living, but as there was no significant reef formation, the fossil record of the epoch is poor. It was thus a radiation of new zooxanthellate genera that populated the tropical seas of the Eocene, the Faviidae being the only Cretaceous family to re-establish numerical dominance. Seventeen Eocene genera are now extant, but as Eocene reefs are sparse in most parts of the world, the record is again poor. The end-Eocene extinctions (p 118) appear to have created a marked reduction in 'species', but not generic, diversity.

The Late Oligocene was an interval of massive reef development, which may have resulted in the highest diversity of corals — Frost's 'cosmopolitan fauna' — for the whole of the Tertiary (Frost 1972, 1977a, 1981; Frost and Langenheim 1974). At least this was so for the Caribbean and Tethys; the Indo-Pacific fossil record is almost non-existent.

Oligocene and Miocene corals provide the first insight into something approaching biologically meaningful species (p 141). Frost's work on them provides a calendar of events about the origins of extant fauna,

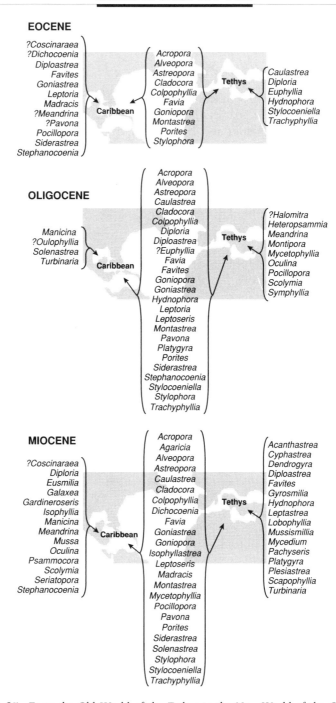

EOCENE

?Coscinaraea
?Dichocoenia
Diploastrea
Favites
Goniastrea
Leptoria
Madracis
?Meandrina
?Pavona
Pocillopora
Siderastrea
Stephanocoenia

Caribbean

Acropora
Alveopora
Astreopora
Cladocora
Colpophyllia
Favia
Goniopora
Montastrea
Porites
Stylophora

Tethys

Caulastrea
Diploria
Euphyllia
Hydnophora
Stylocoeniella
Trachyphyllia

OLIGOCENE

Manicina
?Oulophyllia
Solenastrea
Turbinaria

Caribbean

Acropora
Alveopora
Astreopora
Caulastrea
Cladocora
Colpophyllia
Diploria
Diploastrea
?Euphyllia
Favia
Favites
Goniopora
Goniastrea
Hydnophora
Leptoria
Leptoseris
Montastrea
Pavona
Platygyra
Porites
Siderastrea
Stephanocoenia
Stylocoeniella
Stylophora
Trachyphyllia

Tethys

?Halomitra
Heteropsammia
Meandrina
Montipora
Mycetophyllia
Oculina
Pocillopora
Scolymia
Symphyllia

MIOCENE

?Coscinaraea
Diploria
Eusmilia
Galaxea
Gardineroseris
Isophyllia
Manicina
Meandrina
Mussa
Oculina
Psammocora
Scolymia
Seriatopora
Stephanocoenia

Caribbean

Acropora
Agaricia
Alveopora
Astreopora
Caulastrea
Cladocora
Colpophyllia
Dichocoenia
Favia
Goniastrea
Goniopora
Isophyllastrea
Leptoseris
Madracis
Montastrea
Mycetophyllia
Pocillopora
Pavona
Porites
Siderastrea
Solenastrea
Stylophora
Stylocoeniella
Trachyphyllia

Tethys

Acanthastrea
Cyphastrea
Dendrogyra
Diploastrea
Favites
Gyrosmilia
Hydnophora
Leptastrea
Lobophyllia
Mussismillia
Mycedium
Pachyseris
Platygyra
Plesiastrea
Scapophyllia
Turbinaria

Figure 35 From the Old World of the Tethys to the New World of the Caribbean. Records of now-living zooxanthellate genera of the Eocene, Oligocene and Miocene. Lists of genera are for Caribbean only (left), both Caribbean and Tethys (centre) and the Tethys only (right) (see text).

indicating something of the speed at which macroevolution occurs, and the environmental events that affect it. Of all Oligocene faunas, that of the north-western Tethys, especially northern Italy, have been exceptionally well studied. Frost (1981) listed approximately eighty 'species' (from five hundred and twenty nominal species) and concluded that almost all range in time through the entire epoch (13 million years). He also noted considerable similarity at the generic, if not 'species' level, between contemporary Tethys and Caribbean/Gulf of Mexico assemblages (figure 35), of which he (1972) listed sixty-five 'species'.

Frost, ahead of his contemporaries and foreshadowing comparable findings for the extant corals of the central Indo-Pacific, speculated that longitudinal variation in coral assemblages (between the Oligocene Tethys and Caribbean) was due to variation in the dispersal ability of planula larvae combined with a lack of 'stepping stones' across the Atlantic, and that latitudinal variation was due to differential tolerance of sea temperature.

THE MIOCENE

The Miocene is the Epoch of greatest interest in the evolution of extant corals. It is probably the time of origination of all remaining (post-Oligocene) genera and, probably, homologues (at least) of most extant species. It is also the time of obliteration of the Tethys, of extinction of all zooxanthellate corals from the Mediterranean, and the start of separate evolution of the faunas of the Atlantic and Indo-Pacific. These events did not take place concurrently: they must be considered as part of a succession of very major global climatic and tectonic changes. By the Early Miocene there was probably a high degree of provincial endemism due to isolation of the western Tethys. As the Mediterranean became restricted 'species' numbers gradually declined (Chevalier 1961), fewer than five surviving until the Messinian (very Late Miocene) salinity crisis, which extinguished the last survivors. *Porites* remained the overwhelmingly dominant genus throughout this time (Rouchy et al 1986).

Throughout most of the Miocene, and certainly the Late Miocene, the eastern Atlantic was too cold for reef formation (Esteban 1980), creating a barrier to dispersion to the Caribbean. Only four extant zooxanthellate genera, *Stylophora, Acanthastrea*[1], *Pocillopora* and the cosmopolitan *Porites*, have been recorded in the fossil record of the northeastern Atlantic (Madeira and Porto Santo) (Chevalier 1972; Boekschoten and Wijsman-Best 1981), suggesting that extant eastern Atlantic corals may have had their origins as far back as this time (p 168).

Early Miocene corals of the Caribbean were a much reduced fauna of twenty-five to thirty-five 'species'. These were transitional between the distinctive cosmopolitan corals of the Oligocene and the regional endemics of the Later Miocene, many of which belong to extant genera.

Figure 36
Total now-living
and extinct genera
of Scleractinia.
Data are from
many publications
(see text and fig-
ure 25). High
numbers of genera
during the
Jurassic and
Cretaceous are the
product of rela-
tively long time
intervals.

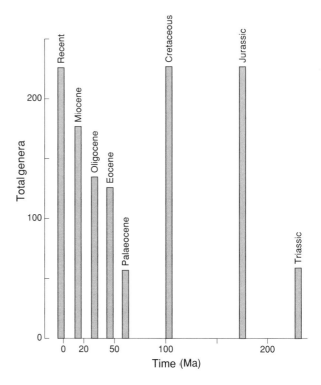

Figure 37
Numbers of new
(solid line) and
inherited (broken
line) zooxanthel-
late and azooxan-
thellate Cenozoic
genera. Newell
(1971), Raup and
S. Stanley (1978)
and G. Stanley
(1981) give
earlier versions of
this figure,
extending back to
the Mesozoic
(all based on
Wells 1956).

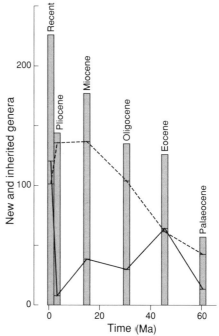

They were dominated by *Porites* and *Goniopora* (Budd et al in press), but also included an array of extinct Oligocene genera and various immigrants from the western Pacific, including *Psammocora* and additional species of *Pavona* (Frost 1977a,c).

The main changes of the Late Miocene/Pliocene of the Caribbean are summarised by Frost (1977a). Whole new groups evolved (eg agariciids and free-living faviids and meandrinids); many modern species in long lineages appeared; distinctive elements of the transitional fauna such as *Psammocora* and (in the Later Pliocene) *Pavona* became extinct; *Goniopora* underwent marked reduction in species diversity (also in the Late Pliocene); and *Stylophora* and *Pocillopora* became abundant.

The Indo-Pacific fossil record is much less interesting, perhaps because of evolutionary stasis, perhaps because of lack of study. Two studies show that corals were distributed much more widely in the Early Miocene than they are today: nine genera, *Stylophora, Goniopora, Alveopora, Goniastrea, Platygyra, Leptastrea, Plesiastrea, Cyphastrea* and *Turbinaria*, reached northern New Zealand (Hayward 1977); and four genera, *Stylophora, Seriatopora, Galaxea* and *Favia* were recovered in the Midway borehole (genera that no longer occur in the Hawaiian chain; Wells 1982; Grigg 1988).

There are no other studies of the Indo-Pacific Miocene of biogeographic interest because most records are from regions of present high diversity and thus do not indicate distribution changes. The oldest records of genera (listed by Veron and Kelley 1988) are mostly from studies carried out earlier this century and are unreliable both stratigraphically and taxonomically.

CENOZOIC STATISTICS

The generic-level database on which much of this chapter, and some of chapter 7, is based is far too extensive to be included in this book other than in summary diagrams focused on extant zooxanthellates. Much of it is badly in need of revision and detailed re-interpretation by specialists (p 108). These data are used to illustrate conclusions that may be drawn from the fossil record rather than specific details of it.

Total generic diversity of Jurassic, Cretaceous and extant corals are very similar (figure 36). Although these data are highly dependent on the different taxonomies, time intervals and preservation potentials (p 138) of the geological intervals they represent, they are compatible with the view (p 218) that there is an evolutionary mechanism limiting absolute global diversity. Closer examination of the generic record of the Cenozoic is summarised in figures 37–41 and as follows[2]:

♦ The total number of genera progressively increases from the Palaeocene[3] to the Recent. This is primarily due to the evolution of new

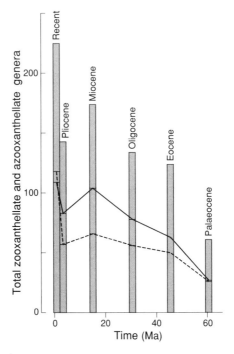

Figure 38 Numbers of zooxanthellate genera (solid line) and azooxanthellate genera (broken line, as inferred from now-living genera), during the Cenozoic. Unattributable genera were apportioned equally.

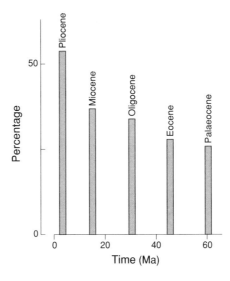

Figure 39 Percentage of total of now-living zooxanthellate genera in Cenozoic fossil record.

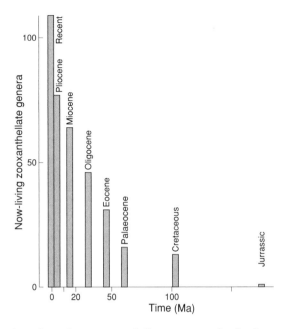

Figure 40 Number of now-living zooxanthellate genera in the fossil record. Mesozoic records are open to doubt (see text).

genera, but is also due to greater chances of fossil recovery (p 139), especially in the Indo-Pacific Plio/Pleistocene (figure 37). The number of inherited genera for each epoch remains approximately stable after the Palaeocene, except for the Recent, where many genera have no fossil record. This stability is primarily due to the geological longevities of genera (see below).

♦ The proportion of zooxanthellate to azooxanthellate genera (based on assumptions from modern corals) is relatively constant (figure 38). The principal reason for this is that the zooxanthellate family Faviidae and the azooxanthellate Family Caryophylliidae are numerically dominant, and numerically uniform, throughout the Cenozoic (figure 25, p 110).

♦ The percentage of extant zooxanthellate genera declines only slowly back in time (figure 39). This is again due to high rates of inheritance and low rates of extinction. The very general principle is that genera have, effectively, an almost indefinite longevity between extinction events.

♦ The number of extant zooxanthellate genera decreases progressively back in time, the oldest genus, *Montastrea*, having been recorded from the Jurassic (figure 40). This diagram is a function of generic longevity, rates of evolution, and preservation potential. Essential points are that the number of extant genera of the Cretaceous and Palaeocene are similar (*Stylophora* being the only point of difference, p 131); that nearly one-quarter of extant genera occurred in the Eocene (mostly in the Caribbean and Tethys, figure 35, p 132) and that nearly half occurred in the Oligocene (Frost's 'cosmopolitan' fauna; p 131 and figure 35).

♦ Figure 41 shows the mean longevity of all Cenozoic coral genera. The fitted curve is created by a very low level of inheritance from the Mesozoic due to mass extinction in the Cretaceous/Tertiary boundary (p 119), and a low rate of subsequent extinction. The detail of the histogram is created by the end-Eocene extinctions (p 121), new speciation in the Oligocene (p 121), and the existence of extant genera which have no fossil record. This figure is not extended back into the Mesozoic because, considered as a whole, the longevity of genera becomes more a product of an incomplete fossil record (ie inaccurately known points of origination and/or extinction) than a reflection of reality[4].

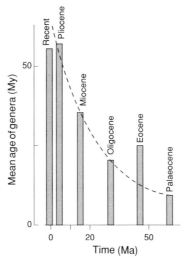

Figure 41 Mean age of Cenozoic genera from the fossil record (see text). The mean age of now-living genera is an artefact of the relatively poor fossil record in the Indo-Pacific; the broken line is a more realistic indicator.

Statistics from the generic fossil record Apart from variations in taxonomic reliability and opinion, as well as issues of spatial and temporal variation, the following caveats seem particularly applicable to corals.

♦ The more widely distributed a genus is at a given point in time, the greater its preservation potential. The fossil record favours genera with high dispersal capability and physical-environmental tolerance.

♦ The more common a genus is at a given point in time, the greater its

preservation potential. The fossil record favours abundant genera.
- ◆ The more recognisable a genus is, the more likely it is to be recorded. The fossil record favours genera with distinctive morphological attributes.
- ◆ Genera with large numbers of species may have greater recovery potential. The fossil record may favour speciose genera.
- ◆ Clearly, the more likely a genus is to be found, the older its oldest records are likely to be.
- ◆ Clearly, also, the more recent a fossil is, the more likely it is to be recovered because it will be in the youngest, therefore the most exposed strata (Raup 1979 appropriately termed this 'the pull of the Recent').

The chance of recovery of fossils from reefs (which are well-defined and mostly well-documented carbonate platforms) is much greater than recovery from non-reef deposits. Thus, the fossil record of zooxanthellate corals should be much more complete than the record of azooxanthellate corals.

Much more is known about the Tethys and Caribbean Cenozoic than about the Indo-Pacific, partly because there are relatively few well-preserved Palaeogene reefs in the Indo-Pacific, and also because Miocene reefs, which are abundant in the central Indo-Pacific, have been little studied.

AGE PATTERNS AND DIVERSITY

The subject of age patterns and diversity has remained overwhelmingly dominated by Stehli and Wells's (1971) seminal paper, which has since become a marine cornerstone of the Darwinian concept of centres of evolution. The principal database of this study was the generic distribution map (reproduced, figure 13, map B, p 46) and the age of genera recorded in Wells (1956) and updated by J Wells (pers comm). This allowed the global contours of generic age (reproduced in figure 14, p 47) to be determined and regressions of generic diversity against average generic age to be plotted for locations within the Indo-Pacific and Atlantic (figure 15, p 48).

Generic age contours have been recomputed here (figure 42) using the present database of generic *ranges* (not records, p 277), and a minimum geological age of 25 million years (see below). Figure 42 indicates the following:

- ◆ The central Indo-Pacific centre of diversity has an average generic age of 30 million years, half that of the Caribbean. The 30 million year contour follows the 70 genera contour of diversity (figure 48, p 158).
- ◆ There is no decrease in average generic age west of the Indo-Pacific centre of diversity, just as there is no attenuation of generic diversity (figure 48).

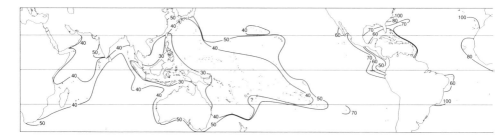

Figure 42 Contours of average age of all now-living genera (million years), generated by multiplying the present ranges of all genera (in Veron 1993) by their maximum age as indicated by the fossil record. All genera were given a minimum age of 25 million years and a maximum age of 100 million years (see text). The computation used an IDRISI Global Information System program.

♦ There is an increase in average generic age east from the Indo-Pacific centre of diversity to the southern Pacific which is inversely correlated with contours of diversity (figure 48).
♦ The far eastern Pacific has average generic age contours that correlate directly with diversity contours rather than inversely (as above).
♦ In all three ocean basins, average generic age increases with latitude.

Several authors have expressed the view that Stehli and Wells's age contours may be a statistical artefact created on the basis that the older a genus is, the more species it will have and the wider its distribution will be by chance alone (eg Vermeij 1978; Jokiel and Martinelli 1992), and that the contours in remote areas are created by a few hardy species (Raup and Stanley 1978). Observations from this study are as follows.
Central regions Because of the poor Indo-Pacific fossil record, 30 per

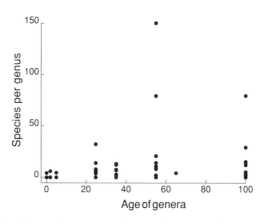

Figure 43 Relationship between the age of genera (million years) and the number of species in them.

cent of all genera within the 30 million year contour have no fossil record earlier than the Pliocene. Thus, the existence of the 30 million year contour becomes partly dependent on the 25 million years minimum age arbitrarily used to generate the contours. Excluding these genera from the computation does not, however, significantly disrupt the basic pattern. It is therefore reasonable to conclude that Caribbean genera are substantially older, on average, than Central Indo-Pacific genera. The most plausible explanation of this is that Tethyan genera are older than non-Tethyan genera. It does not demonstrate faster evolution in the Indo-Pacific (as concluded by Stehli and Wells 1971).

Peripheral regions The average age of peripheral regions in both the Pacific and the Atlantic is the product of a small number of highly dispersed species of (the same) early Cenozoic genera. It is not created by older genera having more species (figure 43), nor is it created by displacement of species. The peripheral pattern is created by dispersion, not evolution.

SPECIES

SPECIES FROM FOSSILS?

Geological research, though it has added new species to extinct and existing genera...has done scarcely anything in breaking down the distinction between species...this is probably the gravest and the most obvious of all the many objections which may be urged against my views. (Darwin 1859)

The species problem in corals in general...has long been notoriously vexatious for no one knows the criteria of species limits among the living forms (which suggests utter confusion in fossil forms). (Wells 1969)

Numerous authors (Wells 1969; Wijsman-Best 1974; Pfister 1977; Beauvais 1992; Budd 1984, 1985; Borel Best et al 1984; Lathuillière 1988; Sorauf and Mackey 1988) have expressed a view that there are 'problems' of various sorts with fossil 'species' of Scleractinia. The same sorts of observations have been made about rugose corals (eg Oliver 1989; Fedorowski 1989) and tabulate corals (eg Scrutton 1989). Yet the rain of species names in the literature, and erroneous inferences made from them, continues unabated. As implied by Wells, the plight of species in coral palaeontology is chaotic. There is, in fact, almost nothing in common between a species of extant coral and most 'species' of fossil coral (p 229). There may be a reasonable basis for transferring information about extant species to very well-preserved fossils of those species or (less confidently) to Tertiary congeners of those species, but further back in time the trail is lost. The above authors have understated the extent of the issue. All Mesozoic and most extinct Palaeogene 'species' have little or no taxonomic status. The reasons for this unhappy conclusion are primarily related to information loss about intra-specific variation, and through preservation.

Intra-specific variation Variation within the vast majority of species can be affected by relatively minor physical-environmental change in geological terms: for example most variation of Great Barrier Reef *Pocillopora damicornis* illustrated in figure 4, p 17, would be expected with a sea level change of about 10 m.

The palaeontological literature abounds with descriptions of sequences supposedly indicating succession or replacement of species with time, or various evolutionary changes within a clade. These processes may indeed have occurred, but a simple alternative explanation is that such sequences, if taken individually, reflect physical-environmental change, especially change in water depth, or prevailing sedimentary regime.

Extant species only become meaningfully defined if physical-environment-correlated variation between contiguous biotopes is known, if principal elements of geographic variation are known, and if the species is meaningfully separable from other closely related species. These are taxonomic necessities, not niceties, and they can only be determined from fossils in the very best sequences. The amount of variation found in most extant coral species is often more than the amount of variation frequently described in several genera of fossils.

Preservation Scleractinia, being aragonitic, usually have much of their skeletal detail destroyed by diagenesis, which normally occurs rapidly in geological terms, depending on exposure, especially to fresh water. Diagenesis causes significant modifications in the coral skeleton: all skeletal elements are thickened and 'pseudo-skeletal elements' are commonly formed. Finding aragonitic skeletons in Mesozoic deposits (associated with aragonite to calcite inhibiting iron ions in the primary matrix according to Roniewicz 1984) helps with the diagnosis of higher taxon levels, but not species.

Identification of most species, extant or fossil, necessarily requires the surface microstructure of calices to be preserved, but these fine skeletal structures are usually the first to be destroyed by diagenesis or erosion. Thin sections provide an alternative means of study, but the information these reveal is seldom adequate for branching species and for most species-rich Indo-Pacific genera.

Essential points about the information value of 'species' in Mesozoic and (to a lesser extent) Palaeogene fossils are:

♦ Fossil species descriptions usually have little information value unless the 'species' has very distinctive characters, or the description relates to some broad body of knowledge about the genus.

♦ Fossil species *may* add value to observations based on co-occurrence. For example, the statement 'the two upper reef slopes had the same genus, but no species in common' could be equally as meaningful for a fossil reef as a modern one.

♦ Numbers of fossil species can convey more information than numbers of genera, but the absolute number has little meaning in most situations other than the aforementioned Caribbean sequences.
♦ Identity of fossil species conveys information only in very restricted circumstances. Species in broad-scale palaeobiogeography are generally not meaningful because they are composites of all the issues and problems associated with extant coral taxonomy, temporal change, lack of environmental information and information loss through fossilisation (as described above).
♦ Fossil species and 'chronospecies' (Stanley 1978) are commonly confused, but are not even vaguely the same thing. The former is an operational unit used in palaeontology; the latter is a conceptual unit in which a temporal dimension is added to some other kind of species (p 66).

THE PLIO-PLEISTOCENE

Depending on state of preservation, the taxonomy of extant species becomes increasingly relevant to Pliocene and Pleistocene corals as we move out of evolutionary time-frames and into ecological ones.

The Caribbean   The Plio-Pleistocene was a time of significant faunal turnover. Budd et al (in press), combining their work with Frost's, records changes in coral composition as follows:

♦ Extinctions were approximately simultaneous in all zooxanthellate families except the Mussidae. The most affected were the Pocilloporidae (*Stylophora* and *Pocillopora*), the Agariciidae (*Pavona* and *Gardineroseris*) and free-living faviids and meandrinids.
♦ Most genera that became extinct in the Caribbean are extant in the Indo-Pacific.
♦ Rates of speciation and extinction were approximately equal.

Pocillopora (resembling *P. eydouxi*) appeared in the Caribbean in the Late Pleistocene (Geister 1977) then became extinct there; *Acropora palmata* first appeared in the Pleistocene (Frost 1977b), and remained. Frost explains the presence of *Pocillopora* by a hypothetical sea highstand, straddling the Isthmus of Panama, allowing immigration from the Pacific. In the absence of corroborative evidence of such a high sea level, a more likely explanation is that the genus was sufficiently rare, or geographically restricted, to remain unrecovered. He explains the existence of the *Acropora* by hybridisation between *A. cervicornis* and the extinct *A. panamensis,* but the same explanation is at least as likely.

The Central American Seaway   Little is known about the physical-environmental effects of the closure of the Central American Seaway (reviewed by Maier-Reimer and Mikolajewicz 1990). Cold eastern Pacific water would have flowed east, seasonally, through the seaway, thus the main effect of the closure would have been an increase in the temperature of the western Caribbean.

Eight Early Pliocene zooxanthellate genera (*Dichocoenia, Diploria, Eusmilia, Porites, Siderastrea, Solenastrea, Madracis* and *?Isophyllastrea*), all now extant in the Caribbean, are known from the Gulf of California (Vaughan 1907; Durham 1947; Budd pers comm); all but *Porites, Siderastrea* and *Madracis* are now extinct in the Pacific. All extant genera and most extant species of the far eastern Pacific occur elsewhere in the Pacific. The origin of this fauna has been the subject of much debate, stemming from the efforts of vicariance-promoting biogeographers to argue against Dana's (1975) hypothesis of long-distance larval dispersal (p 59). The essential points are:

◆ No far eastern zooxanthellate Pacific species is extant in the Caribbean.

◆ Some far eastern species are morphologically close to species of the Plio-Pleistocene Caribbean (eg *Pocillopora eydouxi, Porites lobata* and *Gardineroseris planulata*).

◆ Some far eastern species *may* be endemic (*Siderastrea glynni, Porites panamensis*). The latter was recorded from the Pliocene of Papua New Guinea (Veron and Kelley 1988), indicating a former widespread distribution.

◆ Most far eastern species have wide Indo-Pacific distributions. These include unattached species that cannot be dispersed on floating objects (p 87).

A reasonable conclusion from this evidence is that most species in the far eastern Pacific came from the west, but that some could be relict populations more than 3 million years old. Reys Bonilla (1992) discusses the issues.

The Indo-Pacific The Era beds of the Pliocene epoch of southern Papua New Guinea contain by far the world's richest fossil coral outcrop (Veron and Kelley 1988). The beds contain the oldest Indo-Pacific records of twelve genera (*Stylocoeniella, Pseudosiderastrea, Coscinaraea, Ctenactis, Heliofungia, Herpolitha, Sandalolitha, Podabacia, Oxypora, Australomussa, Merulina, and Duncanopsammia*), as well as a further five genera (*Gardineroseris, Mycedium, Cynarina, Symphyllia and Caulastrea*) that are more-or-less contemporaneous with outcrops in Indonesia. They also contain (according to B Rosen pers comm) the world's oldest occurrence of a major *Acropora*-dominated community (pers obs). The extent of 'oldest records' in the Era beds says little about the beds themselves; what it does do is demonstrate that the Indo-Pacific fossil record is very incomplete, especially when compared to that of the Caribbean.

Interest in the Era beds is, however, in species, because a large number are exceptionally well preserved. Only 144 (more than one-quarter) of all extant Indo-Pacific zooxanthellate species have any fossil record at all. Of these, 70 per cent have been found in the Era beds[5]. These corals lived in inshore conditions similar to those found in the nearby Motupore region of southern Papua New Guinea, where 87 per cent of Pliocene bed species have extant equivalents. Comparison between the fossil and extant outcrops shows that, over an interval of 2 to 3 million years:

- Sixty-five species showed no taxonomically significant change; nine were sufficiently different that they bordered on being called different (probably extinct) 'species'; nine are (as far as is known) now extinct, two belonging to extinct genera.
- As the Era beds are within the Indo-Pacific centre of diversity, few palaeobiogeographic observations can be made, but one species (*Parasimplastrea simplicitexta*), which was believed extinct, has since been recorded in Oman (Sheppard 1985), and another species (*Porites panamensis*) is now known only in the far eastern Pacific.
- The study allows rates of origination to be estimated, but only for a single point in space and time. No other comparable Indo-Pacific studies add further points, and thus macroevolutionary rates for the entire realm are nowhere near as well understood as they are in the Caribbean.

Twenty-one extant Indo-Pacific genera have older records in the western Tethys and/or Caribbean than they do in the Indo-Pacific (Veron and Kelley 1988). *Gardineroseris planulata* and *Leptoseris gardineri* are extant Indo-Pacific species recorded by Budd et al (in press) from the Caribbean. *Diploastrea heliopora* may be the only extant Indo-Pacific species recorded from the Miocene Tethys (pers obs).

THE HOLOCENE

Interest in the Holocene focuses less on evolutionary change and the Caribbean region, and more on distribution change and the Indo-Pacific. The short epoch, 10 thousand years ago to the present, links the last glacial interval to the environments of today. For most places, Holocene coral outcrops are of little interest because they have the same species composition as extant corals.

The corals of Tateyama (near Tokyo, Japan), however, are an extraordinary exception[6] that provides a glimpse of biogeographic change in the 'thousands of years' time-frame, and does so at a place where physical-environmental constraints on distribution and survival are critical. The site is the world's highest-latitude coral outcrop (p 85); it was uplifted and preserved in a single tectonic event 5000 to 6000 years ago, a time sometimes referred to as the 'Middle Holocene High'. At least seventy-two coral species (of which fifty-three have been identified) occurred at Tateyama 5000 to 6000 thousand years ago; by contrast, only thirty-five species have been recorded there this century (Veron 1992b)[7].

Comparison between the Middle Holocene fossils and the present distribution of corals in mainland Japan shows that 85 per cent of the former are extant as far north as Kushimoto, and all except two are extant somewhere in mainland Japan (figure 57, p 176). There have been major changes in abundance however: *Acropora* is rare in the fossil record, *Caulastrea tumida*, *Favia speciosa* and *Echinophyllia aspera* are among the dominants. No such community composition exists in Japan today (Veron 1992a and earlier authors cited therein).

Three important findings are suggested by the Tateyama fossils:

♦ A brief period of regional or global warming can produce an apparently subtropical coral community, even at the latitudinal extreme of extant corals.

♦ Change over ecological time can approximate change over biogeographic space (p 219). Change in the recorded coral composition at Tateyama over 5000 to 6000 years is approximately equal to variation in extant coral communities throughout all mainland Japan.

♦ Detailed comparisons between the temperature regimes and species compositions of extant coral communities of mainland Japan (Veron and Minchin 1992) indicates that the Tateyama fossils had sea temperatures 1.2–1.7°C above present: that is, a temperature rise of less than 2°C can result in a doubling of species richness.

GEOLOGICAL LONGEVITY

Rates of speciation (and in some cases, extinction), as indicated by points of evolutionary divergence, can be estimated by three distinct methods: from the longevity of taxa in the fossil record; by comparing the proportion of extinct to extant taxa at particular points in the fossil record; and by molecular studies of extant taxa. The first and second methods are applicable to Cenozoic genera, and to species in restricted circumstances. The third may dominate most aspects of the subject, but so far has been applied only to the Poritidae (Potts and Garthwaite 1990; Potts et al in press; Garthwaite et al 1994) and to various unpublished estimates of family-level divergences.

All methods and results raise issues of the nature of clade formation. Genera in the fossil record tend to 'appear' or 'disappear' abruptly. This is partly an artefact of taxonomy (reflecting arbitrary points of decision as to what are generic-level divergences, or not), partly the result of genera being monospecific at points of origination if not extinction, and partly an abundance effect where the genus may have evolved from a rare (or azooxanthellate) clade that existed long before the genus 'appeared' as a fossil. Species may originate either as a result of gradual phyletic change within a clade, or by the division of a clade, or by repeated division and fusion of clades. The fossil record may distinguish the second clearly enough, but the first and third may well appear as various forms of uncertainty or noise.

Rates of evolution can only be determined from the fossil record in very restrictive cases because extinct equivalents of extant species are (except as noted above) largely indeterminable in fossils; very good (aragonitic) preservation of surface structure is usually essential; and a high level of chronological continuity is essential. These conditions are met only in the Caribbean region. Elsewhere, rates of evolution can be estimated from fossils only by comparing numbers of extinct species with numbers extant species at specific points in geological time (the second instance noted in the previous paragraph).

The most informative study is that of Budd et al (in press) referred to above (p 143). This showed that Mussids had the highest rates of speciation (over 10 per cent of species per million years over the past 6 million years), while pocilloporids had the lowest (less than 5 per cent of species per million years over the last 22 million years). Speciation and extinction rates are notably uniform during the Miocene and increase during the Plio-Pleistocene. Throughout both intervals, there is a high level of correlation between speciation and extinction rates within families. Both rates are also generally correlated among different families within any given time interval. The authors note that these observations indicate an abiotic control of evolutionary rates, that is, one that acts indiscriminately on different taxa, and that these taxa in the Indo-Pacific are not similarly affected.

No fossil sequences comparable with the Caribbean are known from the Indo-Pacific. The aforementioned study of the Era beds of southern Papua New Guinea presents a single glimpse of evolutionary change. Here, the rate of extinction was exactly the same as the rate of speciation. The combined rate in all families was 4.4 per cent per million years, that is, about half that recorded by Budd in the Caribbean. This outcome, superficially at least, conflicts with the conclusion (p 139) that Caribbean genera, which are poorly speciated, have twice the average age as Indo-Pacific genera, which are relatively well speciated (p 159).

An average speciation rate of 4–10 per cent of species per million years suggests an average longevity of species of more than 10 million years. In the real world, there will have been many influences on this figure, global climatic change not the least.

CONCEPTS OF
MACROEVOLUTIONARY CHANGE

> Critical evaluation of macro-evolutionary theories demonstrates that their validity is not corroborated by evidence. The term punctuated equilibrium may describe some paleontological patterns of morphological evolution, but the main postulates of punctuated equilibrium about patterns of specific evolution have been refuted. (Hecht and Hoffmann 1986)

As the above authors point out, the relationship between microevolution and macroevolution has been among the most contentious of all evolutionary biology, a controversy that was, at least in 1986, 'hotter than ever'. Corals may be destined to make a significant impact on central issues, partly because they have one of the best fossil records of all animal groups, and partly because they are attracting study from so many different points of view, including genetics and molecular biology.

MACROEVOLUTION AND THE FOSSIL RECORD

The literature on this subject is astutely interpreted by Levinton (1988) and has been recently reviewed by Briggs and Crowther (1990). It raises two simple points concerning corals.

The first point can be illustrated by example. The two corals of figure 44 have an age difference of 2 to 3 million years, yet they are so similar that they may have been different parts of the same corallum: are they the same species? In concept (figure 45), species A at time T_0 is the same species as B at time T_1 and D at time T_2. Species J at time T_3 is probably

Figure 44 Modern *Pavona maldivensis* from the Great Barrier Reef (the same specimen as figure 3, p 16) (left) and a 2–3 million year old specimen from nearby southern Papua New Guinea (right).

genetically distinct, but is morphologically similar, having been generated from the same evolutionary forces as species A and the extinctions at time T_2. This simple pattern would be greatly complicated by the inclusion of hybrids. Continuous clades are recognisable in post Late Miocene fossils, but before that time, morphological homologues are not necessarily conspecific. The same argument applies to genera: Jurassic *Montastrea*, to take an extreme example, closely resembles some extant *Montastrea*, but that could be for many reasons that have nothing to do with phylogeny. Thus Budd and Coates (1992) referred to their Cretaceous *Montastrea* as '*Montastrea*-like' corals; and the same reasoning applies equally to Cretaceous *Hydnophora*-like and *Astreopora*-like corals. Names on fossils, no matter how justified, have different meanings in different geological intervals.

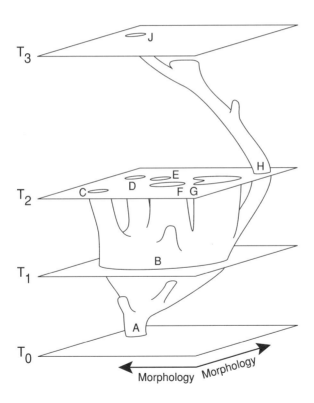

Figure 45 Concept diagram of the phylogeny of a 'chronospecies'. Species A at time T_0 undergoes increase in variation becoming polymorphic species B at T_1. By time T_2 it has evolved into six species (C–H), the first five of which become extinct. By time T_3 species I has evolved in response to the same selective forces as species D. With corals, it is possible for species D and J to be morphologically indistinguishable (eg as in *Pavona maldivensis*, figure 44) yet to have different phylogenies.

The second point is that macroevolutionary change, as observed in the fossil record (mass extinctions excepted), occurs over intervals of ten thousand to one million years. Within the Indo-Pacific, this is much slower than the rate of major distribution changes, which occur at intervals of thousands of years or less (p 85). The inescapable conclusion is that rates of macroevolution and rates of distribution change are not correlated. Distribution changes do, however, underpin most theories of speciation (p 44); the difference in evolutionary rates is the difference between macroevolution and microevolutionary processes, a difference of 1 to 2 orders of magnitude.

In general, macroevolutionary change in corals is very slow, much slower that those reported in most other animal groups (Stanley 1979; Campbell and Day 1987).

PUNCTUATED EQUILIBRIA AND
THE FOSSIL RECORD

The concept of punctuated equilibria, stemming from the classic works of Mayr (1963, 1970), asserts that the history of species is dominated by stasis, with change being minimal and rarely directional, and change, when it occurs, is mostly 'revolutionary', involving marked speciation events. According to the most challenging versions of this view (Eldredge and Gould 1972; Stanley 1979, 1982), speciation by division of clades is the necessary source of evolutionary change, because evolutionary processes within populations (phyletic evolution) is too sluggish to produce the changes observed in the fossil record.

With corals, the weight of evidence is that this may be so only in some families of the Indo-Pacific. Although only a small part of the fossil record (extant zooxanthellate genera) is discussed in this book, this is by far the best-known part, and it shows little evidence of 'revolutionary' evolution. The exceptions may be the Indo-Pacific genera that have little or no fossil record (p 131, 135). *Ctenactis,* for example (which has one of the greatest preservation potentials of all corals), is first recorded in the Indonesian Miocene/Pliocene (Wells 1966, based on very unreliable dating), yet Pliocene *C. echinata* in the Era beds of Papua New Guinea is in all detectable respects identical to those extant. Similar observations apply to other fungiid genera (except *Cycloseris),* as well as those of other taxa that have an equally high preservation potential. Whether this is a result of a weakness of the fossil record or of the existence of a rapid evolutionary event is at present indeterminable. Perhaps the answer will come more from molecular studies than from fossils, but here there must be cautions because molecular methods generally assume constant rates of change, whereas the fossil record demonstrates episodic change.

PHYSICAL ENVIRONMENT VERSUS BIOLOGICAL CONTROL

The 'big picture', as described in the previous chapter, is indisputably controlled by physical environment, by plate tectonics, mass extinction events (discussed for reef organisms in general by Raup and Boyajian, 1988) and episodes of palaeoclimatic change. The species-level fossil record of the Caribbean indicates physical-environmental control of macroevolution (p 147). In chapter 12 it is argued that the finest of details of microevolution are also largely non-biologically controlled. When the big and small scales of evolutionary change are combined, there is strong reason for concluding that most aspects of evolutionary change are primarily mediated by the physical environment. These observations stand in great contrast to the Darwinian concept of evolution being primarily controlled by competition between species (figure 72, p 221).

Red Queen and Stationary Hypotheses Debate over physical-environmental versus biological control of macroevolution, as determined from the fossil record, gained a great deal of momentum when Van Valen (1973) plotted species survivorship curves for different groups of plants and animals and found (to his surprise and that of most other evolutionists) that extinction rates (between intervals of mass extinction) remain constant with time. Van Valen's explanation, which he termed the Red Queen Hypothesis[8], was that species within a community maintain constant ecological relationships with each other, and that it is these relationships, rather than the species, that evolve (much like the 'arms race' between competing countries: new weapons replace older ones in a race to maintain the status quo). With organisms finely tuned to competitive survival, the Red Queen mode may be well-founded; with organisms less finely tuned, or in times of major physical-environmental perturbation, Darwinian selection may take over (ie the arms race becomes a war) until stasis is regained. The alternative 'Stationary model' of Stenseth and Maynard Smith (1984) postulates that evolution is primarily driven by change in the physical environment and ceases in the absence of change. Thus the Red Queen model predicts that rates of speciation, extinction and phyletic evolution will remain constant, even during times of environmental change; the Stationary model predicts punctuated equilibria in response to physical-environmental change.

Hypothetical points of view about the control of evolution appear very different for different taxa, partly because microevolutionary and macroevolutionary processes are usually inextricably interlinked, and partly because time intervals under consideration are often very different. The fossil record of corals does not accord with Red Queen predictions of constant rates of extinction (and its correlate, biological mechanisms of control). Evolutionary change in corals is episodic. Intervals of extinction are associated with sea-level changes and may be short in evolutionary terms. Intervals of origination are ill-defined but appear to be broadly correlated with climatic change. At a macroevolutionary level, the fossil record is perhaps supportive of the punctuated equilibrium ('Stationary') perspective but there is no evidence that the origin of species has proceeded by revolutionary change in time intervals shorter than several glacial intervals. Be that as it may, the fossil record suggests nothing about the mechanism of origin of species. This subject is taken up in chapter 12.

IMPORTED CONCEPTS

Basic aspects of coral speciation (concurrent multi-species mass spawning, mass hybridisation and physical-environmentally controlled long-distance dispersal) are not found in most terrestrial macrofauna, nor is genetic connectivity in terrestrial fauna so directly and continually subject to climatic fluctuations as it is in the shallow ocean. Much of the debate over the relationships between microevolutionary and macroevolutionary processes, punctuated equilibria versus phyletic gradualism, allopatric versus non-allopatric speciation and vicariance versus other means of speciation, has very questionable relevance to corals and probably a wide range of other shallow marine biota.

PART C

MODERN DISTRIBUTIONS

The main conclusions of chapters 9 and 10 are summarised at the end of these chapters. The following is a summary of the main topics.

THE MAIN POINTS

GLOBAL DISTRIBUTIONS AND SPECIES DIVERSITY

- At family level, contours of diversity are dominated by the Eastern Pacific Barrier; the Caribbean is almost as diverse as the Indo-Pacific; there is no Indo-Pacific centre of diversity.
- At generic level, contours of diversity become dominated by the Indo-Pacific centre of diversity, a lack of attenuation across the Indian Ocean, and a relatively low diversity in the western Atlantic.
- At species level, the tropical Indo–west Pacific (from the Red Sea to Fiji) is very uniform. Distinct longitudinal provinces occur only east across the Pacific. The western Atlantic and eastern Atlantic form provinces that are distinct from each other and from those of the Indo-Pacific.
- At species level, contours of diversity are overwhelmingly dominated by the high diversity of the equatorial central Indo-Pacific. Diversity in the Caribbean is similar to the most depauperate provinces of the Indo-Pacific.
- Species compositions are relatively uniform across the Indian Ocean; the Eastern Pacific Barrier creates the most major faunal division of the Pacific. The corals of Brazil are the most distinctive of the western Atlantic.

CENTRAL INDO-PACIFIC SPECIES DIVERSITY

♦ The pattern of species composition in the central Indo-Pacific is dominated by latitudinal attenuations along three continental coastlines from an equatorial centre of species diversity. There is almost no species replacement along these coastlines.

♦ Species composition in equatorial locations is controlled by physical environment more than geographic position.

♦ Species composition in high-latitude locations along continental coastlines is controlled by poleward-flowing boundary currents, which determine the temperature regime and create one-way genetic connectivity with upstream (tropical) communities. Latitudinal dispersion is primarily controlled by 'connectivity ratchets'.

♦ 'Drop-out' sequences vary according to ecological constraints and upstream seeding. Sequences are different for different continental coastlines.

INDIVIDUAL SPECIES DISTRIBUTIONS

♦ The majority of central Indo-Pacific coral species have distribution ranges extending beyond the central Indo-Pacific. Many species span the Indo–west Pacific; a few span the whole Indo-Pacific. Taxonomic issues affect these observations for many species.

♦ Single regions within the central Indo-Pacific have less than 10 per cent endemic coral species; in most regions the level of endemism is less than 5 per cent.

♦ Wide disjunctures in distributions are a common characteristic of corals. These may raise taxonomic issues if they involve extensive tracts of open ocean.

CHARACTERISTICS OF DISTRIBUTION RANGES

♦ Geographic variation in abundance is unpredictable: abundance is an important, though little-studied, aspect of biogeographic patterns.

♦ With most coral species, there is little geographic variation in the range of habitats occupied. Most species have been recorded in a wide range of habitat types and the majority of all species have been recorded in non-reef habitats.

♦ Morphology, colour, and other species attributes vary greatly over very wide geographic ranges. These variations are different for each individual coral species. Different types of geographic variation (figure 67, p 204) occur with different types of disjunct distributions and geographic isolation.

♦ Geographic variation involves intergradations at all taxonomic levels below that of genus. Depending on both taxonomic and conceptual issues, coral species characteristics change over distance in such a way that morphological distinctions, and therefore species boundaries, merge.

9

LONGITUDINAL DISTRIBUTIONS WORLD-WIDE

This chapter is mostly about *longitudinal* (west–east) distributions around the world; chapter 10 is mostly about *latitudinal* (north–south) distributions within the central Indo-Pacific. The two chapters are separated because they have very different data bases, and because longitudinal distributions potentially involve all the complexities of biogeography introduced in chapter 3, set in both ecological and evolutionary time-frames, whereas latitudinal distributions correlate with simple physical-environmental gradients and changes in ecological time-frames (p 269).

Distribution analyses of this chapter are based on the generic distribution maps of Veron (1993, pp 346–400).

GLOBAL DISTRIBUTIONS OF CORALS
FAMILIES

The global pattern of family diversity (figure 46) is entirely created by:

- ◆ the geographic pattern of land masses and reefs;
- ◆ the depauperate faunas of the far eastern Pacific and eastern Atlantic;
- ◆ temperature-correlated latitudinal attenuation;
- ◆ the peripheral distributions of a few species (where these are the only representatives of families).

There is no Indo-Pacific centre of diversity. The Caribbean is almost as diverse as the Indo-Pacific. There is almost no inter-regional variation within the twelve to fourteen family contours. It may thus be concluded that family-level distributions are inherited from the cosmopolitan distributions of the Eocene to Early Miocene (p 131).

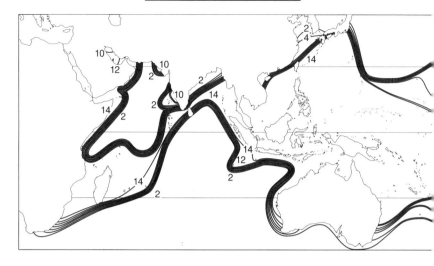

Figure 46 Contours of family diversity generated by combining the distribution *ranges* of all families using an IDRISI Global Information System program. Contours indicate maximum family diversity for large regions; remote or small areas within those regions may have a lower diversity.

GENERA

Figure 47 is a plot of species richness against generic richness for all locations indicated in figure 54 (p 172). This plot shows that for many, but not all purposes (p 48), generic diversity data are meaningful proxy-indicators of species diversity. Historically, most biogeography has been at generic level (p 6); figure 13 (p 46) shows the global patterns of diversity that have been published.

The global generic pattern of diversity generated from the present database (figure 48), shows (as does its predecessors), that there is a well-

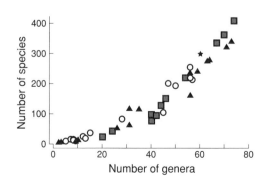

Figure 47 Relationship between number of species and number of genera for all locations in figure 54, p 172. Symbols are as in figure 59, p 182.

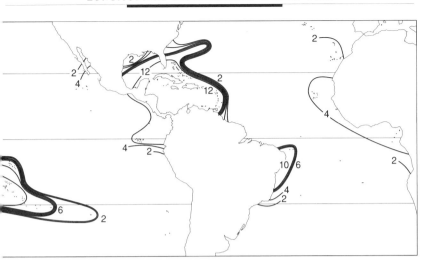

defined Indo-Pacific centre of coral species diversity. The main points of figure 48 are as follows:

♦ West from the centre, there is no attenuation of diversity across the Indian Ocean: the main pattern is created in the north by regional physical-environmental constraints (principally rivers and the sedimentary environment of the Asian continental coast) and in the south by the emptiness of the Indian Ocean (p 47).
♦ The Chagos Stricture is the dominant biogeographic feature of the Indian Ocean.
♦ Along the western border of the Indian Ocean, diversity attenuates with latitude in a manner comparable to the latitudinal attenuations of the central Indo-Pacific (chapter 9).
♦ In contrast to the Indian Ocean, there is an attenuation of diversity east across the Pacific, mostly for the same reasons that created the pattern of attenuation of families (above) for the same area.
♦ Distance of particular locations in the Pacific from the Indo-Pacific centre of diversity is the overriding parameter determining generic diversity.
♦ The pattern in the far eastern Pacific and the Caribbean remains almost unchanged from the pattern of family distributions, but differences in diversity between these provinces become marked.
♦ Some of the generic pattern at peripheral locations is created by the distributions of individual species (p 141).

GENERIC PROVINCES

The longitudinal provinces of genera (figure 49) illustrate, more than any other analysis undertaken for this book, the world's major biogeographic *divisions* of corals. The main points of figure 49 are as follows:

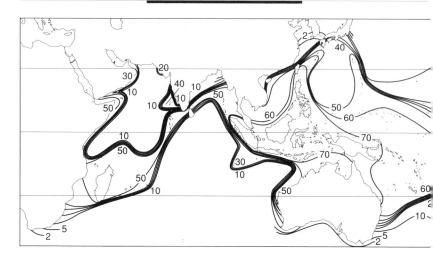

Figure 48 Contours of generic diversity generated by combining the distribution *ranges* of all genera (after Veron, 1993), using an IDRISI Global Information System program. Contours indicate maximum generic diversity for large regions; remote or small areas within those regions may have a lower diversity.

Figure 49 Regional affinities of the major provinces of Indo-Pacific corals, based on the geographic ranges of genera. Finer division in the dendrogram (not shown) separate off high-latitude regions as well as successive layers of the western Pacific. The region enclosed by the 50 genera contour in figure 48 is very strongly conserved. The classificatory procedure was as for figure 55, p 174.

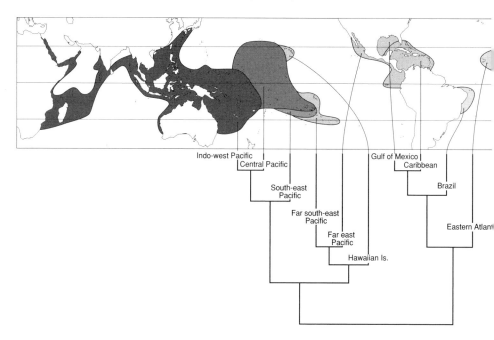

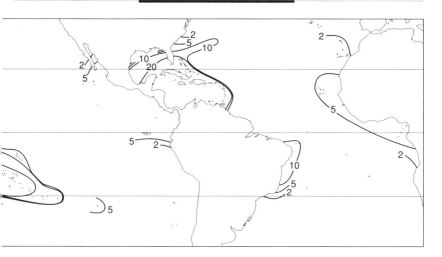

- There is a single tropical Indo–west Pacific province extending from the central Red Sea to east of Fiji. This is the dominant feature of global generic distributions. At increasing levels of resolution (not shown), the algorithm that generated figure 49 generates more than 100 nodes, representing areas of 5° latitude × 5° longitude, before it separates the central Indo-Pacific from the high-diversity band that crosses the Indian Ocean[1] (ie separates the 50 genera contour from the 60 genera contour, figure 48). In short, the whole Indo–west Pacific is a single, highly conserved province.
- Hawaii is a distinct province in the northeast Pacific.
- The southeast Pacific has two provinces of increasing dissimilarity from the central province. The far southeast province has greater affinity with the far eastern province than with its geographic neighbour. These affinities are re-examined at species level (p 165).
- The four provinces of the Atlantic reflect generic contours of diversity. Affinities within the Atlantic are re-examined at species level (p 168).

GENERIC VERSUS SPECIES CONTOURS

Generic contours, familiar to biogeographers, tend to be accepted as being representative of the real world. But in this they are flawed, the worst flaw being that all genera have equal weight. This has no major effect on Atlantic diversity contours, where genera are poorly speciated, but in the Indo-Pacific, where the mean number of species per genus is 8.8, the distortion is major. At extremes, *Acropora,* with about 150 species, contributes the same amount of information as each of 40 genera that have only one species (figure 81, p 265). A second flaw is that generic contours are distorted by higher-level taxonomic boundaries; thus, the family Acroporidae, with approximately 250 species, contributes 4 genera, while the family Faviidae, with 133 species, contributes 24.

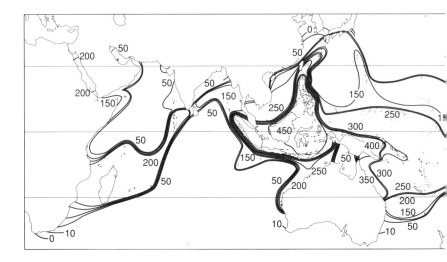

Figure 50 Contours of species diversity derived by multiplying the distribution ranges of all genera by number of species in each genus; multiplying all matrices thus produced (n=833) by the distribution pattern of figure 48; reducing the values of the thus generated contours by a consensus of known values in the central Indo-Pacific (figures 55, 56 and 57); and adding the 450 species contour. The first two procedures were undertaken using an IDRISI Global Information System program. (The first gives weighting to the species diversity of genera; the second gives weighting to spatial variation in diversity.) Contours indicate totals of distribution ranges (as for figures 46 and 48, see text).

At present, these flaws cannot be redressed by creating a contour map of species distributions. The quantity and quality of information required to compile distribution ranges is available for less than 25 per cent of species (60 per cent of Caribbean species, species of paucispecific Indo-Pacific genera, and some of the common species of the other genera). Figure 50 is an alternative that may contain errors within regions, but is unlikely to contain significant error at the level of inter-regional comparisons. Unfortunately, it is not possible to generate meaningful regional classifications (as in figure 49) with such data.

SPECIES

The derived species-level contours of figure 50[2] are the most informative portrayal of the global *distribution* of corals possible with present data. This figure indicates a substantially higher number of species than has generally been recorded in non-central-Indo-Pacific studies. This is partly because it is based on ranges and not records, but also because within-region studies are seldom comprehensive. The main points of figure 50 are as follows:

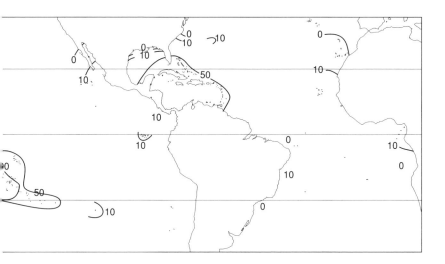

- Within the tropical Indian Ocean, species diversity is so uniform that figure 50 has a pattern of contours similar to those generated from family-level and generic-level data (figures 46 and 48, respectively). Finer contouring (not shown) indicates a slightly higher species diversity in the central Red Sea than elsewhere in the province.
- The boundaries of the Indo-Pacific centre of diversity are mostly well defined: Sumatra and Java in the southwest; Sabah and the Philippines in the northwest; the Philippines, eastern Indonesia and Papua New Guinea in the northeast.
- Species diversity of the Great Barrier Reef is intermediate between that of the Indonesia/Philippines Archipelagos and the western tropical Pacific.
- East of the Great Barrier Reef, the tropical Pacific is sharply divided, by the 200 and 250 species contours, into a western province (the eastern part of the Indo–west Pacific province of figure 49), and an eastern province.
- A central and southeast Pacific province is contained by the 150 and 50 species contours.
- Other Pacific provinces, including Hawaii and the far eastern Pacific, are very peripheral at species level.
- The species diversity of the Caribbean equates with that of the most depauperate provinces of the Indo-Pacific.

REGIONAL PATTERNS

Neither space nor data permit accounts of corals within regions, nor any account of the environmental conditions that may have led to within-region distribution patterns. The following are inter-regional comparisons, focused as far as possible on broad-scale patterns of species diversity[3].

THE GULF, THE RED SEA

AND WESTERN INDIAN OCEAN

The Gulf The (Arabian or Persian) Gulf is partly isolated from the Indian Ocean by an enduring (in geological time) zone of cold upwelling against the coast of southern Oman and by the narrow Strait of Hormuz. Depending on details of global sea level change, local hydro-isostacy and tectonics, most of the Gulf has been aerially exposed for approximately 70 per cent of the past 10 million years. The 'reefs' of the region are relatively poorly consolidated, only superficially resembling wave-resistant reefs of open oceans. The Gulf is presently occupied by a small suite of temperature-tolerant species (p 91), which may include at least one endemic (a *Porites*). An endemic species in the region would presumably be the result of extinction over a former greater distribution range, as the Gulf has been flooded for less than 18 000 years.

The Red Sea and western Indian Ocean The Red Sea is biogeographically divisible into northern, central and southern regions, of which the central region has the greatest concentration of reefs, and the highest species diversity of coral. The entrance to the Red Sea is narrow and shallow and this, combined with the depth of the central and northern regions (more than 3000 m maximum), makes physical oceanographic parameters rather autonomous from those of the adjacent Indian Ocean.

The Red Sea has been periodically isolated from the Indian Ocean, possibly for intervals long enough (depending on tectono-eustacy as much as glacio-eustacy) to have created salinity crises and perhaps regional extinctions, especially in the Late Miocene.

At generic level the corals of the Red Sea are almost indistinguishable from those of the tropical Indian Ocean. At species level the region is separated from the Indian Ocean, but this separation cannot at present be evaluated quantitatively. Sheppard (1987) attempted a species-level classification of intra-regional relationships of the Red Sea and the Indian Ocean by synonymising species names obtained from within-region studies. The outcome (as it would have been if the same procedure were applied to early central Indo-Pacific intra-regional studies) is more a classification of who-worked-where than a reflection of either taxonomic or biogeographic order.

Recent studies do, however, give an indication of similarity between Red Sea/Indian Ocean corals and those of the central Indo-Pacific (table 1). These data as a whole are not suitable for numerical analysis, but they are useful indicators of similarity, and they *do* approximate, as far as morphological taxonomy allows, the author's own conclusions from study of collections (p 8) and some *in situ* observation[4]. Figure 50 predicts a substantially higher number of species than has so far been recorded anywhere in the realm, a gap that is likely to be closed by more detailed studies.

Table 1 Number of species and percentage of them in common with the central Indo-Pacific, recorded in the Indian Ocean and adjacent seas.

	Region	Number of species recorded	Central Indo-Pacific species %
1	Northern Red Sea	139	88
2	Central Red Sea	150	89
3	Southern Red Sea	115	90
4	Gulf of Aden	59	86
5	Gulf of Oman	75	89
6	The (Arabian/Persian) Gulf	46	87
7	Kuwait	25	92
8	Gulf of Kutch	29	97
9	Lakshadweep	88	97
10	Gulf of Mannar	88	94
11	Andaman and Nicobar Islands	108	96
12	South Africa and South Mozambique	118	92

Data sources: regions 1–5 Sheppard and Sheppard (1991); region 6 Hodgson (in press); regions 7–11 Wafar (1986 and pers comm); region 12 Riegl (1993).

CENTRAL INDO-PACIFIC

Biogeographically, the eastern Indian Ocean is the western border of the Indo-Pacific centre of diversity (the subject of chapter 10). Increase in species diversity east of India appears to be abrupt, but just how real this is is not clear. The small and isolated Cocos (Keeling) Atoll alone has ninety-nine species (Veron 1990c). The atoll's species are mostly Indonesian; only one species there is not found elsewhere in the central Indo-Pacific.

NORTHERN AND EQUATORIAL PACIFIC

The complex of issues of historical biogeography come to the fore in the Pacific. The *corals* themselves have the same heritage as those of the Indian Ocean (the cosmopolitan fauna of the mid-Cenozoic, p 131). The *geography* of the Pacific is as old as the oldest Scleractinia. The *coral biogeography* of the Pacific is, essentially, the blending of the fauna with the geography.

The distribution of species within the province is poorly known. Taxonomic and hence biogeographic issues are complex, involving both distributional and evolutionary issues over very variable ranges of space and time. Spatial variation is not simply latitudinal as in the central Indo-Pacific (chapter 10) but has very different latitudinal and longitudinal components. Temporal variation is not simply in an ecological time-frame as in the central Indo-Pacific, but has a major evolutionary component, especially at within-species taxonomic levels. A taxonomic database adequate for comprehensive inter-regional comparisons at species level, comparable to that underpinning the corals of the central Indo-Pacific, seems remote and largely beyond the reach of morphological taxonomy.

Marshall Islands The results of a substantial study of the corals of the Marshall Islands, undertaken in 1976 by many coral taxonomists, have never been published. Unpublished personal observations, however, accord with figures 49 and 50, that is, they demonstrate that the Marshall Islands have a lower level of species diversity than the central Indo-Pacific, but a high level of affinity with it.

Hawaii and Johnston Atoll Hawaii is a mountainous island chain with a very high habitat diversity. Reefs are mostly restricted to localities protected from strong wave action, either in the very protected Kaneohe Bay, or elsewhere in subtidal depths. Johnston Atoll, 800 km to the southwest of Hawaii, is a small atoll with a simple physiography. Both regions are very geographically isolated.

Hawaii has the highest percentage of endemism of any major coral region of the Indo-Pacific, and several general hypotheses, with varying relevance to corals, have been proposed to explain it (p 57). Johnston Atoll has no endemics (Maragos and Jokiel 1986).

The corals of Hawaii have a distinctive composition, with a high diversity of *Porites* (nine species) and *Psammocora* (4 species) and a low diversity of *Acropora* (three species), and faviids (four species altogether)[5]. This contrasts strongly with the corals of Johnston Atoll, which has a faunal composition essentially similar to other depauperate areas of the western Pacific (except for faviids of which only two species are recorded). Johnston Atoll is likely to receive propagule immigration from the Marshall and Phoenix Islands at a rate sufficient to suppress formation of endemic species. The level of endemism in Hawaii may be the result of some allopatric speciation (p 235), but it is more likely that it is a relict coral fauna that rarely receives immigrants from anywhere. Thus at species level Hawaiian corals have closest affinity with those of Johnston Atoll and the far southeast Pacific (figure 51). At below-species levels, some species show no region-specific characters, but the majority are regionally distinctive.

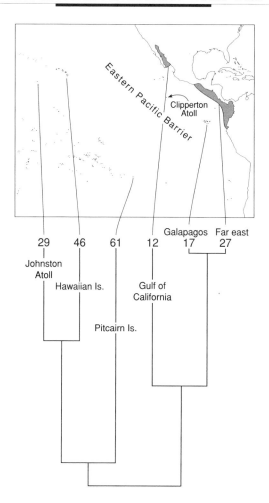

Figure 51 Geographic patterns of affinity of coral species of the far eastern Pacific, based on the geographic ranges of species. The classification overrides a wide range of taxonomic uncertainties. Data are from Veron (1993), updated (see text). The classificatory procedure was as for figure 55, p 174.

THE SOUTHEAST PACIFIC

The corals of the southeast and far east Pacific have little in common at the species level (figure 51). The Line and Marquesas Islands have, by far, the highest species diversity of the eastern Pacific (figure 50), but these have not yet been adequately studied for analysis of affinities.

The generic affinities of the Pitcairn Island corals (figure 49) are repeated at species level, that is, they have primary affinities with the central Pacific (figure 51). The level of endemism is not precisely known, as detailed taxonomic comparisons with the corals of French Polynesia have yet to be undertaken. Approximately 40 per cent of species are morphologically distinct from their western Pacific equivalents. Eight species form geographic subspecies, which are probably distinct from French Polynesian equivalents. There appear to be at least two endemic species (a *Pavona* and a *Cyphastrea*, both undescribed)[6].

The corals of Easter Island, despite ease of access, are largely unknown. Wells (1972) lists six species, all collected before the advent of scuba diving.

THE FAR EASTERN PACIFIC

Glynn and Wellington divide the far eastern Pacific into a series of biogeographic regions, the equatorial regions of the Galapagos Islands and Panama having highest species diversity. The latter region lies in the path of the eastward-flowing Equatorial Counter Current (figure 23, p 103); the former lies in a warm gyre from that current. The whole province has, essentially, a common coral fauna, overwhelmingly dominated by *Pocillopora* (?six species) and *Pavona* (seven species) and devoid of faviids and mussids[7]. There is only one (perhaps temporary) acroporid (*Acropora valida*) (p 85). This fauna attenuates abruptly towards the cold south, but gradually towards the north, with the most northerly coral outcrops occurring in the Gulf of California and Clarion Island (six species).

Pavona gigantea, Siderastrea glynni and possibly two undescribed species (a *Pocillopora* and a *Psammocora*) are restricted to the southeast and far east Pacific. The *Siderastrea* may be the only species endemic to the far eastern Pacific.

Clipperton Atoll, at latitude 10°S, south of the Gulf of California (figure 51), is a true atoll well to the east of all others. Its coral fauna consists of seven species only[8].

The biogeographic emphasis placed by several other authors on the origins of the far eastern Pacific (p 58) seems somewhat misplaced. The province has intrinsic interest in being the most remote of the whole of the Indo-Pacific, but it is so depauperate and distinctive that its composition sheds little light on dispersion mechanisms of coral *in general* and no light on mechanisms of speciation.

THE CARIBBEAN, GULF OF MEXICO AND BERMUDA

The history of the movement of corals from the Old World (the Tethys Sea) to the New World (the Caribbean region), is one of the best documented pieces of all palaeobiogeography (p 131). The extant corals of the Caribbean region are relicts of a cosmopolitan fauna that may have been as diverse (Acroporidae excepted) as that of the equatorial Indo–west

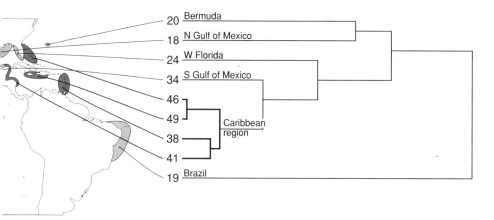

Figure 52 Geographic patterns of affinity of coral species of the western Atlantic, based on the geographic ranges of species for which data appear comprehensive. The actual number of species for each region is greater than indicated. Data are from Veron (1993), updated (see text). The classificatory procedure was as for figure 55, p 174.

Pacific today. At generic level it is old (p 139), and genera have low numbers of species (p 159).

The corals of the western Atlantic are divisible into two regions: the first extending from Bermuda to the southern Caribbean, the second being Brazil (figure 52).

Biogeographically the corals of the Caribbean appear more uniformly distributed than in most comparable areas of the Indo-Pacific (figure 52, which includes only those regions studied by the author). This is probably due to a lack of inter-regional isolation, uniformity of physical-environmental conditions (especially temperature), and low levels of terrigenous sediment throughout most of the province. Some *very subjective* comparisons between the corals of the Caribbean and those of the Indo-Pacific are as follows[9]:

♦ Caribbean corals are distributed relatively uniformly *within* zones (such as along reef fronts at constant depth). This is comparable, perhaps, to the uniformity of community structure found in northern Red Sea corals (Loya 1972).

♦ Intra-specific variation in Caribbean corals is generally greater as that of Indo-Pacific corals *within a given location.*

♦ Geographic subspecies, such as are common in high latitudes of the Indo-Pacific, are (for obvious geographic reasons) generally absent in the western Atlantic as a whole.

♦ Intra-specific variation in the Caribbean appears to be much less physical-environment-correlated than in the Indo-Pacific and may therefore be under a more direct genetic control.

♦ Some species are geographically restricted, or appear to be restricted, by general rarity, but the province has few species (perhaps *Agaricia tenuifolia*, T Hughes pers comm) which are common in one region and absent elsewhere, ie there are few, if any, intra-regional endemics.

The western Gulf of Mexico has a reduced but similar fauna compared with that of the Caribbean. The eastern and northern regions of the gulf have a progressively lower species diversity, and a progressively greater dissimilarity. This is correlated with sedimentary environment.

Bermuda has a very distinctive coral fauna, primarily characterised by the absence of major genera (*Acropora, Mycetophyllia, Mussa, Colpophyllia*) and paucity of others (*Agaricia and Porites*).

BRAZIL AND THE EASTERN ATLANTIC

The corals of Brazil are separated into two regions by the Rio Sao Francisco, a major barrier to dispersion. Both regions have the same species complements (Belém et al 1986) because the southern region includes the Abrolhos Islands, which lie east of the freshwater barrier. Of the nineteen species currently recognised in the region, six appear to be endemic (a *Siderastrea*, a *Meandrina*, a *Favia* and all three species of the genus *Mussismilia*[10]).

The corals of the eastern Atlantic include the endemic monospecific genus *Schizoculina*, but no other endemics. Thiel (1928) records *Acropora* in the Gulf of Guinea, but there has been no subsequent confirmation of this. The eastern Atlantic corals have affinities with those of both the Caribbean and Brazil (Laborel 1974), leading to speculation that they have had both a northern and southern route of immigration.

CORAL AND CORAL REEF DISTRIBUTIONS

> The problem of accumulating quantitative data concerning the sea floor favourable to zooxanthellate coral growth in any region is insurmountable. (Stehli and Wells 1971)

There is no straightforward way of comparing patterns of coral distributions and species diversity with patterns of reef distribution, as the latter are associated with many continuous variables of bathymetry, reef morphology and community composition. Maps depicting the distribution of reefs (eg Wells 1988) show an overall correlation between the regional concentration of reefs and the regional diversity of corals: that is, reefs are primarily concentrated in the Indo-Pacific centre of coral diversity. The reefs of the Caribbean are, however, as concentrated as they are in most regions of the Indo-Pacific. High regional concentration of reefs, therefore, is not necessarily correlated with high species diversity.

Length of reef fronts, which can be used as proxy indicators of area of reef, can be determined much more objectively by satellite[11]. When this is done globally (figure 53), it is seen that there is a clear correlation between concentration of reef per unit area of ocean and generic diversity, and that there are no major developments of reefs in regions where corals have low generic diversity.

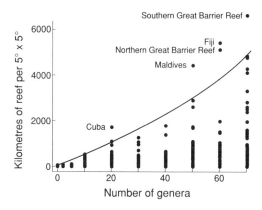

Figure 53 Relationship between concentration of reefs (in 5° latitude × 5° longitude pixels), as determined from satellite data (see text) and generic diversity. The distribution of points under the curve has no meaning as they may reflect the proportion of land, rather than the concentration of reef, in pixels.

SUMMARY

Distribution data on which this chapter is based are summarised in a separate volume (Veron 1993).

- At the family level, global diversity contours are generally uniform.
- At the generic level, global diversity contours are uniform in the Indian Ocean. The central Indo-Pacific province is well defined. East of this province, there is a progressive attenuation of diversity across the Pacific. The Caribbean centre of diversity is also well defined.
- At generic level, the main longitudinal divisions of the fauna are a highly conserved equatorial province extending from the central Red Sea to the western Pacific; a central Pacific province; a northeast and two southeast Pacific provinces; a far east Pacific province; a Caribbean province; a Brazilian province; and an eastern Atlantic province.
- At species level, the central Indo-Pacific centre of diversity is very pronounced, and a high diversity is retained across the Indian Ocean. Hawaii, the southeastern Pacific, the far eastern Pacific and the Caribbean provinces all have much lower diversity than is indicated by generic-level data.

♦ Species compositions of adequately studied biogeographic regions to the west of the Eastern Pacific Barrier (Johnston Atoll, Hawaii and the Pitcairn Islands) are very dissimilar. Hawaii has a high level of endemism; that of the Pitcairn Islands is lower; and Johnston Atoll has no endemics.

♦ Species compositions of regions east of the barrier are relatively uniform, except for the depauperate Gulf of California. The level of endemism is low.

♦ Species of the Caribbean appear to have intra-specific levels of variation similar to species of the Indo-Pacific within regions of comparable area, but lack the well-defined geographic subspecies found in the Indo-Pacific.

♦ Species compositions of regions within the Caribbean are generally uniform compared with regions of similar area of the Indo-Pacific. Species compositions of the western, northern and eastern regions of the Gulf of Mexico show substantially increased dissimilarity from each other and from Caribbean regions. Bermuda has a lower number of species than any reef of the Indo-Pacific. Brazil has a distinct biogeographic province, also with very few species, and with a high level of endemism.

♦ On a global basis, the within-region concentration of coral reefs per unit area is broadly correlated with generic diversity, but reefs and even whole atolls can be formed by as few as six species of coral.

10
LATITUDINAL DISTRIBUTIONS IN THE INDO-PACIFIC CENTRE OF DIVERSITY

This chapter is a companion to the preceding one, but longitude-correlated issues are now replaced with latitude-correlated ones: scales of space and time are smaller; physical-environmental gradients are better defined; and questions tend to be less complex and more readily answered. The database is the species-level distribution compilation of Veron (1993, pp 23–340) (p 8). The primary data, derived from original field-work, are extensive and taxonomically detailed, covering all known coral species throughout the full length of both the east and west coasts of Australia, as well as the east coast of Asia from the Philippines to mainland Japan (figure 54). The reader is referred to the aforementioned monograph for original data sources, and supplementary biogeographic and taxonomic references.

THE THREE LATITUDINAL DIVERSITY SEQUENCES

The central Indo-Pacific has three of the world's four principal latitudinally contiguous distribution sequences[1] from the high diversity of the tropics to the low diversity limits of coral distributions (figure 75, p 235). The province as a whole is the world's centre of zooxanthellate coral diversity, has long been believed to be the principal centre of zooxanthellate coral evolution (p 45), contains 37 per cent of the world's area of coral reefs[2], and it has the world's greatest concentration of tropical shorelines. It is the 'catch-bag' of all westward-flowing currents of the Pacific.

Presence/absence data of principal coral localities within each of the three regions indicated in figure 54 are summarised in figures 55, p 174,

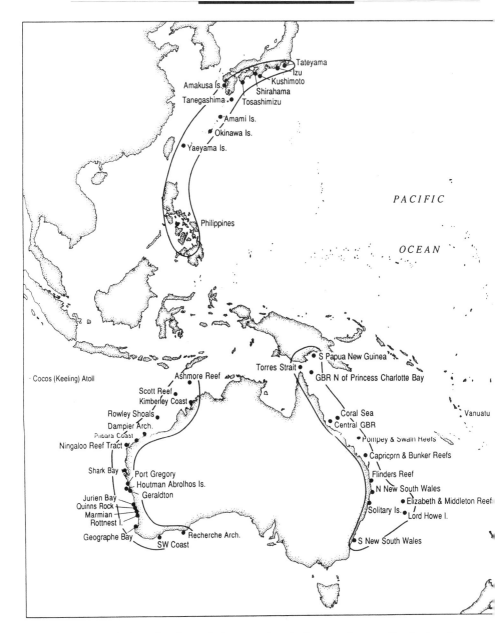

Figure 54 Central Indo-Pacific locations for which comprehensive distribution data are available (see text). Most locations are contained within three regions, each extending from low-latitude locations of high diversity to high-latitude locations of low diversity.

56, p 174 and 57, p 176[3]. These figures are very similar in principle. They are all numerically 'robust'[4] (and are not significantly altered when recalculated by different numerical fusion strategies in the PATN package of Belbin 1987), they are all strongly uni-dimensional (in non-metric multi-dimensional scaling procedures, Kruskal 1964), and all have unbranched 'minimum spanning trees' (Gower and Ross 1969) connecting localities in geographic sequence[5]. Although computer-generated purely from presence/absence data, these figures strongly reflect subjective impressions, that is, they are intuitive, and are thus also likely to be indicative of broader ecological patterns and physical environments.

Brief sketches only of relevant points about regional geography, palaeoclimates and surface currents are included here; cited references give further detail. Correlations between coral species distributions and latitudinal changes in physical environment are discussed in chapter 6.

SOUTHERN PAPUA NEW GUINEA AND EASTERN AUSTRALIA
CORAL DISTRIBUTION AND GEOGRAPHY

Principal localities studied can be initially clustered into three groups: tropical reefs; temperate reef and non-reef communities; and high-latitude outlying populations (figure 55). The region is dominated by the Great Barrier Reef, the largest existing structure ever made by non-human organisms. Patterns of species diversity within the Great Barrier Reef are little known, but it is known that they are complex, being determined by both cross-shelf and latitudinal physical-environmental gradients. Cross-shelf patterns are noted below (p 183); latitudinal patterns primarily reflect the fact that the region is large enough to extend from the warm tropics to temperate regimes. Within this latitudinal range, there are very different climates (wind patterns and rainfall), tidal regimes, water qualities, bathymetry, island types, substrata and even geological histories.

There are only two other reef areas south of the Great Barrier Reef: the Elizabeth/Middleton reef group and Lord Howe Island. Both are probably relicts of a wider Plio-Pleistocene reef development (Slater and Phipps 1977). Elizabeth and Middleton reefs (Hutchings 1992) are large platform reefs, the southernmost platform reefs in the world. Lord Howe Island (Veron and Done 1979) is mountainous, with a reef lagoon on its western side, which was probably once much more extensive than it now is.

Flinders 'Reef' (figure 55) is a sandstone platform surrounded by a rugged terrain offering great habitat diversity; neither this 'reef' nor the Solitary Island reefs (Veron et al 1974) have any limestone accretion. There are no other high-latitude offshore islands off the eastern Australian coast.

The southern latitudinal extremity of the Great Barrier Reef is determined by deep water and soft substrates; it is not temperature-limited.

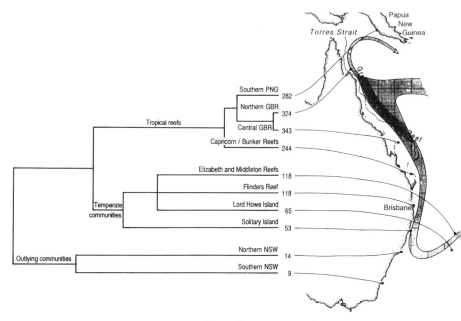

Figure 55 Geographic patterns of affinity of coral species of southern Papua New Guinea and eastern Australia. Each point represents a location, or several locations (p 270), within which the species complement is known in comprehensive detail from extensive *in situ* studies. Veron (1993) lists the species found at each point. The classification is an agglomerative hierarchical component of the PATN package of Belbin (1987), using the Bray-Curtis dissimilarity coefficient and flexible unweighted pair mean averages. Total numbers of species in each location are indicated.

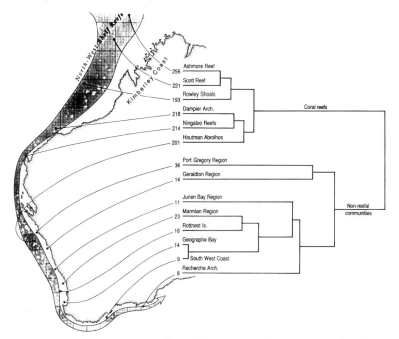

Figure 56 Geographic patterns of affinity of coral species of western Australia (after Veron and Marsh, 1988). Geographic characteristics, database and classificatory procedure are as for figure 55.

PALAEOCLIMATIC BACKGROUND

The age of the Great Barrier Reef varies geographically: both the far north (eastern Torres Strait) and far south (Capricorn Basin) are Middle to Early Miocene in origin (Davies et al 1989). The Queensland Plateau and Marion Plateau, east of the central and southern Great Barrier Reef (respectively), are both of this age. The rest of the Great Barrier Reef is probably younger.

Virtually nothing is known of pre-Pleistocene Great Barrier Reef corals, as no well-preserved fossil deposits have been found. The entire region has been repeatedly submerged and has re-emerged throughout the Plio-Pleistocene (figure 34, p 126): during intervals of low sea levels, coral communities would have been predominantly non-reefal, although a variety of submerged relict reefs have been recorded in the Coral Sea (Harris and Davies 1989) and may have been common there. Continental slope refuges, devoid of islands and major areas of topographic complexity, would have had little of the habitat diversity of the present Great Barrier Reef, or the archipelagos of eastern Indonesia and the Philippines (at either high or low sea levels). This may be a reason why the Great Barrier Reef has a lower diversity of species (especially of species restricted to shallow, low-energy environments) compared with Indonesia and the Philippines.

SEA SURFACE CURRENTS

The South Equatorial Current (figure 23, p 103) brings tropical water westward into the Coral Sea between the Solomon Islands and New Caledonia, then branches near the Queensland Plateau (east of the central-north Great Barrier Reef). The northward-flowing branch forms a clockwise coastal current and partly closed gyre in the northern Coral Sea, exiting into the Solomon Sea. The southward-flowing branch forms the East Australian Current. The point of the north/south split fluctuates seasonally. Some of the East Australian Current is returned north by an offshore counter-current south of the Great Barrier Reef; the main stream forms eddies and meanders off the New South Wales coast before ultimately separating from the coast south of the Solitary Islands (figure 55). The current then crosses the Tasman Sea and returns to the western Pacific via the northern tip of New Zealand (Pickard et al 1977; Burrage 1993).

Coral propagules north of the central Great Barrier Reef have uni-directional transport to the northern Coral Sea and thence to Vanuatu; propagules from the south are entrained in uni-directional southward-flowing currents. No zooxanthellate corals survive the journey to New Zealand under present conditions (p 135).

WESTERN AUSTRALIA
CORAL DISTRIBUTION AND GEOGRAPHY

Coral assemblages (Veron and Marsh 1988) can be initially clustered into two groups: tropical reefs, and temperate non-reef communities (figure

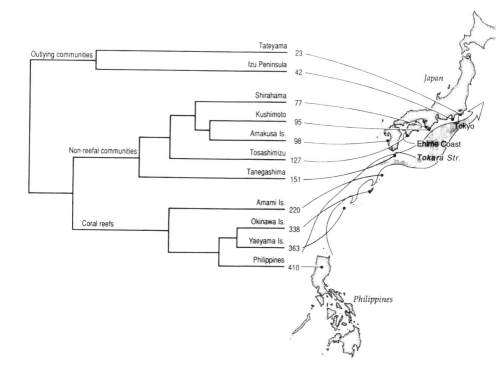

Figure 57 Geographic patterns of affinity of coral species of the Philippines and Japan (after Veron and Minchin, 1992). Geographic characteristics, database and classificatory procedure are as for figure 55.

56). Tropical reefs form two distinctive groups: the offshore reefs of the North-west Shelf, and coastal islands and reefs, including the Houtman Abrolhos Islands. Non-reef high-latitude coral communities are all onshore, or around nearshore islands. All coral communities of the entire coastline are under the influence of the southward-flowing Leeuwin Current.

The Kimberley coast, with its myriad small islands, might be conducive to coral growth but for its very high tidal range (more than 10 m) and turbid waters. Further south a small number of island groups, all onshore, are all associated with shallow reef communities. Australia's longest fringing reef, the Ningaloo Reefs, occur at the continent's western extremity, where the Leeuwin Current comes closest to the coast. The Port Gregory region and all locations further south, except the Houtman Abrolhos Islands, are non-reef coral communities, generally associated with onshore islands and rock outcrops. The Houtman Abrolhos Islands are the highest-latitude reefs of the Indian Ocean and are of particular biogeographic interest as extensive coral communities occur adjacent to macroalgae communities, with which they compete (p 92).

PALAEOCLIMATIC BACKGROUND

The affects of palaeoclimatic changes are seen on the southern coast: abundant deposits of *Acropora* from the last inter-glacial occur in the mainland coast at the latitude of the Houtman Abrolhos Islands, and *Acropora* and *Montipora* of this age have been recovered as far south as the southwest coast (G Kendrick et al 1991, and pers comm). There are extensive carbonate platforms of reef origin along the southwest Australian coast although none now have the morphological characteristics of active reefs.

SEA SURFACE CURRENTS

The Leeuwin Current (Pearce and Walker 1991) is the only south-flowing eastern boundary current of the southern hemisphere. It brings warm, nutrient-poor equatorial waters to high latitudes and consequently may ultimately be responsible for the world's only occurrence of reefs on a western continental coastline. Biogeographic interest in the Leeuwin centres on the connectivity ratchet it generates (below); temperature, which permits coral growth at high latitudes; and nutrients, which may give corals sufficient competitive advantage over macroalgae to allow their continuing the existence at the Houtman Abrolhos Islands (Hatcher 1991).

PHILIPPINES AND JAPAN

CORAL DISTRIBUTION AND GEOGRAPHY

Coral assemblages (Veron 1992b) can be initially clustered into three groups: tropical reefs, temperate non-reef communities, and high-latitude outlying populations (figure 57).

The Philippines-to-Japan island arc provides the world's best geographic setting for studying latitudinal dispersion of corals, as reef then non-reef coral communities occur sequentially from the global maximum species diversity of equatorial southeast Asia to the highest-latitude coral communities in the world. As the region has also produced the world's most comprehensive oceanographic data, the relationship between distribution and physical environment can be studied in detail.

The Ryukyu Island arc exhibits a sharp demarcation between reef and non-reef communities. Extensive reef formation occurs north to the Amami Islands (figure 57), but only very minor limestone accretion occurs at Tanegashima, 225 km further north.

PALAEOCLIMATIC BACKGROUND

Deep water on both the western and eastern sides of the Ryukyu Islands chain, together with the absence of a continental shelf, would have made the effects of Plio-Pleistocene glacio-eustatic changes small by comparison with the Philippines and the Great Barrier Reef. However, glacio-eustacy and tectono-eustacy combined, during the Pleistocene, has seen the East China Sea change dramatically from a vast sedimentary basin fed by the Palaeo-Yellow River, into an inland sea (illustrated by figure 33, p 125) (Kimura 1991). The reefs of today are mostly Pleistocene in origin, although there are older deposits, reflecting sub-continuous shorelines back through the Cenozoic and into the Mesozoic Era (figures 26–30, pp 111–122).

SEA SURFACE CURRENTS

The Kuroshio (figure 33, p 125), a major boundary current, originates in the northern Philippines, enters the East China Sea through a strait between Taiwan and the Yaeyama Islands, and flows northward to the west of the Ryukyu Islands. It then veers to the east and flows along the south coast of mainland Japan to the furthest extremity of coral dispersion at Tateyama (figure 57, p 176). East of the Tokara Strait, the Kuroshio has five separate meander paths, each of which may last for several years. The large meander path is the more prevalent, having been present over thirty-five years between 1895 and 1984 (Kawabe 1985). It forms a loop between Shikoku and southeast Japan and greatly increases transport time to that region. The loop is often associated with a cold surface current, which has had catastrophic effects on coastal biota.

The geographic setting of the Kuroshio in Japan and the wealth of data about it allow detailed comparison between species diversity and physical environment. Dispersion time for propagules varies seasonally; it also varies in scales of decades or centuries as the path of the Kuroshio varies (figure 33, p 125). Detailed records of the speed of the Kuroshio over a four-year time span show that dispersion time has a small influence on observed distributions where it is less than ten days, and no influence at all where it is more than ten days (Veron and Minchin 1992); that is, although there are some local effects of proximity, the dispersal ability of corals is sufficiently great to not limit intra-regional dispersion. What is limiting is sea surface temperature and factors correlated with temperature (p 90).

PRINCIPAL CONCEPTS OF LATITUDINAL ATTENUATION

Environmental aspects of latitudinal attenuation of corals have been discussed in chapter 6; morphological and other changes within species are the subject of chapter 11. The following observations, of general significance to coral biogeography, are based on studies within the three regions described above.

CONNECTIVITY RATCHETS

The concept of connectivity ratchets (p 99) is of primary biogeographic importance (figure 75, p 235) and underpins the patterns of affinity of figures 55–57. In all three regions, boundary currents flow poleward, taking entrained propagules towards higher latitudes and offering very reduced options for return dispersion. This effect is strongest on the west Australian coast where the Leeuwin Current (p 177) provides no significant northern transport south of mid-coast latitudes. At the Ningaloo Reefs (figure 56), the continental shelf edge, hence the core of the Leeuwin Current, comes close to the reef edge, leaving no room for any significant northward counter-current. These reefs, and all those further south, receive annual imports of propagules from the north and can therefore be considered under genetic influence from the north. This is likely to restrict local genetic drift and retard establishment of high-latitude-adapted genotypes.

On the east coast of Australia, the East Australian Current has a predominantly northward flow in the far northern Great Barrier Reef; in the southern Great Barrier Reef region it generates semi-continuous gyres, which allow both northward and southward dispersion at all latitudes within the Great Barrier Reef. South of the Great Barrier Reef, however, transport is one-directional and has endured, perhaps off and on, at least as long as the period of evolution of modern coral species. All coral communities south of the Great Barrier Reef depend for their existence on the continuity of the East Australian Current for maintaining sea surface temperature, and they are certainly affected by periodic immigrations from the north, and not only immigration of corals[6].

In Japan, as the core of the Kuroshio flows well to the west of the Ryukyu Islands, there is a substantial return flow, providing transport in both directions along the Ryukyu arc. To the south and to the north of the arc, however, transport is one-directional. Corals of the Ryukyu Islands cannot, under the present current regime, reach the Philippines; those of mainland Japan cannot reach the Ryukyu Islands. This may be the reason why the mainland islands have twice the proportion of endemics as the Ryukyu Islands (p 91). The 'connectivity ratchet' is similar to that of the two Australian coastlines; the effects are much more readily observed at a subspecies level (chapter 13) than among species.

EQUATORIAL CONNECTIONS

The reefs of southern Papua New Guinea and northeast Torres Strait (figure 55) are separated by only 300 km. Nevertheless, their coral faunas are significantly different. Surface circulation is predominantly clockwise within the Coral Sea (Hughes et al in press) and this has probably been a long-standing barrier to migration of Papua New Guinea corals to the Great Barrier Reef. More importantly, major rivers of the Gulf of Papua create a low-salinity barrier, particularly prevalent during the early-summer spawning of most coral species. In a geological context, the rivers are responsible for the absence of reefs anywhere in the northwestern Gulf of Papua, and this would have been the case since the rivers of southern Papua New Guinea came to occupy their present positions.

Although there is no comprehensive account of the corals of Indonesia, it is clear that species diversity is very high, equal to (Borel Best et al 1989) or more likely greater than, that of the Philippines. Maximum species diversity is likely to be centred around the complex coastlines of Sulawesi and the Banda Sea rather than the southernmost reefs of Indonesia bordering the Java Trench. Nevertheless, the origination of the Leeuwin Current in Indonesia is reflected in the species composition of the corals of northwest Australia, especially Ashmore Reef, only 150 km from Indonesia.

RATES OF LATITUDINAL ATTENUATION

In all three latitudinal diversity sequences, coral species attenuate latitudinally and are not replaced by other species. In chapter 6 it was shown that species attenuation is correlated with latitudinal change in physical environment, especially temperature and correlates of temperature.

In two latitudinal sub-sequences, temperature probably has no influence, and thus these sub-sequences can be used to estimate attenuation attributable to isolation under present surface circulation vectors:

Ashmore Reef, Scott Reef and Rowley Shoals Each of these coral reefs in northwest western Australia (figure 56) is, in effect, a stepping stone along the southward-flowing path of the 'Indonesian through-flow' which is part of the origin of the Leeuwin Current. At Scott Reef sixteen species only (7.2 per cent of the total) have *not* been recorded upstream at Ashmore Reef. At the Rowley Shoals, nine species only (4.6 per cent of the total species complement) have not been recorded at upstream Scott or Ashmore Reefs. The attenuation is 7.9 per cent of species per 100 km between Ashmore Reef and Scott Reef, and 3.6 per cent of species per 100 km between Scott Reef and Rowley Shoals. Surface current velocities are not yet sufficiently well-known to express these attenuations in terms of dispersion time.

The Philippines and the Yaeyama and Okinawa Islands of Japan At the Yaeyama Islands (figure 57), 7.5 per cent of species only have *not* been recorded in the Philippines, giving an attenuation of 1.6 per cent of species per 100 km. Similarly, there is an attenuation of 2.3 per cent per 100 km between the Yaeyama and Okinawa groups. The minimum transport time between the Yaeyama and the Okinawa Islands under present surface circulation velocities is about three weeks (Veron and Minchin 1992). There are no adequate data to indicate minimum transport time between the Philippines and the Yaeyama Islands, but if the maximum speed of the Kuroshio is assumed to be the same, minimum transport time would be forty-nine days.

Clearly, there are too many assumptions in these dispersion data (not the least that present faunas have reached a state of equilibrium[7]) for them to be generally applicable to other geographic situations. They are, however, consistent with the hypothesis that, although temperature acting in ecological time-frames controls dispersion to high latitudes, dispersion from equatorial refuges of the last glaciation is still an ongoing process.

ENVIRONMENT VERSUS GENETIC CONNECTIVITY

An obscure but fundamental aspect of coral biogeography (at least within the central Indo-Pacific) is that, although boundary currents drive latitudinal dispersion in ecological time-frames, widely separated locations with comparable physical environments have a greater similarity than have adjacent locations with different physical environments. Corals of the North-west Shelf reefs of western Australia have a greater similarity to those of the Great Barrier Reef[8] than they have to the adjacent mainland coast (figure 58). This similarity is underscored by a 92 per cent overlap in species[9]. It is also seen in similarities of relative

Figure 58 Geographic patterns of affinity of coral species of all Australia (ie Figs 55, p 174 and 56, p 174 combined). Geographic characteristics, database and classificatory procedure are as for figure 55.

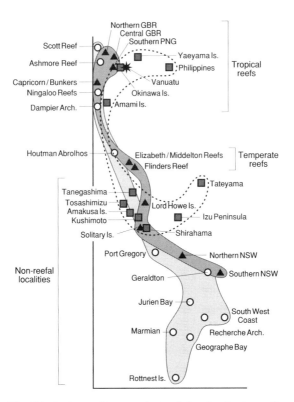

Figure 59 Principal coordinate analysis of the distributions of corals of Papua New Guinea/eastern Australian sites (triangles), western Australian sites (circles), Philippines/Japan sites (squares) (ie the same data used to generate figures 55, 56 and 57) and Vanuatu. The classification procedure is a component of the PATN package of Belbin (1987).

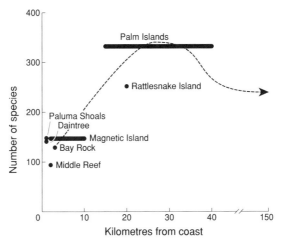

Figure 60 Cross-shelf distribution of total number of species in the central Great Barrier Reef in an area of 1 km² or less, recorded from comprehensive *in situ* studies. Onshore sites are unconsolidated sedimentary accumulations; Bay Rock, Magnetic Island, Rattlesnake Island and the Palm Islands are all high islands. Data are from many sources. The number of species of offshore reefs is an estimate only.

abundances, and in intra-specific variations. The reason is environmental: offshore reefs of both the west coast and east coast are carbonate platforms with clear-water, high-energy environments. These contrast with onshore reefs, which have turbid environments dominated by terrigenous substrates. Connectivity between eastern and western Australia has been via Indonesia, not via the shorter route of Torres Strait[10]. Thus, physical environment dominates both latitudinal change (by driving it) and conversely, longitudinal change (by being more important than connectivity).

This observation is reinforced by figure 59, which shows that the species composition of all tropical reef localities, all temperate (high-latitude) reef localities, and all non-reef localities, are grouped together (in principal coordinate analysis) irrespective of geographic position. The species composition of Vanuatu corals (figure 54) on the eastern border of the Coral Sea is a case study[11]: of the 296 species recorded, only 24 have *not* been recorded from the Great Barrier Reef. Of these, eighteen occur in the Philippines and all but four (all undescribed) have been recorded elsewhere in the central Indo-Pacific (two only, appear to be endemic, figure 64, p 191). Not only are the species complements usually similar, but so also are their relative abundances, colours and habitats (chapter 11). This similarity is so striking that some upper reef slope communities of Vanuatu are virtually indistinguishable from their equivalents on the Great Barrier Reef.

CROSS-SHELF VERSUS LATITUDINAL GRADIENTS

There is a comparatively weak latitudinal gradient within the Great Barrier Reef with species numbers decreasing (and intra-specific changes occurring) only towards its southern end (Capricorn and Bunker Reefs) (figure 55)[12]. The reason is, again, mostly environmental: the Capricorn and Bunker reefs have no high islands, or terrigenous substrates and the reefs themselves are morphologically simple and surrounded by deep water. In general, there is much more variation across (east–west) the Great Barrier Reef than down its length. The relatively shallow, turbid, terrigenous coastal waters, protected from strong wave action and subject to seasonal river flooding, with attendant pulses of low salinity, silt and organic nutrients, support a reef and inter-reef fauna and flora of a very different character to that found offshore. Also, high (continental) islands mostly occur inshore, and it is these islands that provide much of the Great Barrier Reef's habitat diversity and probably house the greatest number of species (figure 60).

A pseudo-latitudinal distribution effect is often seen in inshore communities, where coral species compositions tend to resemble those of higher-latitude locations rather than adjacent offshore locations. The resemblance between inshore and high-latitude communities of eastern

Australia is seen in the relative abundance of some major groups of species (eg most *Turbinaria* and *Acanthastrea* and many *Acropora*)[13], and in the similarities of ecomorphs, especially explanate and encrusting morphologies. There are certainly multiple reasons for this: the primary one is physical-environmental (shared terrigenous substrates and turbid water as opposed to carbonate substrates and clear water), but relative lack of *Acropora* dominance may also be relevant.

HIGH-LATITUDE REEFS AND NON-REEF COMMUNITIES

Aspects of general biogeographic interest are as follows:

♦ Species compositions of subtropical *reefs* are relatively similar, irrespective of geographic location because, on the one hand, they are populated by species that have very broad ranges in all directions, and on the other, they are not severely depleted by latitudinal attenuation (figure 61, p 189).

♦ Species composition of increasingly high-latitude *non-reef locations* are increasingly dependent on geographic position (figures 58 and 59) because of upstream seeding and connectivity ratchet effects.

♦ There are major discontinuities in species richness between reef and non-reef locations.

♦ Species compositions of very high-latitude locations are very dissimilar (figures 55–57) primarily because they involve a relatively small number of species and may be the outcome of stochastic founder events.

EASTERN AUSTRALIA

There are substantial species differences (notably in *Montipora*) between Flinders Reef and Elizabeth/Middleton Reefs. No species are known to be endemic to either region, although the latter has a wide range of geographic subspecies and other species (eg *Acropora glauca, A. lovelli* and *Scolymia australis*) that are so rare elsewhere that these reefs are probably acting as high-latitude refuges.

Elizabeth/Middleton Reefs and Lord Howe Island are only 200 km apart, yet have very different numbers of species (118 and 65 respectively)[14]. The reason appears to be temperature: Lord Howe Island is periodically inundated by cold water from the Tasman Front, whereas Elizabeth/Middleton Reefs lie well within the path of the East Australian Current as it is deflected away from the coast of southeastern Australia by eastward-flowing water from the Tasman Sea (figure 55, p 174). As a probable result, most of the benthos of Lord Howe Island, including much of the reef lagoon, is dominated by macroalgae, not coral.

With the exception of three high-latitude endemics (*Coscinaraea mcneilli, C. marshae* and *Scolymia australis*) (p 192) and a tropical species that extends to southern Australia (*Plesiastrea versipora*), the corals of coastal New South Wales are, presumably, among the most low-temperature-tolerant of all southern hemisphere species. However, only ten species (the aforementioned plus *Pocillopora damicornis, Cycloseris costulata, Favites abdita, F. flexuosa, Cyphastrea serailia* and *Turbinaria mesenterina*) exist in habitats not commonly occupied by other tropical invertebrates and fish.

Comparison between the presence/absence records of Veron et al (1974) and Smith and Simpson (1991) for the Solitary Islands region (latitude 30°S, figure 55) indicates that coral populations are generally stable over intervals of decades but further south they appear to be unstable, with occasional sightings of species well outside their normal range (eg *Pavona decussata* from Wollongong, latitude 34.5°S).

WESTERN AUSTRALIA

The Houtman Abrolhos Islands are topographically complex, with an exceptional environmental diversity and an exceptional range of coral communities, many of which are found nowhere else. These communities are frequently dominated by species that are rare elsewhere, and by monospecific stands or species combinations unknown elsewhere, or by unique combinations of corals and macroalgae (pp 92, 177) (which themselves dominate most high-energy reef faces). Seasonally changing aspects of water quality add a further dimension to variation, including variation in nutrients, temperature, and extremes of turbidity, alkalinity and salinity.

The morphological distinctiveness, hence taxonomic complexity, of Houtman Abrolhos Island corals is exceptionally great. This is primarily correlated with physical environment, but is far from entirely so. The level of dissimilarity between these corals and those of the Ningaloo Reefs (figure 56), the occurrence of endemic species (figure 64, p 191), of disjunct populations (figure 65, p 193), and of many location-specific species attributes (chapter 11), all point to genetic isolation. It seems probable that most species have had a long residence time and that Holocene immigrations are rare[15]. Perhaps more indicatively, *Acanthaster planci,* with its great dispersal capability, has not yet reached the islands.

Species attenuate in an orderly sequence along the coast of southwest Australia, with only minor disjunctures. The attenuation is temperature-correlated, except for the aforementioned widely dispersed *Plesiastrea versipora,* and the same high-latitude endemics that occur on the southeast Australian coast (*Coscinaraea mcneilli, C. marshae* and *Scolymia australis*). *Symphyllia wilsoni,* the only described endemic, occurs primarily in high-energy, kelp-dominated or *Sargassum*-dominated platforms and is thus ecologically distinct from all other corals.

JAPAN

There is a 31 per cent decrease in species between the northernmost reefs of Japan (the Amami Islands) and Tanegashima, a distance of only 225 km. This attenuation correlates with temperature, but appears to be primarily due to loss of reef habitats. Most of the species diversity at Tanegashima is restricted to small parts of the island; there are two conspicuous endemics (*Acropora tanegashimensis* and *Euphyllia paraglabrescens*).

The role of temperature in the distribution of mainland corals is discussed elsewhere (p 90), as is the temporal stability of these distributions (p 85).

SUMMARY

Distribution and taxonomic data on which this chapter is based is summarised in a separate volume (Veron 1993).

◆ The central Indo-Pacific has three of the world's four principal latitudinal contiguous sequences from very high to very low species diversity. These sequences occur along the east Australian coast; the west Australian coast; and the east Asian coast (from the Philippines to mainland Japan). Coral distributions are known from comprehensive presence/absence data of nine to fourteen principal locations along each coastline.

◆ Figures 55, 56 and 57 (pp 174–176) summarise the distribution of coral along the aforementioned sequences. Within each sequence, coral species diversity attenuates in an orderly manner: principal locations are connected by the East Australian Current, the Leeuwin Current and the Kuroshio (respectively).

◆ Along all three sequences, boundary currents flow poleward, taking entrained propagules towards higher latitudes and offering reduced options for return dispersion. In each case, this system operates as a connectivity ratchet.

◆ There is almost no replacement of species with other species along these contiguous sequences.

◆ Widely separated tropical locations (from the North-west Shelf of western Australia to the eastern Coral Sea) have a greater similarity than have contiguous locations along either Australian coastline, indicating that physical environment, rather than geographic position, is the primary determinant of coral species composition.

◆ At higher latitudes along all three sequences, coral species compositions become increasingly determined by geographic position because of upstream seeding and the connectivity ratchet effect.

◆ There are major discontinuities between reef and non-reef communities.

11

GEOGRAPHIC CHARACTERISTICS OF INDIVIDUAL INDO-PACIFIC SPECIES

'Those forms which possess in some considerable degree the character of species, but which are so closely similar to some other forms, are so closely linked to them by intermediate gradations, that naturalists do not like to rank them as distinct species, are in several respects the most important to us.'
(Darwin 1859)

Geographic patterns of coral species distributions contribute little to our understanding of evolutionary processes because (the author concludes) of the lack of correlation between place of occurrence and place of origin (p 233). Geographic patterns *within* species are another matter, and it is here, and here alone, that patterns suggest process, not because the place of occurrence and place of origin are now correlated, but because patterns *within* species indicate that the majority of species are geographically undefinable.

This chapter focuses on what is known of the geographic distribution of *individual* operational species, and then on patterns of variation *within* individual operational species. The former subject has received some attention in the literature, the latter has received almost none, for good reason, as indicated here:

♦ Intra-specific variations are matters of detail, most of which are trivial on intra-regional scales.

♦ Every coral species has a unique morphology, and thus a unique pattern of morphological variation. These cannot be combined to produce a quantitative outcome (such as patterns of distribution described in the previous two chapters), nor can they be encapsulated by simple summary statements.

♦ The taxonomic value of observations usually decreases with distance (figure 66, p 203). Within regions, species are generally cohesive units that can be described and recognised; among increasingly distant regions, this cohesion becomes increasingly less recognisable.

♦ Molecular methods, which have the potential to shed much light on geographic variation, have yet to be used for any Indo-Pacific-wide species[1].

The chapter is divided into four parts. The first two are about the size and the characteristics of individual species distributions; the latter two are about geographic variation within species. As with the two preceding chapters, observations based on the central Indo-Pacific and those based on the whole Indo-Pacific are separated, and for the same reasons:

♦ *Across the whole Indo-Pacific* (chapter 9). Biogeographic variations are poorly known, are not supported by published studies, and are not adequately supported by taxonomy.

♦ *Within the central Indo-Pacific* (chapter 10). Biogeographic variations are relatively well-known, are at least partly based on published studies, and are adequately supported by operational taxonomy.

Geographic patterns, whatever their nature, are generated by patterns of genetic connectivity and physical environment. Within the central Indo-Pacific, where levels of genetic connectivity are relatively great, patterns are strongly latitudinal (chapter 10) and are mostly generated by physical-environmental gradients in biogeographic, rather than evolutionary, time-frames. Within the whole Indo-Pacific, the reverse is predominantly true: weak connectivity and evolutionary time-frames create patterns that are sufficiently great for most species to make their boundaries arbitrary. Thus, the onus of proof of reticulate evolution falls primarily on distributions across the whole Indo-Pacific, that is, on the weak point of coral biogeography. The reason is circular: Indo-Pacific-wide biogeography is weak because its operational taxonomic base is inadequate; the taxonomy is inadequate because of geographic variation.

This chapter, as with all observations of patterns that are the immediate outcome of evolutionary processes, is unavoidably based on thousands of details about hundreds of species. Considered singly, most of these details are trivial; it is only when they are put together that they suggest the process by which they were formed. To avoid this chapter being a catalogue of 'trivia', but at the same time to present its factual basis, summary points are listed at the end, and the chapter itself is organised into categorised observations illustrated by examples. Specific examples (given in endnotes) for the central Indo-Pacific can be sourced through references in Veron (1993). Examples for the whole Indo-Pacific are, for reasons given above, much less well supported (see p 203).

THE SIZE OF DISTRIBUTION RANGES

Patterns produced by combined distribution ranges were the subject of chapters 9 and 10; this chapter deals with individual species.

LATITUDINAL AND LONGITUDINAL BOUNDARIES

Latitudinal boundaries The bimodal pattern of figure 61 is due to the majority of species having a wide latitudinal range, and a substantial group of species being restricted to equatorial regions (with the addition of a small number of high-latitude endemics) (p 192).

Longitudinal boundaries If, for present purposes, all taxonomic complexities that arise over geographic variation within species are set aside and the distribution boundaries given in Veron (1993) are accepted 'at face value' (p 4)[2], then 21 per cent of central Indo-Pacific species have a longitudinal range of 10° or less and, at the other extreme, nearly 3 per cent of species span the entire Indo-Pacific (figure 62)[3]. The remainder have boundaries within these extremes. These boundaries show no conspicuous pattern, that is, no geographic areas which have an especially high frequency of distribution boundaries[4]. This strongly indicates a lack of distinctive faunal provinces within the Indo-Pacific[5], which accords with the findings of chapter 9 (p 9).

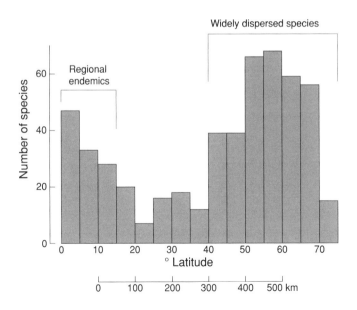

Figure 61 Frequency histogram of the length of latitudinal distributions within the *whole* Indo-Pacific, of *central* Indo-Pacific species. Some widely dispersed species have doubtful geographic boundaries in high latitudes.

LARGE AND SMALL DISTRIBUTION RANGES

LARGE RANGES

A total of 62 per cent of all central Indo-Pacific species have distribution ranges extending beyond the central Indo-Pacific region. Individual examples of taxonomically straightforward species with large ranges are illustrated in figure 63. Large distribution ranges are also found in some Atlantic species[6] which, of geographic necessity, exist in non-Caribbean/ Gulf of Mexico regions as disjunct populations.

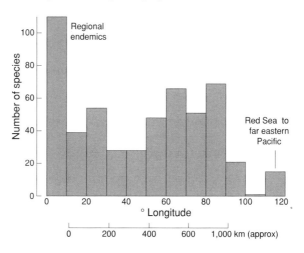

Figure 62 Frequency histogram of the length of longitudinal distributions within the *whole* Indo-Pacific, of *central* Indo-Pacific species or supposed species. Most widely dispersed species have doubtful geographic boundaries: data compilations are maximum ranges irrespective of taxonomic and/or identification problems.

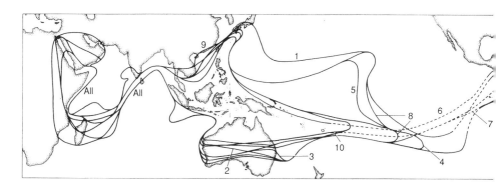

Figure 63 The distribution ranges of some widespread, common, and easily identified Indo-Pacific species. 1 = *Pocillopora damicornis*, 2 = *Seriatopora hystrix*, 3 = *Stylophora pistillata*, 4 = *Montipora verrucosa*, 5 = *Acropora valida*, 6 = *Porites rus*, 7 = *Gardineroseris planulata*, 8 = *Fungia scutaria*, 9 = *Favites abdita*, 10 = *Diploastrea heliopora*.

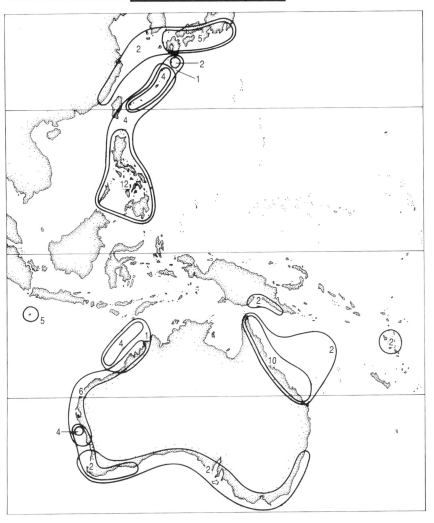

Figure 64 The number and location of species with restricted distribution ranges (endemics) within the central Indo-Pacific (see text).

SMALL RANGES

Several major issues are associated with the determination of restricted distribution ranges, among which are abundance, recognisability and taxonomic status. Clearly the boundaries of abundant, conspicuous and taxonomically straightforward species are much more reliably ascertained than those that are rare, cryptic or doubtful.

Approximately 22 per cent of all central Indo-Pacific species are restricted to less than half the area of the central Indo-Pacific. Approximately 17 per cent of these species are known to occur outside the central Indo-Pacific. The remainder have varying degrees of uncertainty associated with their taxonomy or distribution records, but of these, approximately seventy species (13 per cent of the central Indo-Pacific total) can be said, with reasonable certainty, to be restricted to a single continental coastline or region of comparable geographic size. Further studies are likely to extend the range of some of these species, but they are also likely to reveal more species with restricted distributions. Figure 64 indicates the extent, if not the absolute values, of endemism in the central Indo-Pacific.

Because endemism is difficult to determine, it is not uncommon for species that were once thought to be endemic to a particular region to be later discovered elsewhere[7]. There are also some well-established records of endemics outside the central Indo-Pacific, as demonstrated most convincingly by distribution maps of monospecific and paucispecific genera[8].

In principle, criteria for determining endemism are different for high-latitude (non-reef) and low-latitude (reef) locations.

High latitude, non-reef localities Endemism in high latitudes is of particular interest because spatial patterns become compressed into small areas. Southern Australia has four endemics[9], high-latitude localities of Japan and neighbouring east Asia have nine (probable) endemics[10].

With these small numbers of species, it can be concluded that *endemism* in the high-latitude central Indo-Pacific is more a product of the distribution of a small number of individual species than a biogeographic property of corals in general. These endemics appear to be relicts of former tropical distributions, restricted to high latitudes because of the effects of connectivity ratchets (p 99). The fossil record supports this conclusion (p 145). Except in the case of the two Australian *Coscinaraea*, which are morphologically similar, there is no evidence that high-latitude endemism results from speciation within their region of occurrence. Endemism of high-latitude *geographic subspecies* is far more common and is easier to determine (p 22).

Low-latitude, reef localities Relatively few species can be considered, with complete confidence, to be endemic to any particular region within the reef area of the central Indo-Pacific. Ten species only are endemic to the Great Barrier Reef[11], eleven species are known only from tropical western Australia[12], twelve mostly distinct species are known only from the Philippines or the Philippines/Indonesia region[13], four species are known only from tropical Japan[14]. It may be concluded that single regions in the tropical central Indo-Pacific have fewer than 10 per cent endemics, and in most major regions the level of endemism is less than 5 per cent.

Few studies indicate levels of endemism in isolated regions of the central Indo-Pacific. The only isolated region adjacent to the Indo-Pacific centre of species diversity studied by the author is Cocos (Keeling) Atoll, which has five possible endemics[15].

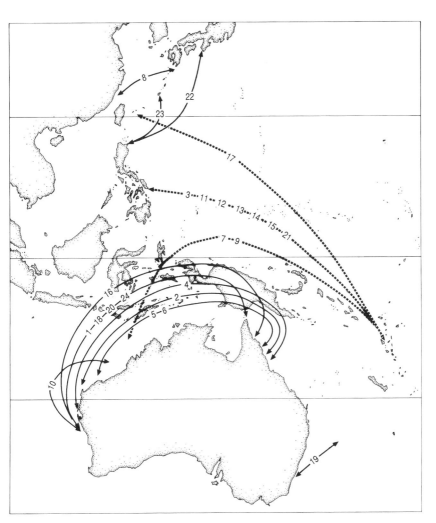

Figure 65 Examples of disjunctures in the distribution of central Indo-Pacific species absent from contiguous regions. 1 = *Palauastrea ramosa*, 2 = *Montipora capricornis*, 3 = *M. altasepta*, 4 = *Acropora bushyensis*, 5 = *A. lovelli*, 6 = *A. listeri*, 7 = *A. abrolhosensis*, 8 = *A. pruinosa*, 9 = *A. exquisita*, 10 = *A. donei*, 11 = *A. rambleri*, 12 = *A. insignis*, 13 = *Porites attenuata*, 14 = *P. deformis*, 15 = *P. horizontalata*, 16 = *Alveopora gigas*, 17 = *P. vaughani*, 18 = *Diaseris fragilis*, 19 = *Pavona decussata*, 20 = *Blastomussa wellsi*, 21 = *Montastrea multipunctata*, 22 = *Cataphyllia jardinei*, 23 = *Turbinaria peltata*, 24 = *T. radicalis*. Arrows indicate well-established separations between adjacent records. Broken arrows indicate recorded disjunctures that may well be shortened by further study in equatorial regions, but not in Australia.

CHARACTERISTICS OF DISTRIBUTION RANGES
DISJUNCT AND CONTIGUOUS DISTRIBUTIONS

The presence of wide geographic disjunctures within the distribution ranges of species are of great biogeographic interest because they demonstrate the capacity (in the absence of other explanations, p 51) of corals to disperse over long distances.

Establishing the presence of geographic disjunctures is often not a simple matter: it must be established that the disjuncture does, in fact, occur within a species and not between 'sibling species', and that no intermediate populations exist. Molecular techniques have the potential (as yet unused for any coral species) to aid the first requirement. The second must always be dependent on the thoroughness of *in situ* studies: requirements are different for contiguous and non-contiguous regions.

Incomplete records are potentially the reason for all disjunctures indicated by Veron (1993, tables 1–3), as it is never possible to prove that a particular species does *not* occur in a particular region. However, some major disjunctures appear well substantiated (figure 65). Although it is likely that further research will show that some of these are not as long as presently indicated, it is clear that these sorts of gaps in distributions are a common characteristic of coral distributions, not a sampling arte-fact. Potentially, they may be due to migrations, regional extinctions, or regional formation of 'sibling species'. They may also be temporary or semi-permanent.

A second aspect of disjunctures is seen in high-latitude, non-reef populations, which form geographic subspecies downstream of reef pop-ulations. Continuity between these subspecies and their parent popula-tions can usually be established by morphological taxonomy. The genet-ic component of this variation has not yet been established for any species.

Disjunctures within species also occur where populations are sepa-rated by wide open ocean. In these circumstances (ie the certain absence of intermediate populations), the distinction between species and sub-species (or 'sibling species') is frequently unclear. In general, the greater the distance of separation, the greater the taxonomic uncertainty, and if one of the populations of the disjuncture occurs in a high-latitude region (eg Hawaii), or in an otherwise different physical or ecological environ-ment (eg the far eastern Pacific), taxonomic uncertainty increases.

Wide disjunctures are not confined to the Indo-Pacific: although there are major taxonomic uncertainties, approximately thirteen 'species'[16] have been recorded from both sides of the Atlantic.

PATTERNS WITHIN SPECIES RANGES IN
THE CENTRAL INDO-PACIFIC

Unequal evolutionary change within the geographic range of individual species creates geographic patterns (chapter 12). These patterns occur in almost all species at all systematic levels from the population up, and occur increasingly with distance. The outcome is that most species are adequately definable within any given region, but over increasing space their definition becomes progressively less straightforward. They also become progressively less well defined in terms of other species, creating geographic patterns in inter-specific differences.

The following account unavoidably omits the detail on which the examples given are based; it also contains errors of omission. Veron (1993) suffers the same problems for the same reasons. Taxonomic publications referred to by Veron (1993) contain mostly intra-regional (and thus primarily environment-correlated) detail, not biogeographic detail. With the exception of solitary and a few colonial species, probably no more than 10 per cent of the morphological variation of any widespread Indo-Pacific species has been fully described.

Geographic variations in reproduction and the physical environment are discussed in chapters 5 and 6 (respectively).

MORPHOLOGY

Even the most casual acquaintance with corals *in situ* provides the observer with ample evidence that different species of corals can vary in size, shape and growth form in much the same way that forest trees do. Further observation reveals that this variation is repeated in different ways in different biotopes and that it is also found in details of skeletal structure (chapter 2). The polymorphic nature of corals is simple enough to observe and may be simple to determine in detail for species that are distinct and have at least some conservative skeletal characters. It is less simple to determine for groups of species that are difficult to separate *in situ* (such as some species of *Platygyra*, *Porites* and *Turbinaria*), or for species that have few taxonomically useful characters, for example, most *Astreopora* and *Galaxea,* or those which have highly variable characters, such as many *Montipora* and *Porites*. The task of separating these species may be complex; the addition of a geographic component may make this complexity substantially greater.

If morphological characters are used for taxonomy without supporting co-occurrence, ecological and geographic data, the result (as taxonomists of old have repeatedly demonstrated), is mostly chaos (p 19). If such studies are extended over a wide geographic range, the result is still greater chaos[17]. In order to establish meaningful species boundaries over wide geographic ranges, significant *geographic* components of variation within species must be separated from all other forms of variation, including those that are correlated with physical environment, have a genetic basis that does not vary geographically, and involve taxonomic uncertainties. This requires a lot of information about a species, and for this reason the number of species that can used for detailed observations of biogeographic variation is much less than the total central Indo-Pacific complement.

SKELETAL MORPHOLOGY

Only a few species (most *Fungia* and a scattering of colonial species such as *Porites rus, Gardineroseris planulata* and *Diploastrea heliopora*) are stable in both morphology and microstructure over great distances; the vast majority show at least some geographic variation. This occurs in gross morphology or microstructure or both, may be masked by environment-correlated variations, may indicate (or clearly demonstrate) geographic subspecies or sibling species, and varies greatly according to geographic region and/or taxon. There are so many aspects to the subject that individual species, at least those that are adequately known, tend to become case examples for very few other species. The following condenses the main points:

♦ *Major geographic variations* Perhaps the most complex problem of operational coral taxonomy is resolving the significant, geographically continuous, morphological changes observed in some form or other in most species. The absence of species boundaries (except in isolated regions) is not the only issue as these geographic variations are usually interlinked with various aspects of environment-correlated variation[18].

♦ *Subtle geographic variations* 'Subtle' geographic variations emphasise the point that minor geographically contiguous morphological changes occur in almost all colonial species. The consequences are usually trivial for regional taxonomy or identification, but become important, by accumulation, over the full range of a species. In some instances accumulation of subtle changes, as interpreted by taxonomists, delimit that range. Many of the examples listed below potentially represent different geographic components of species complexes, each of which is a subset of a continuum of minor variations[19].

♦ Morphological variation in some species appears to be due to an undefined combination of local physical environment and geographic factors, reflecting particular genetic controls acting in particular environments in particular regions[20].

♦ Some species have one or more ecomorphs that are only found in a particular region while others have ecomorphs that are absent or rare in a particular region[21]. In all cases, observations are biased by the amount of the species' range that has been studied. As with abundance, the presence/absence of ecomorphs is difficult to determine if there are changes in morphology at the species' distribution boundary, or if that boundary is not associated with an environmental discontinuity or physical barrier. There are few examples of specific geographic ranges of ecomorphs[22].

♦ A group of species in one region may have minor points of similarity retained by the same group in another distant region. This is usually readily observed *in situ* by divers who, after becoming familiar with location-specific points of identification within a particular species group in one region, then see the same points repeated in quite a different region[23].

♦ Morphological characters used to separate very similar species may vary geographically. Distinctions between species, such as can be seen in co-occurring colonies (p 20) and which are often very subtle, may vary geographically[24].

♦ Morphologically distinct geographic subspecies commonly occur in high-latitude environments (p 7); there are only rare instances where these are not contiguous[25]. Similar groupings of either geographic subspecies or 'sibling species' are also found within regions of high diversity, or occur in isolated regions, or occur gradually over distance[26].

♦ *Variations that are latitude correlated* Species may display no geographic variation in morphology over most of their range, but do so in high-latitude parts of it[27].

♦ Geographic variation is often masked by physical-environment-correlated variation. This is an interpretive problem in which geographic variations that have a genetic basis may be obscured by ecomorph variations or geographic patterns in habitats. This may defeat attempts to resolve geographic variations, especially where particular environment-correlated variations are very dominant[28].

CALICES AND SKELETAL MICROSTRUCTURE

It is hardly surprising that most species commonly display correlated geographic variation in both morphology and skeletal microstructure. There are exceptions: some species display geographic variation in calice size[29], or skeletal microstructure[30], but not in gross morphology.

It is common for most species in the same region to have similar levels of skeletal density[31]. It appears less common for individual species to display unusual levels of skeletal density independently of most other species of the same region[32].

COLOUR

Colour variations in corals are almost as difficult to generalise about as abundances (below), primarily because they involve so many different categories, but also because they appear to have so many different causal relationships, involving not only the corals themselves but also their zooxanthellate symbionts. Some species have specific colours, or colour ranges, that can be used as an identification aid; sometimes these vary geographically. Interest here is not so much in geographic variation in colour as such but in colour as a geographically-variable characteristic that may have a genetic basis.

By far the most common variations in colour are correlated with physical environment, especially light. Colonies, or parts of colonies, exposed to strong sunlight are usually relatively pale; thus massive colonies in shallow water often have pale tops and darker sides, whereas colonies of the same species in deeper water are uniformly coloured.

Another very common source of colour variation, especially in *Acropora*, occurs as zooxanthellae progressively infect growing branch tips or corallite margins. Some species of other genera display such a wide range of colours that two colonies of the same colour are seldom seen together[33]. Geographic variation in colour may be difficult to separate from these sorts of non-geographic variations, yet colour is a species characteristic that is constantly observed and (consciously or not) used as an identification aid. It is sometimes also a taxonomic character, or, more frequently, signals a taxonomic issue.

The observations presented below are from the central Indo-Pacific. In general, Caribbean corals appear to show relatively little intra-specific geographic variation in colour, perhaps owing to the relative paucity of *Acropora* species (and because all three Caribbean *Acropora* have similar colour), *Favia* and pectiniids and, perhaps, the relative clarity of Caribbean waters. Lack of colour variation may be primarily a function of area and limited latitudinal range. It may also relate to a relative lack of habitat variation, especially in the sedimentary environment. These observations would not equally apply to the corals of the Gulf of Mexico.

Lack of variation in colour Lack of variation is just as important as variation in colour from a biogeographic point of view, as it may reflect genetic homogeneity within the species. It may also indicate a lack of genetic control, or a lack of environmental control, or other factors, including limited habitat range. Almost all coloration in corals is attributable to soft tissues, the odd instances where this is not the case are mostly geographically conservative[34].

♦ Some species have distinctive colours, with little environment-correlated or geographic variation throughout wide distribution ranges. In most of these, colour can be a reliable identification character[35].

♦ A high proportion of species show wide colour variation correlated with physical environments, but little or no geographic variation except in high-latitude locations.

♦ A few species have wide geographic variations in colour, but in some locations, only a small proportion of this range expressed[36].

♦ Very few species have ubiquitous colour morphs[37].

Variation in colour Regional variations in colour may be strikingly obvious or quite subtle, may be restricted to some colonies (that are distinctively coloured) but not others, and may be more apparent than real, for example where a species is restricted in a particular region to a particular habitat.

♦ Many species have well-defined colour variations from one tropical region to another. Of these, *Acropora* species often have regional colours that are independent of physical environment[38].

♦ Other species have no region-specific colours; inter-regional variations, if any, are due to differing proportions of the same colours[39].

♦ Some species have geographically restricted exceptions to otherwise predictable ranges of colours[40].

Colours of geographic subspecies   Corals in high-latitude locations usually have more intense or darker colorations than their tropical counterparts. This is best observed in Japan, where species are spread in a continuum from the tropics to high latitudes[41]. High-latitude-specific colours are particularly common in the Faviidae[42]. Very few observations reliably associate high-latitude colours with a particular environmental variable[43].

Although colour is commonly used as an identification character, it is seldom used as a taxonomic one. Many species occurring in high latitudes, including most of those mentioned above, have distinctive colours as well as distinctive morphologies. Most of these colorations are likely to be at least partly induced by physical environment, but some are very likely to have a genetic basis[44]. Whether these can reasonably be called geographic subspecies or 'sibling species' remains doubtful.

SPECIES ABUNDANCES

Geographic variation in the abundance of species is very unpredictable. The first dives on high-diversity reefs of a country or region not previously visited repeatedly reveal the unexpected: species thought to be always rare (in the type of habitat under observation) are often found to be common, and vice versa for common species. Presence/absence observations are, in general, much more predictable. This may simply be due to a multiplier effect (presence/absence × abundance) giving enhanced variation of information to the observer. Alternatively, variation in abundance may well be a fundamental aspect of species' distributions, especially distributions within different regions of high species diversity. If so this requires more detailed data to verify than presently available: considering the amount of information required, such a study would be long in the making.

Determination of species abundances is a complex and time-consuming task. Terrestrial botanical ecology has provided blueprints of methodologies that are potentially applicable to coral communities, but there are obviously major differences between studies of (say) terrestrial forests and of corals, foremost amongst which are the problems of working underwater, and the capacity of observers to identify species *in situ*. Historically there has tended to be an inverse relationship between sophistication of quantitative methods used in field studies and level of identification skill, so that highly quantified studies tend to be taxonomically weak and vice versa. For some purposes, species identification is not important; for the present purposes it clearly is, and this greatly restricts the sorts of quantitative data useful to biogeography.

Lack of abundance data is unquestionably a weakness of the primary biogeographic database of this book. Binary (presence/absence) data, which equates rare species with common ones, raises the question: have records of rare occurrences distorted distributed patterns? In the central Indo-Pacific it is commonplace for the abundance of a species to vary several orders of magnitude between one biotope and another (or even within single biotopes), and it is this that makes meaningful quantification of abundance a tedious undertaking. Nevertheless, as with most types of macrofauna and macroflora, relative abundances can be estimated visually. Very general estimates of relative abundance of species have been recorded in many countries, but not always for different locations within those countries. Perhaps the most useful dataset for biogeographic analysis is that of Japanese species, divided (by Veron 1992b) into three abundance categories: 'common', 'uncommon' and 'rare', comprising 129, 151 and 119 species (respectively). Deletion of 'rare' species from the classification of Japanese localities (the dendrogram of figure 57, p 176, repeated without rare species) does not, in this case, change the pattern of the dendrogram, but this may not hold true for other regions.

As previously emphasised (Veron 1993, p 21), abundance estimates referred to in this book are, of necessity, general indications only (p 86), and are not sufficiently well established to be used in numerical analyses. No estimates of geographic variation in relative abundance have been recorded for at least half the species of the central Indo-Pacific. Even crude estimates may be unobtainable for species that are always uncommon, or are always common but in very restricted habitats (eg inter-tidal mud flats), or that are restricted to seldom-visited habitats (eg very deep or inter-reef habitats). Present records reveal no consistent geographic patterns in species abundances, but they do indicate the sorts of variation that can occur. Veron (1993) gives brief summaries of geographic variation in abundance of the examples indicated below; further details are in intra-regional studies.

Regions of high diversity Species may be:

♦ Uncommon or rare throughout their range, except in one or two regions where they are common[45]. Those regions may be environmentally very dissimilar[46].

♦ Uncommon or rare throughout their range, except in specific physical environments in specific countries, where they are common[47].

♦ Common in major parts of a wide distribution range and uncommon or rare in other major parts[48].

♦ Common throughout most of their range, but uncommon or rare in one or two major regions[49].

♦ Common throughout their range except in one or more regions, where they are restricted to a particular habitat[50].

Regions of low diversity Although several forms of diversity indices have been used to correlate abundance with species diversity within particular regions, this has not been done on a comparative biogeographic scale. Again, observed estimates of intra-regional abundance only can be used. This has been done in high-latitude regions (above), and clearly other physical-environmental factors may cause low diversities independently of latitude. Of more biogeographic interest are low diversities associated with isolation.

Regions that have a low species diversity, but do not have major physical-environment limitations, frequently have a very patchy distribution of species, resulting in great abundance in limited areas[51]. Apart from such extremes, most species in isolated areas are prone to having unusual abundances[52].

A high proportion of species have their latitudinal boundaries within major reef areas and thus do not, apparently, have their ranges delimited by physical environment. These species usually appear to be relatively uncommon at the margins of their distribution ranges[53].

Latitudinal variation and geographic subspecies A wide range of species' attributes, including abundance, change with latitude, especially where ranges extend into non-reef environments. There have been no quantitative studies of this.

♦ Species may be uncommon or rare throughout their distribution range except in high-latitude locations, where they are common[54].

♦ Species may be common throughout their distribution range except in high-latitude locations, where they are uncommon or rare[55].

♦ Most species that form geographic subspecies do so in high-latitude or isolated regions, where relative abundance appears to be strongly correlated with physical environment. However, within the tropics some species forming geographic subspecies do so with no apparent correlation with physical environment; and these subspecies may have associated variations in abundance[56].

Colony size Species that are rare in a particular region appear to be just as likely to form large colonies as in other regions where they are common. The converse is also true[57]. Many species, especially of *Acropora*, which seldom form large colonies in the tropics, do so in high latitudes[58]. In general, a wide variety of species form exceptionally large colonies in high-latitude locations[59], but others do the opposite.

HABITATS

A habitat may be considered to be the summation of all physical and biological environmental parameters, any one or combination of which may be limiting for a particular species. Principal physical parameters that are correlated with zonation appear to be of overwhelming importance, but biological parameters, especially competitive exclusion, may well be a major determinant of habitat occupancy. For the present purposes, however, habitats can be considered as 'black boxes', the workings of which need not be known (p 267): the focus is on the distribution patterns that are influenced by habitat availability, and intra-specific geographic variation in habitat occupancy.

In contrast to abundance, the habitats in which a species is most commonly found vary little from one region to another. Whatever variation there is in habitat occupancy between different regions is, for most species, primarily a sampling effect reflecting variation in abundance. Principal exceptions are non-reef habitats, especially those of high latitudes; specialised habitats (including inter-tidal habitats and those dominated by macroalgae); and geographically restricted habitats primarily determined by substrate type and tidal range.

Distribution patterns and habitat occupancy One of the primary ecological characteristics of zooxanthellate corals is that most species are generalists, that is, most species can occupy a wide range of high-diversity habitats[60]. Within the central Indo-Pacific, 67 per cent of all corals have been recorded in non-reef habitats[61], 95 per cent have a depth range extending from the subtidal to depths where species diversity starts to decline[62], 48 per cent tolerate a water temperature of 14°C (4°C below the threshold for development of consolidated reefs, p 91). There is little geographic pattern in these habitat limitations: ecological constraints on species' occurrence, though important for individual locations, impose few inter-regional constraints; the main exceptions are habitat-dominating sedimentary and tidal regimes (p 102)[63].

Intra-specific variation and habitat Variations in abundance between habitats is a vague subject, made so by spatial variations in relative abundance and the 'generalist' nature of most coral species.

♦ Species occurring in a particular habitat in one region may be absent from similar habitats in another[64].

♦ Species that are almost always associated with a particular type of habitat in one region may be found in a completely different habitat in another[65].

♦ Species that usually occur in a wide range of habitats may, in some regions, be restricted to one habitat type[66].

Despite the number of clearly defined geographic subspecies in corals, none has a consistently different habitat occupancy from other geographic subspecies of the same species[67].

BEHAVIOUR

Species-specific behaviour Apart from variations in reproductive behaviours, corals have wide intra-specific variations in polyp behaviour

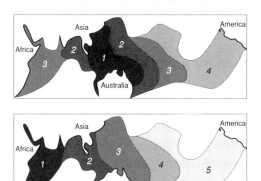

Figure 66 Concept diagrams of levels of taxonomic uncertainty due to geographic distance (see text). The two taxonomies are hypothetically based on the study of Great Barrier Reef corals (above) and Red Sea corals (below). In regions marked 1, operational taxonomy is directly supportive; in 2, taxonomically significant uncertainties arise in many species; in 3, most identifications involve substantial uncertainties; in 4, most species are uncertain; in 5, only a few species are identifiable. The diagrams exclude differences due to presence/absence, and it is assumed that the hypothetical taxonomist does not have a knowledge of geographic variation of the species under study.

(including retraction and extension of tentacles, vesicles and/or the polyp trunk) in response to external physical or chemical stimuli. In the case of free-living species they also vary in methods of mobility. There are also intra-specific variations in aggressive behaviour.

There appears to be no geographic variation in polyp behaviour among widespread species that have large, semi-permanently extended polyps that fully retract only after persistent stimulation[68], or those that have middle-sized (5–20 mm diameter) or small polyps that are also semi-permanently extended but that retract in a few seconds after physical stimulation (presumably as a defence mechanism)[69].

Some species show clear geographic variations in diurnal polyp and/or tentacle extension[70], particularly in high-latitude locations[71]. This may be a learned response resulting from reduced fish predation, as it is also commonly seen in corals kept in aquaria[72].

PATTERNS OVER THE WHOLE INDO-PACIFIC

With corals, as with plants, a species that is readily identified in one region or province may be much less readily identified in another, that is taxonomic distinctions between species become less reliable with distance. Figure 66 illustrates in concept the 'reach' (relevance or, conversely, level of doubt), of taxonomy originally based on studies on the Great Barrier Reef and that based on studies in the Red Sea. The reason that the reach of the two taxonomies differ is purely geographic variation. For the same reason, reference collections and identification guides relevant to one region may be of limited value for another.

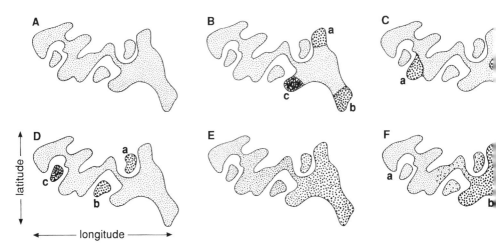

Figure 67 Concept diagram of common types of geographic variation within widespread Indo-Pacific species and species complexes. *A* No taxonomically significant variation occurs throughout the range. *B* Morphologically distinct disjunct races or geographic sub-species (a, b and c) occur in high latitudes. *C* Morphologically distinct races occur in geographically marginal (a) or central (b) regions. *D* Morphologically distinct disjunct races or geographic subspecies/sibling species occur in isolated regions. These groups may be similar (a and b) or individually distinctive (c). *E* Continuous variation within a species becomes increasingly marked over distance so that coralla from opposite ends of the longitudinal range are very distinct, but at no point are there two overlapping or distinctive groups. *F* As with E except that there is a region of overlap, or hybridisation, between geographic subspecies/sibling species a and b (see text).

Indo-Pacific-wide 'biogeographic' taxonomy

Most taxonomic studies of Indo-Pacific corals are based on original studies made within a single region. A few are based on original studies over a whole province. None are Indo-Pacific-wide. The examples given below are, of necessity, unpublished personal observations, based on a combination of limited original field work, studies of major museum collections and the taxonomic literature.

The relevance of taxonomic decisions made from original studies in one region (eg the Great Barrier Reef) to those of other regions (eg the Red Sea) (figure 66) raises important issues, not least that single taxonomic descriptions are of geographically-limited use. Taxonomic descriptions covering geographic variation are equally valid and important as those covering environment-correlated variations (p 17). It follows that the synonymy applicable to a species in one region may not be equally applicable to the same species in another region, that is, synonymies and nomenclatorial priorities (in theory at least) can vary geographically.

Geographic variation is normally described as being inter-specific or intra-specific. In practice, the two go together and are not mutually exclusive or fundamentally dissimilar: presence/absence (inter-specific) differences between the corals of two regions are *indicative* of intra-specific differences.

Figure 67 illustrates the sorts of geographic patterns commonly found over very large ranges. As with most human endeavours in taxonomy, this figure suggests categorisations where none, in reality, exists. The categories (below) form a continuum; they are not mutually exclusive, and many species can be placed into two or more categories. Most species that have very wide geographic ranges (p 190) are 'species complexes' over that range.

Species that have little or no geographic variation Fewer than ten species in the Red Sea are virtually indistinguishable from those on the Great Barrier Reef[73], but at least forty are recognisable without raising any taxonomically significant issue. Approximately similar proportions of the total species complement apply to comparison between the southeast Pacific and the Great Barrier Reef. Most of these species are those that have the greatest longitudinal distribution ranges (figure 62, p 190).

Species that have disjunct distributions in high latitudes High-latitude occurrences of species may be in one of the Indo-Pacific's four major latitudinal sequences or isolated from the tropics. In the former case, upstream seeding and connectivity ratchet effects will ensure one-way genetic connectivity with tropical populations. In the latter case, upstream seeding is a relatively rare event under present surface circulation regimes, and genetic isolation may continue into evolutionarily significant time.

High-latitude localities on the east African coast are (sequentially) downstream of Red Sea and western Indian Ocean tropical locations. The fauna is isolated from the central Indo-Pacific by the combined effects of a western Indian Ocean connectivity ratchet and Indian-Ocean-wide longitudinal displacement across the Chagos Stricture[74].

Species that form geographic subspecies in isolated regions Depending on level of detail, the majority of species in any given region have unique points of distinction from the same species in other regions. Occasionally the same variant of a species found in one region appears in another very distant one, suggesting long-distance founder dispersion[75]. Because intra-specific dissimilarities appear to reflect inter-specific ones[76], combined measures of inter-specific affinities (eg figure 59, p 182) usually suggest a measure of the effect of isolation on individual species[77].

Species that have increasing variation with distance By its very nature, taxonomy is binary: two corals are either given the same name, or they are given different names. By necessity this gives false information if corals from two regions are separated by a continuum. Thus, figure 67E might be called a single species if the full continuum were known, but would be called two species if only the extremes were known. Whether or not operational taxonomy tends to favour the former or the latter position depends entirely on point of view and information available. One member of a geographic 'species complex' may have one name on one side of the complex, and another name on the other side. Where it occurs in intermediate regions, it might be given the one name or the other (perhaps with a 'cf' or '?'); it cannot be given *both* names, although it is, in effect, both 'species'. Conversely, the 'complex' may be given one name, but the characters of the species in one region may be very unlike its characters in another[78].

Hybrid patterns When is a species not a species? Conceptually, hybrid zones may occur as illustrated in figure 67F, but more commonly species pairs or groups appear to have differing levels of taxonomic separation in different regions[79]. The answer to this question is not just a matter of biogeographic taxonomy, it is also a matter of concept, taken up in the following two chapters.

SUMMARY

The distribution and taxonomic data on which this chapter is based are summarised in a separate volume (Veron 1993).

♦ Distribution ranges are mostly so large that they can only be produced by long-distance dispersal.

♦ Approximately two-thirds of all central Indo-Pacific species have distribution ranges extending beyond the central Indo-Pacific region.

♦ Approximately 10 per cent of central Indo-Pacific species are limited to single continental coastlines or geographic regions of similar size.

♦ Endemism in the high-latitude central Indo-Pacific is more a function of the distribution of particular species than a characteristic of corals in general.

♦ Many species form discrete geographic subspecies in high latitudes, or in isolated regions. In such cases, distinctions between species and geographic subspecies may be unclear.

♦ There is no taxonomically significant morphological variation in approximately 25–50 per cent of species *within* the central Indo-Pacific province. The remaining species have a wide variety of geographic variations that can be categorised into different types of subspecific taxa. These variations may be subtle or pronounced, occur in growth form and/or fine skeletal structure, have different geographic characteristics and, as a result, create a wide range of conceptual and operational issues for taxonomy.

♦ Colour predominantly varies with light availability, but there is a wide range of geographic variations in colour, some of which co-occur with variations in other species attributes, especially in high latitudes where they are commonly associated with geographic sub-species.

◆ Relative abundance (as determined from visual estimates) is very unpredictable. It varies greatly in regions of both high and low species richness. Species that have distribution boundaries not apparently limited by physical environment tend to have reduced abundance at their distribution boundaries.

◆ Habitat occupancy tends to vary little within distribution ranges. Most species are generalists with respect to habitat: 67 per cent of central Indo-Pacific species presently occupy both reef and non-reef habitats; 95 per cent occur over a substantial depth range.

◆ It is relatively common for species to form very large colonies, or monospecific stands in high latitudes.

◆ Unequal evolutionary change within the distribution range of a species creates patterns of geographic variation. Biogeographically important variations may be subtle; indicate discrete geographic sub-species; be sufficiently great to raise questions about the validity of species; or make particular species undefinable.

◆ Within the central Indo-Pacific, variation in most species is contained within recognisable species boundaries. The existence of geographic subspecies in high latitudes is the most common and pronounced type of intra-specific geographic variation.

◆ When expanding the area of interest to the whole Indo-Pacific, there are many inadequacies in available data and there are no directly applicable publications. Latitudinal variations are largely replaced by less well-known longitudinal ones. Categories of geographic patterns illustrated in figure 67, which involve most species, are common types of geographic continua.

◆ Variation in very widely dispersed species may not be accommodated by operational taxonomy and binomial nomenclature. This has raised taxonomic issues parallel to many groups of vascular plants.

PART D

EVOLUTION

The main topics of chapters 12 and 13 are listed at the beginning of each chapter and the main conclusions are listed at the end of each chapter. The following is a summary of these topics and conclusions.

Although these concluding two chapters are based on the factual content of the rest of this book, they are a fundamental departure from it in being speculative.

THE MAIN POINTS

The following points are not necessarily self-explanatory. They are statements of concept with varying degrees of factual basis.

- ♦ Surface circulation vicariance, a function of both divergence and hybridisation, is driven by fluctuations in surface circulation patterns creating fluctuations in rates and amounts of genetic connectivity.
- ♦ Major fluctuations in surface circulation patterns are probably driven by palaeoclimatic cycles.
- ♦ Surface circulation vicariance is a continuous, benign process, that is not necessarily dependent on major palaeoclimatic upheavals.
- ♦ Reticulate patterns are seen over wide geographic ranges; the outcome is also seen in single locations where different colonies

of the same coral species may have very different phylogenies in space and time.

♦ Surface circulation vicariance acts equally and simultaneously on all coral species. The outcome is a form of mass synchronous evolution.

♦ Reticulate evolution occurs within metaspecies. A metaspecies is the highest taxon that never hybridises.

♦ Reticulate patterns at present include coral species that are syngameons; coral species that are parts of syngameons; races; and populations. These are all potentially taxonomic units which can be operational species.

♦ Operational species are the most clearly identifiable discontinuities in continuous variation from the population to the genus. They are essentially human-created and are not qualitatively different from other taxonomic levels.

♦ Operational taxonomy requires morphological distinctions. Because evolution is not primarily biologically controlled, systematic (genetic) units, which do not necessarily have clear morphological distinctions, may be common. If so, these units will create discrepancies between the results of molecular taxonomy (which will recognise them) and those of morphological taxonomy (which will not recognise them).

♦ Indo-Pacific patterns of biodiversity are created by attenuation (dispersion) from an equatorial band of maximum species diversity. The ultimate control of biodiversity is the relationship between broad-scale sources (which are dependent on genetic connectivity and habitat) and local sinks (which are ecologically dependent).

12
VICARIANCE AND RETICULATE EVOLUTION

'Evolutionary biology enjoys the peculiar status of being that subject which clearly unites all biological endeavours, while seeming to be nearly as remote from complete understanding as when Darwin brought it within the realm of materialistic science'.
(Levinton 1988)

The subject of evolution is at the one time the centrepiece of our understanding of life and the focus of most of our conceptual ideas about it. The reasons are clear: all fields of biology can raise questions of evolution, and questions of evolution relate to all fields of biology. The spectrum of evolutionary concepts that have been proposed for corals by a multitude of authors were introduced in part A of this book. The last two chapters are largely restricted to this author's concepts. The word 'concept' is stressed because evolutionary processes cannot be observed in most organisms: they can only be inferred from taxonomy, observations of distributions and genetic studies. Because these two chapters are largely conceptual, they differ in principle from parts B and C of this book, which are, to the maximum extent possible, empirical.

The central points linking concepts of coral evolution to observations about their distribution are their capacity for rapid, long distance dispersal (p 87); and the immense geological longevity of their families (p 108), genera (p 131) and species (p 146). When combined, these two factors have long obliterated all evolutionary tracks and have done so, perhaps, hundreds of times over. Observed patterns do, however, demonstrate that corals have not evolved by Darwinian displacement of species from centres of origin (p 36). They also discount speciation by classical vicariance (p 41), because of the lack of correlation between their present place of occurrence and any determinable place of origin.

Editorial note: taxonomic and systematic terminology used in this chapter is explained in the next chapter.

THE MAIN TOPICS

◆ Palaeoclimatic cycles driving surface circulation vicariance.
◆ Surface circulation vicariance creating reticulate evolution.
◆ Non-reticulate evolution; mass synchronous evolution.

SURFACE CIRCULATION VICARIANCE

The concept of vicariant speciation, introduced in chapter 3 (p 40) in the context of biogeography, must now be extended to include vicariant hybridisation as well as vicariant speciation. This departure from previous usage of the term 'vicariance' is, at least, in keeping with dictionary definitions of it.

THE CONCEPT

Figure 68 shows a pattern of reefs and surface circulation currents[1] in the hypothetical coral distributions illustrated in figure 67 (p 204). In an imaginary situation where all surface currents cease, there would be no genetic communication between reefs for fauna (like corals) that rely on passive dispersal. Over time, every reef would develop its own distinctive coral species complement through genetic drift and local selection. This complement would be but a tiny fraction of the world's total coral species diversity, which would, consequently, be very great. The reverse imaginary situation would be where surface circulation currents were so strong, and so variable, that propagules were repeatedly dispersed en masse back and forth between one region and the next. Over time, every reef would develop a similar coral species complement to every other reef, and each would have a major proportion of the world's total coral species diversity, which would, consequently, be relatively small. These imaginary conditions have never happened, but intervals of fluctuations in surface circulation over time would have created ever-changing patterns of genetic communication between reefs. These in turn would create complex changes in patterns of distribution and diversity. Potentially, this is a vicariance mechanism where evolutionary change is driven by the formation and removal of ocean currents that alternate between being barriers to dispersal and vehicles of transport. In this case, the barriers are relative (not absolute) and cyclical (not isolated events).

Surface circulation vicariance *must* produce environmentally-determined biogeographic patterns (figure 67, p 204). These patterns

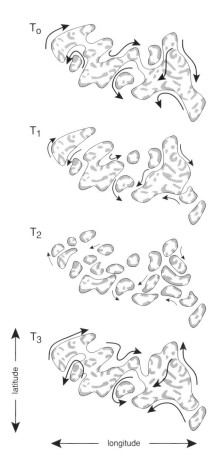

Figure 68 Concept diagram of changing genetic connectivity in a gene pool over a single palaeoclimatic cycle. Individual regions are indicated by patches of grey; regions in genetic communication are enclosed by a line. At time T_0, strong surface currents are maintaining genetic connectivity throughout most of the gene pool. At T_1, surface currents are weakening and the gene pool is becoming partitioned. At T_2, surface currents are very weak and the gene pool is partitioned into many isolated regions. At T_3, strong surface currents are restored, but the pattern of genetic connectivity is not the same as at T_0.

will also be *biologically* determined because they will vary according to the endurance of propagules, levels of post-migration reproductive isolation, and the demographic success of hybrid mixtures.

CHARACTERISTICS

Unlike previous concepts of vicariance in coral evolution, which generally require specific intervals of major tectonic upheaval, or environmental or sea level change, surface circulation vicariance operates continuously, forming and removing bridges and barriers to genetic communication without physical disruptions, and doing so simultaneously on all geographic scales from the global to the local. Major palaeoclimatic events (p 120), especially sea level changes, would cause major change in surface circulation vectors, but these are only a part of the process, albeit a major part.

Surface circulation vicariance provides an explanatory basis for the high level of variation found in most coral species in most localities (because it produces unending founder events, p 43), as well as geographic variation among localities. It explains the principal characteristics of coral species distribution patterns, and it underpins reticulate evolution (below). General characteristics (to which there are likely to be numerous specific exceptions) are as follows:

♦ Surface circulation vicariance drives both speciation and hybridisation (and perhaps localised extinction) at different times in the cycle. Speciation will occur if formerly allopatric taxa do not successfully hybridise after becoming sympatric.

♦ Surface circulation vicariance primarily impacts on any shallow-water marine life where dispersion occurs by the passive transport of larvae. It is a purely physical mechanism, although it must act in association with natural selection and biological mechanisms of species cohesion.

♦ Surface circulation vicariance will have varying regional impacts. Relatively remote faunas (in terms of currents) will tend to have less frequent genetic contact with other faunas and more prolonged periods of genetic isolation. Whether or not the mechanism operates in an area as uniform as the Caribbean is a moot point.

♦ Successive vicariance cycles will continuously modify genetic patterns created by previous cycles, resulting in alternating periods of genetic isolation and potential genetic contact. The more cycles that a species survives without hybridisation with another species, the more genetically cohesive and isolated it will become, and the more taxonomically distinct it is likely to be over a wide geographic range. Species that are well-defined over wide geographic ranges have probably had a history of stability over many cycles. Species complexes (whether allopatric, clinal or sympatric) are probably caught in unending cyclical change.

♦ The effects of a vicariant event will vary among species, depending primarily on their reproductive characteristics. The more widespread and the more common a species is, the greater its production of propagules will be, and the shorter and weaker will be the impact of a surface circulation vicariant event. This applies within species in different regions as well as among species in the same region. The more common a species is, the greater will be its capacity to restore genetic connectivity, and (perhaps) to migrate, after periods of isolation. Conversely, the less common a species is, the less success-ful it will be in recolonising reefs in the next vicariance cycle. One might surmise that, for a widespread species, abundance is its most essential stabilising attribute.

◆ Opportunities for sympatric recombinations may be temporary (eg occur because of a temporary change in current pattern). Most recombinations will be geographically restricted by surface currents at any given point in time. Hybridisation and stasipatric speciation may spread geographically from areas of sympatry by a range of genetic mechanisms, including outcrossing and transilience.

◆ Importantly, surface circulation vicariance, being an evolutionary mechanism, externally driven primarily by environment and not by natural selection[2], may produce genetically distinct hybrids that need not be morphologically distinct. From this it may be concluded that complexes of 'genetic species', which have no morphological distinc-tion, may well be abundant.

◆ There is no geometric increase in the number of species (figure 69), as predicted by classical vicariance.

SPATIAL AND TEMPORAL PATTERNS

We can safely assume that the evolution of all coral species has taken place in *some* dimension of space and time. What happens when space and time are continua, driving both speciation and hybridisation?

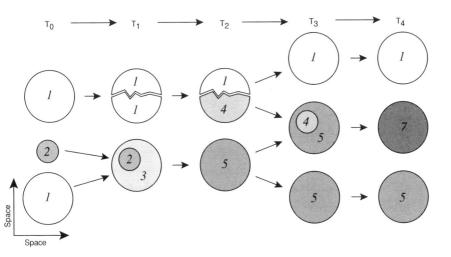

Figure 69 Illustrating evolutionary change over sucessive time intervals T_0 to T_4, due to surface circulation vicariance (using the same icons as for classical vicariance, figure 10, p 40) and resulting reticulate evolution. Circles represent the distribution ranges of individual species. Hybridisation occurs where circles overlap. The rate of species orig-ination equals the rate of extinction. Species do not have discrete phylogenies (see text).

Spatial patterns The present-day outcome of surface circulation vicariance may be illustrated by the following simple hypothetical example.

Co-occurring races of a species (perhaps indicated by different colorations) may have substantially different geographic origins if these races are maintained by reproductive or genetic isolating barriers. Thus, for example, three races of a species on the Great Barrier Reef might have 'originated' from Fiji via the southern Coral Sea; from Vanuatu via the northern Coral Sea; and from Fiji via Vanuatu. A host of factors will determine the level of cohesion each race will have with each other race: two races may be effectively isolated in one region, but reproductively compatible in another, perhaps because one contains some genetic component of yet another race with another place of origin.

If this species has a distribution extending to the Red Sea, it may have a very large number of races, each with varying levels of potential or actual reproductive compatibilities with every other race. Most coral species in any given region are likely to be composed of unique mixtures of such races. Over large areas, these will create reticulate networks. These networks can be observed in geographic variation from the smallest scale to the largest, and involve gradations from groupings of races to groupings of species. Some simplified large-scale outcomes are illustrated in figure 67 (p 204).

Temporal changes In the example above, the three immigrations to the Great Barrier Reef may have occurred at different times. These times may vary from recently to many millions of years ago. Given that coral species may exist over periods of millions of years (p 146), the composition of any given species in any given place is likely to be the outcome of a long and complex history spanning hundreds of Milankovitch cycles and several glacial intervals.

The combination of space and time creates an almost limitless capacity for pattern variation. At present, past variation can only be imagined, but spatial variation is at least indicative of temporal variation, because it is the product of temporal change. Figure 70 illustrates the fate, in concept, of a taxon over a major palaeoclimatic cycle.

RETICULATE PATTERN FORMATION

Concepts of *intra*-specific reticulate pattern formation are commonplace in plant and animal population genetics. *Inter*-specific reticulate patterns, that is, patterns generated through hybridisation, have also frequently been described in plant genetics and taxonomy[3] (reviewed Riseberg and Wendel 1993). In animals, they have only been referred to in occasional genetic studies (eg Benzie, 1987), and not at all in general concepts of speciation. Speciation in animals is always envisaged either as gradual change within lineages, or as the branching of lineages. As a result, of the hundreds of

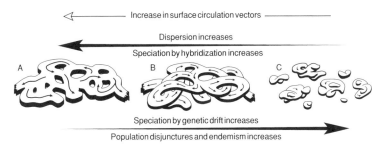

Figure 70 Concept diagram of a metaspecies or syngameon over its full geographic range, illustrating the effects of changing genetic connectivity over a single palaeoclimatic cycle. In *A*, strong surface circulation vectors provide a high degree of homogeneity. In *B*, weakened genetic connectivity has created some regional reproductive isolation (see figure 73, p 227). In *C*, very weak surface circulation has created geographically isolated regions.

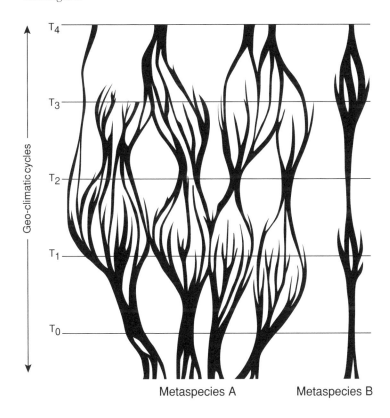

Figure 71 Concept diagram of evolutionary changes in two metaspecies. Metaspecies A, a syngameon, is caught-up in palaeoclimatic cycles of reticulate evolution. Metaspecies B survives these cycles. Phylogenies have varying patterns of spatial separation. For A, high levels of genetic communication during periods of strong surface circulation (at times T_0, T_2 and T_4) produce small numbers of well-defined species, while low levels of genetic communication during periods of weak circulation (at times T_1 and T_3) produce large numbers of ill-defined species complexes.

phylogenetic trees of various sorts that have been constructed to illustrate change with time in a major taxon, none (to the author's knowledge) has branches of operational species that fuse. The reason for this is probably the general view that hybridisation between species is a genetically marginal phenomenon restricted to hybrid ('suture' or 'tension') zones[4] (Barton and Hewitt 1985; Harrison 1993). Secondary reasons may be that concepts of animal speciation are mostly terrestrial, and are commonly cladistic. Cladistic studies can readily fail to distinguish between numerical procedure and phylogenetic analysis by forcing a hierarchical descent where there may be none (cf McDade 1992). If artificial hierarchies of attributes and actual hierarchies of descent are to be congruous, most attributes must be transmitted to descendants, not intermittently added or subtracted on a large scale. Conditions under which phylogenetic attributes can be given to cladogram nodes are far more restricted than commonly practised (Davis and Nixon 1992). This applies to any study involving any level of hybridisation[5].

THE CONCEPT

Figure 71 illustrates taxonomic repercussions of reticulate evolution driven by surface circulation vicariance in two metaspecies (p 228). Metaspecies A at time T_0 is composed of four distinct species; at time T_1 it is a single 'species complex' composed of many geographically isolated units; at time T_2 it appears to be five species, two of which are indistinct; at time T_3 it is a single 'species complex' plus a sibling species; at time T_4 it is four species again, but not the same four species it was at time T_0. Metaspecies B remains largely intact during these particular palaeo-climatic cycles, but may be affected by other cycles that are broader-scale or create other surface circulation patterns.

CHARACTERISTICS AND CONSEQUENCES

Reticulate evolution requires a very different concept of species, and predicts very different biogeographic patterns than previously proposed mechanisms of evolution. Most of the following points, *which* are *speculative only,* are potentially applicable to the majority of Indo-Pacific coral species.

♦ In figure 71, the total genetic heterogeneity of metaspecies A may be greater during intervals of weak connectivity (T_1 and T_3) than during intervals of strong connectivity (T_0, T_2 and T_4), but not necessarily so. What the vicariant cycles are doing is moving genetic diversity up and down systematic levels by alternately packaging it into larger and smaller units. During major cycles of increased connectivity, metaspecies A may (for example) become a single taxonomic unit, perhaps something equivalent to an operational genus. At the other extreme metaspecies A may become hundreds of discrete races (p 229).

These are microevolutionary changes that will occur at much higher frequencies than most other microevolutionary processes because they do not create heterogeneity, they 're-package' it.

♦ 'Reservoir species' may develop, which dominate the evolution of other species. For example, *Porites lobata* in Hawaii is genetically close to the morphologically distinct *Porites compressa* (D. Potts, pers comm): they may have undergone genetic exchange at the time of apparent divergence.

♦ Allopatric pseudo-speciation (during weakening effects of palaeo-climatic cycles) will alternate with sympatric pseudo-extinction (during strengthening effects of palaeoclimatic cycles) (figure 71). The rates of both speciation and extinction will appear to be similar over geological intervals spanning many cycles. Otherwise, allopatric, parapatric and sympatric forms of speciation have no basic distinction. Pseudo-speciation will occur after mass extinction until some sort of provincial genetic saturation is reached. After that, populations will be allopatric, or non-allopatric, according to phases of palaeoclimatic cycles (Huntley and Webb 1989). Similarly, both dispersal (or founder) and classical vicariance mechanisms of speciation have no fundamental distinction as dispersal and non-dispersal intervals will alternate.

♦ Hybridisation within a group of taxa may occur more-or-less contemporaneously (and to varying extents) over a wide geographic range: resulting species are unlikely to have specific places of origin. Hybrids are likely to be geographically restricted to particular niches if parental species are niche-specific, but not otherwise. Coral polymorphism, and consequent lack of niche specificity, will be enhanced by reticulate evolution.

♦ Just as reticulate patterns become more apparent with increasing space, they are also likely to become more significant with increasing time (eg a reticulate pattern observed across the whole Indo–west Pacific is likely to have a longer history than one extending across the central Pacific). In such examples, variations in space will be proxy indicators of variation in time.

♦ Taxonomic criteria tend to become artificial on biogeographic scales. By the simple process of giving names, taxonomy *must* override biogeographic patterns, because it *always* produces discrete units. The finer the taxonomic unit, the more artificial it will be if it is part of a reticulate pattern, that is, numbers of discrete taxa are inversely proportional to levels of reticulate hybridisation: taxonomy artifically enforces binary decisions on natural continua.

♦ Most major extinction events, as determined from the fossil record, do not occur at specific points in time, but occur over geological intervals commonly associated with major eustatic fluctuations.

These intervals may be long enough for evolutionary modifications to take place, that is, for systematic heterogeneity to be restricted to a minimal number of very discrete species. There will certainly be extinctions (as illustrated in figure 71 by the terminating lineages), some of which may be recognised in the fossil record, but genetically, most will be pseudo-extinctions in that they are not complete terminations of lineages.

THE EVIDENCE

Few would argue that reticulate pattern formation occurs *within* species. The question is 'does reticulate evolution occur at significant levels *between* operational species?' Any conclusion that it does, must beg the question 'what, then, are operational species?' Reticulate evolution in operational coral species is incompatible with the biological species concept, as the latter is based on reproductive isolation (p 64). It also denies most aspects of other concepts that have been derived from the biological species concept. The nearest equivalent to 'biological species' in corals are syngameons (p 228), which are hypothetical units, equivalent, for example, to *Quercus* (Burger 1975), *Eucalyptus* (Potts and Reid 1985) and *Iris* (Arnold et al 1990) syngameons in plants. A high proportion of operational species in corals, as in these plants, are at lower systematic levels.

At present there is no proof of reticulate evolution in corals any more than there is proof of dichotomous evolution, but the former is much more explanatory of a wide spectrum of observations made throughout this book.

◆ There must be a direct relationship between surface circulation patterns and genetic connectivity. Changes in the former must result in changes in the latter because passive dispersion is the only option open to corals. Reticulate patterns must develop to the point where genetic isolating mechanisms, space or survival, prevent hybridisation.

◆ Morphological variation of most 'species' over their full distribution range is much greater than their variation within single regions. This variation produces the patterns illustrated in figure 67 (p 204); these are mixtures of clines within 'species' and reticulate patterns among 'species'. The overall outcome is a decreasing applicability of species-specific characters with increasing distance (figure 66, p 203); the diagnostic characters of an operational species in one region are not necessarily the same for the same species in another region.

◆ Very different species of corals can hybridise (p 83). This does not necessarily mean that they do hybridise naturally or at a high frequency, or that hybrids are viable. However, considering the dimensions of space and time involved in coral species evolution, the levels of variation within species and the extent and frequency of global physical-environmental change, it is reasonable to suppose that if two species *can* hybridise, they probably will at some point in space and time. Whether this occurs frequently, or rarely, is a secondary consideration: rare events probably play a dominant role in many aspects of biogeography and evolution.

♦ Syngameons exist, as proposed in the next chapter (p 228). Syngameons among operational species are a necessary outcome of reticulate evolution among those species.

♦ Operational species have very great geological ages: rates of macroevolution are perhaps one to two orders of magnitude slower than rates of major distribution change (p 85). As the latter is correlated with microevolutionary change, a widespread species at present must be the outcome of a very large number of changes in genetic connectivity.

♦ Reticulate evolution in oceans will occur most commonly in organisms that have passive larval dispersal (p 86) (ie have patterns of genetic connectivity created and altered by ocean currents); exhibit concurrent multi-species spawning with external fertilisation (facilitating hybridisation); have wide distribution ranges (providing opportunity for geographic patches of both isolation and hybridisation); and have patchy or disjunct distributions (allowing genetic drift or natural selection to occur in isolated patches). Most corals have most of these 'attributes', some being causes of distributions, others being outcomes.

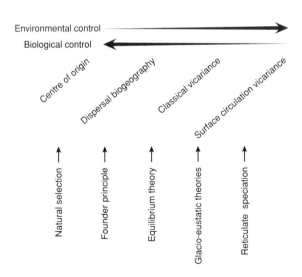

Figure 72 The changing proportions of physical environment versus biological control of evolution implied in the biogeographic concepts of evolution indicated.

♦ As emphasised in the previous chapter, the outcome of evolutionary changes is seen in within-species details. Reticulate evolution implies:

1 Constant genetic mixing. The outcome is seen, for example, in small genetic distances between widespread species.

2 Uniformity of biodiversity. There is a high degree of uniformity of levels of biodiversity (as indicated by taxonomic diversity) across the whole Indo–west Pacific (figure 49, p 158).

3 Equal rates of speciation and extinction. The Miocene to Recent fossil record strongly implies that there is a high degree of constancy in species numbers (p 143) (discussed in general by Stanley, 1990).

4 A primarily physical-environmental control of evolution (figure 72). Species distributions and species morphologies are both primarily controlled by physical environment. Biological mechanisms are primarily responsible for limiting biodiversity and hence rate of evolution; physical mechanisms are primarily responsible for creating biodiversity and biogeographic patterns[6].

5 The macroevolutionary predictions outlined in the next chapter (p 236).

RETICULATE EVOLUTION IN OTHER ORGANISMS

Reticulate evolution is now widely recognised in plants (Arnold 1992), but only in isolated studies of animals.

The prerequisites of reticulate evolution may be absolute or relative, depending on a host of detail. In principle, oceans and extensive arid regions in which pockets of favourable habitat expand and contract over time (creating gene-flow corridors), offer the best conditions for reticulate pattern formation because in both cases changes in physical environment (facilitating changes in genetic connectivity) may be dominant over hybridisation-inhibiting natural selection. Potentially, reticulate evolution may be dominant in the following organisms.

♦ Marine invertebrates, notably groups of micro-Crustacea, molluscs, polychaetes and echinoderms, which broadcast gametes for external fertilisation to produce larvae that are then subject to surface circulation vicariance.

♦ Terrestrial animals, notably Amphibia and soil fauna that are confined to isolated pockets for long intervals of time, but are capable of hybridising.

♦ Terrestrial plants that are seed-dispersed and have ranges that are periodically divided and united by palaeoclimatic cycles. A wide variety of mechanisms is available to plants for perpetuating favourable gene combinations that have arisen from hybridisation, for example, strong restrictions on gene recombination (Grant 1981).

NON-RETICULATE EVOLUTION

Not all corals potentially demonstrate reticulate patterns. The Caribbean and Gulf of Mexico may be too uniform an area for reticulate patterns to develop. Monospecific and most paucispecific genera do not appear to be the outcome of reticulate evolution; conversely, reticulate patterns appear to be dominant in the Pocilloporidae, Acroporidae, Poritidae and Faviidae.

Although the outcome of reticulate evolution is very different from non-reticulate evolution, differences are probably ones of degree. Reticulate interactions have probably long been concentrated in the Indo–west Pacific centre of diversity. With increasing remoteness from the centre, the frequency of hybridisation is likely to be progressively reduced and evolution by allopatric genetic drift more prevalent.

Reticulate patterns are not necessarily dependent on hybridisation, or the formation of clines. Fixation of mutations within populations may itself create reticulate patterns. Thus, somatic mutations in the growing parts of corals can be incorporated into sex cells through budding in much the same way as mutations in the meristem of plants are incorporated into buds (Whitham and Slobodchikoff 1981; Grant 1981). 'Neoplasms', which presumably arise from somatic mutations, are common in corals (pers òbs) and are likely to reflect high mutation rates.

SYNCHRONOUS EVOLUTION

Zooxanthellate coral taxa at all systematic levels have synchronised evolution. In the next chapter, it is postulated that metaspecies behave as single genetic units in evolutionary time, but even metaspecies have highly synchronised evolution because all are simultaneously subjected to the same surface circulation vicariance at all scales of space and time. They also share the same 'evolutionary reward': the control of their physical environment by the building of reefs. The very existence of reefs is the outcome of enormous coordinated productivity on the part of their builders. This coordination is the outcome of synchronous evolution which may have no equivalent in other ecosystems, marine or terrestrial.

THE MAIN CONCLUSIONS

◆ Palaeoclimatic cycles cause continual change in surface circulation vectors. These, in turn, cause continual change in genetic communication between one geographic region and another. This process, termed surface circulation vicariance, is postulated to be the principal mechanism driving evolution in corals.

◆ Surface circulation vicariance drives both speciation and hybridisation, and perhaps localised extinction, in response to successive palaeoclimatic cycles.

◆ Surface circulation vicariance primarily occurs in shallow-water marine life. It has varying regional impacts that underpin geographic patterns. Successive vicariance cycles continuously modify genetic patterns created by the previous cycle, creating continual genetic mixing.

◆ Different operational species will be differently affected by vicariance cycles. Some are caught in unending cycles of change, and may take the form of species complexes. Others have broken free (by the evolution of genetic isolating mechanisms) of all (or all but the most major) cycles, and take the form of reproductively isolated units (metaspecies).

◆ Surface circulation vicariance, or any other evolutionary process primarily driven by non-biological processes, may result in very large numbers of genetic 'species' that have no distinguishing morphological characters.

◆ Surface circulation vicariance generates reticulate patterns of genetic connectivity at all systematic levels within metaspecies.

◆ The more abundant a coral species is, the more resistant to vicariance cycles it will be: abundance affecting reproductive capacity may be the most important stabilising attributes of species.

◆ Evidence supporting reticulate evolution largely comes from morphological variation within coral species over very wide distribution ranges; the capacity of corals to hybridise; aspects of their reproductive behaviour; their geological longevity; and biogeographic patterns of distribution, distribution change, and diversity.

◆ Reticulate evolution predicts that there may be little distinction between central and peripheral origin of species, that the evolution of a new species would be a rare event, and that rates of speciation and extinction would be similar.

13

THE
NATURE
AND ORIGINS
OF SPECIES

The question 'what are species?' has been raised by many hundreds of authors. Most that have done so have sought a general concept applicable to all species. This has resulted in a progression of species concepts including the 'biological species concept', and the 'evolutionary', 'isolation', 'recognition' and 'cohesion' concepts (p 63), as well as methodological concepts, notably the phylogenetic species concepts of cladists (p 24). Many authors have asked 'is there one kind of species or many?' (p 66) and 'do species have a special status relative to other taxonomic levels?' (p 67).

The term 'species', meaning 'operational species' (p 3), has been used throughout this book with only occasional reservation. For most uses of taxonomy, species are adequately defined as the most clearly identifiable morphological discontinuities in nature, and what problems remain for operational taxonomy are with individual species, not with species in general (p 30). This concept, however, breaks down over great geographic distances (p 4), is liable to break down in the face of molecular taxonomy (p 32), and is demonstrably artificial when dealing with evolutionary processes (p 216).

The basic issues that confront this book, and most others that deal with evolution, concern spatial and temporal continua on a very wide range of scales. What most authors (including the present one) attempt to do about the complexities this creates is to take pieces of continua, put them into word-packages (which are intended to have meaning in undefined space and time), label them and tell stories about them. We sometimes complain that this process is somehow a misrepresentation of an underlying truth we don't really know about, but we get on with it just the same, usually by avoiding the hard bits. These hard bits are never items of taxonomy (which are opinions), or geography (which can be plotted), or of physical environment (which can be measured), or even of the geological past (which can be guessed): they are almost always to do with *processes* of evolution.

There are, and always will be, unendingly complex issues of space and time, pattern and process, observation and explanation (p 35). The essential need in all this complexity is to recognise the simple fact that the complexity does, in fact, exist. Only when this is done is it possible to create an environment where labels, concepts and generalisations can have real meaning. So many descriptors of the biological world have, in fact, no meaning other than a particular one given them (often by default) by a particular author (or reader), and the most ambiguous word in the history of biology is 'species'.

THE MAIN TOPICS

◆ Taxonomic and systematic hierarchies in corals.
◆ Operational taxonomic units.
◆ The control of biodiversity through vicariance re-packaging and sink/source relationships.
◆ The origins of species.

THE NATURE OF SPECIES

Local racial variation grades into geographic variation. The distinction between the two is arbitrary but practical. Local races are the focus of population biological studies, whereas the broader racial groupings are of interest in systematics...The geographical variation of patterns within plant species are exceedingly diverse. Indeed each plant species is unique in a descriptive sense. Nevertheless, the diversity falls into a limited number of general variation patterns. We can classify the major (e.g., non-local) races of species into three categories: continuous geographical races, disjunct geographical races, and ecological races. (Grant 1981)

Grant's summation of the nature of species in plants is more applicable to corals than anything in the zoological literature. This description implies a lot of taxonomic fuzziness created by the uniqueness of individual species, variation in levels of genetic communication at different spatial scales and related variation in morphological and reproductive characteristics.

HIERARCHICAL LEVELS IN TAXONOMY AND SYSTEMATICS

Editorial note: The meanings of terms used to describe levels in taxonomic and systematic hierarchies vary greatly among authors. They especially vary among the separate disciplines of taxonomy, genetics and biogeography. Terms used in this book have been selected to minimise ambiguities, but readers who apply concepts from other sources are very likely to misinterpret present usage.

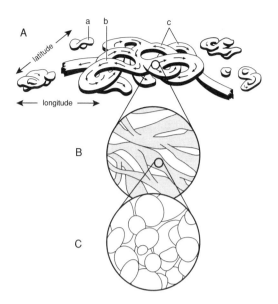

Figure 73 Concept diagram of the three-tiered structure of a hypothetical species at present point in time (see figure 70). The upper tier (A) indicates the full geographic range of the species. Arrows indicate surface circulation vectors (connectivity): some regions are geographically isolated (a); some are reproductively isolated (b); others are not isolated from the main gene pool (c). The middle tier (B) indicates the genetic structure of the species at a single location, such as a group of reefs. The branches are races that have come from different places at different times. The lower tier (C) are individual populations. The ratio of A:B:C, their shapes, and their complexity, will vary among different species and depend on spatial variation, local abundance and dispersion capability (see text). Over its full geographic range, the species may have some or all of the components illustrated in figure 67, (p 204).

Families and genera are the most arbitrary, yet straightforward of all taxonomic levels: they represent groupings of lower taxa according to convenience and the opinion of taxonomists. Most biological interest starts with taxa below the level of genus.

Figure 73 illustrates the author's general concept of the components of a species. Not all species have all these components: a non-varying species endemic to a single location may consist of only a single population whereas a widely distributed 'species complex' may consist of several tiers of complexity. To describe these levels systematically and taxonomically in space and time, the terms *metaspecies, syngameons, species, races, geographic subspecies, ecomorphs* and *populations* (summarised in figure 74) are the most useful. These terms are not just ones of hierarchical level, they also have varying morphological and genetic implications. They are also as essential to the understanding of the biogeography and evolution of species in corals as they are in plants.

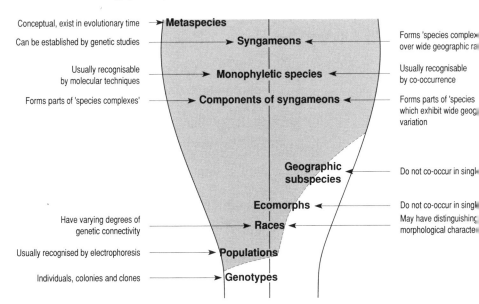

SYSTEMATICS	TAXONOMY
Taxa are based on molecular characters and are largely conceptual	Taxa are based on observable chara and serve a practical purpose

Conceptual, exist in evolutionary time → **Metaspecies**

Can be established by genetic studies → **Syngameons** ← Forms 'species complex over wide geographic ra

Usually recognisable by molecular techniques → **Monophyletic species** ← Usually recognisable by co-occurrence

Forms parts of 'species complexes' → **Components of syngameons** ← Forms parts of 'species which exhibit wide geog variation

Geographic subspecies ← Do not co-occur in singl

Ecomorphs ← Do not co-occur in singl

Have varying degrees of genetic connectivity → **Races** ← May have distinguishing morphological characte

Usually recognised by electrophoresis → **Populations**

Individuals, colonies and clones → **Genotypes**

Figure 74 Summary of the relationship between systematic and taxonomic hierarchies of corals. The shaded area indicates those parts of these hierarchies which have been justifiably described 'species'.

METASPECIES

A metaspecies[1] is the lowest taxon level that never hybridises, that is, metaspecies are permanently reproductively, hence genetically, isolated. Metaspecies are conceptual, not operational, as they exist in time as well as space. Reticulate evolution occurs within, but not between, metaspecies. Syngameons (which exist in space only) may be metaspecies or parts of metaspecies. The existence of metaspecies depends entirely on genetic isolating mechanisms; they are therefore the lowest taxon level that are monophyletic (the product of unfused branches of a phylogenetic tree).

SYNGAMEONS

A syngameon is a group of species that are capable of hybridising[2]. Syngameons are reproductively isolated in the present. Their component species may be allopatric (isolated geographically) or sympatric (isolated by genetic or reproductive mechanisms, or lack of hybrid fitness). Species in corals may be whole syngameons, or individual components of syngameons. At present, the only well documented examples of syngameons are in plants (p 226).

SPECIES

Species in corals are the most clearly identifiable discontinuities in continuous morphological and/or genetic variation from the population to the genus. The morphological boundaries (within which all morphologies of the species occur) of the majority of species are clear within regions, but become artificial over very large geographic ranges (p 203). Minor discontinuities (which are sometimes called 'microspecies', 'sibling species', 'cryptic species', 'subspecies', 'races' etc, with widely varying author-specific or situation-specific implications) are nested within larger discontinuities. Over very large geographic ranges, the majority of individual species become parts of syngameons. In so doing, these 'species' become increasingly non-operational and are sometimes referred to as 'species complexes'[3]. Such 'species', or their component parts, may be sympatric, clinal, or allopatric, according to origin.

Reticulate patterns will not necessarily yield congruence between morphological units and genetic units identified by molecular techniques. Morphological boundaries are likely to be wider (tier A of figure 73) than molecular boundaries, unless the latter are derived from techniques that are sufficiently broad-ranging to detect syngameons. Molecular techniques may identify morphotaxonomic units more definitively than any techniques based only on morphology. Both morphological and non-morphological units may exist semi-independently as nested continua.

SUBDIVISIONS OF SPECIES

Races Races are systematic taxa with both spatial and temporal dimensions. The middle tier of figure 73 are races; they may have recognisable colours or minor skeletal characteristics, but these are mostly masked by environment-correlated variations. Races are nested within species. Conceptually, they are the highest taxon level of continuous reproductive cohesion. 'Varieties', 'ecotypes' and 'subspecies' are commonly used subspecific labels that may be equivalent to races, with different author-specific or situation-specific implications.

Geographic subspecies Geographic subspecies[4] describe a combination of geographic and morphological variation, and they are mostly allopatric. Because most corals are polymorphic, geographic subspecies are difficult to recognise as discrete units except in remote or high-latitude locations, where they may be as well-defined as species (p 194).

Ecomorphs Ecomorphs are morphological units resulting from combinations of environment-correlated and genetic variations, that is, they are the morphological outcomes of intra-specific variation, whatever its origin. Ecomorphs exist primarily because coloniality allows very great morphological variation within species. Ecomorphs may be largely genetically determined (ie equivalent to races), or they may be largely environmentally determined (the outcome of genetic plasticity).

Ecomorph diversity (the number of ecomorphs that may be found in a particular region) is inversely related to species diversity (the number of species that may be found in a particular region) (p 74); this may be an evolutionary strategy[5] that is enhanced by reticulate evolution.

Populations Populations are systematic taxa, used (often with different meanings) in genetics and ecology. They are the highest level within which frequent (although not necessarily unrestricted) genetic exchange takes place. Populations in corals are nested within races, from which they are separated only in concept.

OPERATIONAL TAXONOMIC UNITS

Recent reviews about the nature of species in general (eg Mishler and Donoghue 1982; Levinton 1988; Nelson 1989) lean towards the view that they have no special status relative to other taxa and that, along with other taxa, they are human constructs. An increasing number of authors (both botanists and zoologists) have also questioned or abandoned the biological species concept (well reviewed by Cracraft 1989). In groups where evolution is primarily driven by natural selection, the products of evolution are likely to have morphological distinctions; in groups where evolution is primarily driven by physical environment (with little or no natural selection), the products need have no morphological distinctions. With corals, if surface circulation vicariance is primarily responsible for driving evolution, species essentially become units of identification. Although these may be distinctive and fully operational, they need not have a well-defined systematic basis, nor conform to any general concept.

Taxonomic (morphological) order can be conceived as the outcome of an underlying systematic (genetic) order. The latter order obviously controls the former, but does so indirectly. The taxonomic outcome (species) may represent different levels in a systematic hierarchy. Some coral species may be neo-Darwinian species, the outcome of biological natural selection; others may be reticulate species, the outcome of environmental changes. Most will be mixtures.

SPECIES THAT ARE SYNGAMEONS OR COMPONENTS OF SYNGAMEONS

The sorts of biogeographic patterns indicating the presence of syngameons are illustrated in figure 67 (p 204). Most 'species complexes' are probably syngameons where individual components, sometimes referred to as 'sibling species', are either in the process of diverging or hybridising. The majority of taxonomically straightforward species are likely to be components of syngameons. The issues may be explored with any operational coral species for which there is adequate systematic and taxonomic information. The following examples are of single operational species of different families that have been studied using molecular methods. They are illustrations only of the sorts of evidence that may be used for or against the existence of syngameons.

1) *Pocillopora damicornis* is the best understood of all coral species, is the most taxonomically straightforward of all *Pocillopora,* and is seldom misidentified in any but extreme physical environments. It is very widespread (figure 63, p 190), and has a great dispersal capability, both through rafting and teleplanic phototrophic larvae (p 87). In high latitudes, it forms distinct geographic subspecies (p 83). In the tropics, it forms races that appear to be distinguishable by details of growth and colour. Within single localities, it can form a wide range of distinctive ecomorphs (figure 4, p 17). The species displays a wide range of reproductive options, some of which occur in the same colony, others being characteristic of particular locations, others of whole regions (p 83). Various geographic subspecies, ecomorphs, and perhaps races, are likely to be substantially genetically distinct, and many may be reproductively isolated. Thus *Pocillopora damicornis* is a well-defined operational species, yet it is also a syngameon, composed of many smaller units, many of which are likely to be genetically distinct and reproductively isolated, at least within single regions. Many of these smaller units may qualify for the status of 'species' (whether operational or not), yet for most there is likely to be little congruence between genetic units and morphological ones.

2) *Montipora digitata* is a 'species complex' over its full geographic range. On the Great Barrier Reef, however, it is a single distinctive morphological unit occurring in great abundance in a very restricted habitat (inter-tidal mud flats). However, two subdivisions of it are reproductively incompatible and have the properties of separate species (figure 8, p 32). *Montipora digitata* is likely to be a syngameon, the components of which may have varying levels of isolation in different regions, and perhaps different locations within regions.

3) *Montastrea annularis* is considered to be a single species, or a group of species, in different studies in different places of the Caribbean (p 33). The reason may be primarily a matter of interpretation of taxonomic data, but it may be the outcome of incomplete separation of species. Conceptually, the component 'sibling species' may be reproductively isolated in one region, but not in another[6].

4) *Porites lobata* is a very widespread species complex. In the far eastern Pacific, it is a single species that has several of the characteristics of *P. lutea.* In Hawaii, it is genetically close to *P. compressa* (p 219). It may be a reservoir species that has persisted through time, periodically hybridising with other *Porites* species.

MONOPHYLETIC SPECIES

Many operational coral species are not syngameons, or components of syngameons, and are not the outcome of reticulate evolution. These are likely to include species of most monospecific genera, most Atlantic species, and others that are well defined and do not display very wide geographic variation.

The logical question that arises is 'should not all "species" be this type of species: monophyletic and reproductively isolated?' This question can also be examined with *Pocillopora damicornis* as an example. This operational species is a syngameon that may have component units that have identifying morphological characters, a knowable distribution range and a known level of genetic isolation. Examples might be a race in Panama, an ecomorph on the Great Barrier Reef, and a geographic subspecies at Rottnest Island. Each may qualify for species status equally with other named species, and therefore be usefully 'split off' from the *P. damicornis* syngameon. The central issues, however, will remain: *P. damicornis* will remain a syngameon, albeit one with reduced variation; the level of 'splitting off' will be a continuum of ever-finer scale; and at fine levels, different splits will be indicated by different taxonomic criteria and/or methods. This is a problem that *cannot* have a common solution for any fauna or flora that has not evolved hierarchically. The simple process of raising small units to the status of 'species' is not an operational solution as it simply creates 'species' that are only components of clines, which hybridise and (if not absolutely genetically based) are products of environmental variation.

SPECIES WHICH ARE SINGLE RACES OR POPULATIONS

A small number of widespread coral species show no geographic variation and thus effectively represent a single race, for example, *Diploastrea heliopora* and *Gardineroseris planulata*, both of which have Indo-Pacific-wide distributions. A small number of highly endemic species may be no more than a single population, for example, *Euphyllia paraglabrescens*, known only from one bay of one island in Japan. Species that are single races or populations in one region may be many races over their full range, for example, most geographic subspecies are single races or populations restricted to single (usually environmentally discrete) localities.

'SPECIES' IN OPERATIONAL TAXONOMY

The author and his colleagues have used the term 'species' many thousands of times as a taxonomic unit, have determined how hundreds of coral species can be distinguished, how they vary with physical environment, and what their distribution ranges are. This work is primarily based on field studies, and its taxonomic basis is widely used by others, that is, it is 'operational'. The issues raised have been presented throughout this book. The focal question, 'are the species of operational coral taxonomy correct, or is there another reality?', has been asked of the author hundreds of times in the context of identification problems, taxonomic methodologies, specific issues to do with particular species, general issues where taxonomy interfaces with other sciences, in fossils and in concept. It has also been raised several times in publications in the same contexts by way of implicit or explicit complaint or to promote a particular taxonomic technique or alternative perspective.

Most coral researchers want identifications from coral taxonomy, they do not want to confront species issues unless the issues affect their results or (if they are taxonomists or geneticists) are the object of their study. For most researchers, morphological taxonomy is, and will long remain, the most meaningful basis of species identification. There will come a time, however, when DNA sequencing will become readily available as an identification tool, allowing species units to have a more systematic basis. Future interactions between morphological and molecular taxonomy are largely unpredictable, but are certain to involve a wide spectrum of issues. The time of the 'all-purpose' species serving the needs of studies ranging from palaeontology to genetics is running out. There is already a need to recognise multiple kinds of species (p 1); that need will never change as there can be no common solution to species issues.

The selection of taxonomic units should depend on the aim or purpose of individual studies. Field identification guides must necessarily be based on morphological characters, and species based on morphological distinctions will continue to serve the needs of coral ecology, just as they do in plant ecology. Experimental biologists, however, should be aware that the species they study may be syngameons (and so they should select replicate colonies that are as morphologically uniform as possible). Studies of reproduction, including hybridisation, should ideally be based on taxa that are identifiable systematically as well as morphologically. Any study involving species over wide geographic ranges must recognise the fact that a species in one region is not likely to be systematically identical to the same species in another region: it may have different morphological, physiological, reproductive, behavioural and ecological attributes.

THE ORIGINS OF SPECIES

This remains one of the most elusive but intriguing subjects in all biology. It is elusive because the evolution of a new species is the outcome of a succession of events each of which originally had an extremely low probability of success. Thus, every new species is likely to have had a highly individual history in space and time. It is this individuality that has defeated, and perhaps always will defeat, attempts to satisfactorily explain the ambit of mechanisms and processes that lead to the origin of species.

The central concept of species evolution suggested by the present study can be divided into hypotheses of microevolutionary processes (considered below in terms of the origins of biodiversity) and macroevolutionary processes (considered below in terms of the origins of species). These hypotheses do not necessarily exclude competing hypotheses: peripheral allopatric speciation, in particular, may well have been the basis of origination of at least some coral species. Most of the following observations concern the Indo-Pacific only. They are, of necessity, only speculations.

THE EVOLUTION OF BIODIVERSITY

Vicariant re-packaging generates continual evolutionary change
In evolutionary time-frames, surface circulation vicariance continually re-packages biodiversity (p 219). On the finest scales of space and time, races are re-packaged into each other. On larger scales, races are re-packaged into species. On larger scales still, species are re-packaged into metaspecies. The process occurs at rates that vary with the frequency and magnitude of palaeoclimatic changes. Within a single metaspecies, the outcome today may be a single 'monophyletic' species or a syngameon. Both are composed of reticulate patterns of races.

Natural selection limits the rate of evolutionary change The mechanisms of vicariant re-packaging generate the spatial patterns of potential evolutionary change. Over evolutionary time, this process creates almost infinite potential for genetic recombinations within metaspecies. The *number* of successful recombinations (total biodiversity[7]) is primarily limited by natural selection (see the following paragraph). The *identity* of those recombinations is individually determined by spatial separation and/or the evolution of isolating mechanisms.

Biodiversity is ultimately limited by limitations to sources in source/sink cascades (p 75) which limits the rate of evolutionary change The mechanisms that ultimately limit biodiversity in corals are not (in the author's view) understood, but they appear to be positively correlated with the range of available habitats (the greater the range of habitats, the greater the diversity that is supported, p 75) and temperature-correlated ecological functions (highest diversity occurs in the tropics, p 161), and negatively correlated with intra-regional genetic connectivity (the greater the connectivity, the lower the diversity, p 76).

THE EVOLUTION OF SPECIES

Phylogenies down to the level of metaspecies can be considered as dichotomous branches of branching trees. Below the level of metaspecies (eg at the level of operational species), there are few 'branches', but rather reticulate patterns of spatial and temporal continua at all systematic (and taxonomic) levels. It is the interactions of these continua of space, time and level that suggest principles of evolutionary change.

Observations on the evolution of biodiversity (as above) may also be applied to species: the final outcome of surface circulation vicariance (an essentially microevolutionary process) is the reticulate evolution of species. The ultimate limits to biodiversity are also the ultimate limits to species diversity. With species, however, evolutionary changes can be put into some sort of geographic context as follows:

♦ Most Indo-Pacific 'speciation' occurs within the province of maximum diversity and is exported from this province to peripheral regions. The province is very strongly characterised by uniformity at

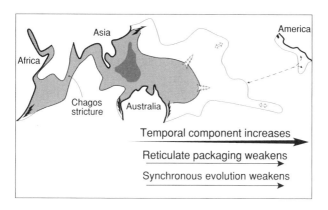

Figure 75 *Summary diagram.* The tropical Indo-west Pacific (pale grey) contains, effectively, a single suite of species which attenuates away from its Indonesia/Philippines centre of diversity (dark grey). Diversity within the (pale grey) high diversity band is (in concept) due to a combination of local extinctions and cascades of source/sink relationships. *Latitudinal* attenuation beyond the band is the product of connectivity ratchets (solid half arrows, where boundary currents provide pulses of dispersion), temperature and the presence/absence of reefs, and occurs in (mostly) non-evolutionary time-frames. *Longitudinal* attenuation east across the Pacific occurs in a combination of biogeographic and evolutionary time-frames, driven by palaeoclimatic pulses (open arrows). Isolation leading to allopatric speciation may occur in the remote north-east and south-east Pacific (return open arrows) and (perhaps), in the far eastern Pacific. Patterns of affinity reflect an increasing temporal component across the Pacific.

species level (p 161). The Indonesia/Philippines Archipelago and other regions extending west to the central Red Sea probably act as long-term net sources of species diversity. Adjacent regions may alternate in time as net sources and sinks according to prevailing surface currents. More peripheral regions are probably always net sinks. Species diversity, but not individual species, is thus continually exported from central sources to peripheral sinks (p 75)[8].

♦ The boundaries of all but the eastern margin of the province of maximum diversity is determined, on a biogeographic scale, by bathymetry (either by continental shorelines and deep ocean), or by continental boundary currents driving connectivity ratchets (pp 99, 178) (figure 75). It is the eastern margin of the province that is not thus defined and which has consequently attracted the attention of coral biogeographers.

♦ The existence of connectivity ratchets along continental margins is unlikely to have evolutionary consequences because the time intervals of genetic change are much too short (p 181). A combination of dispersion and evolutionary mechanisms, however, creates biogeographic patterns (of both species diversity and affinity) across the Pacific. It is these patterns that are probably created by orbital forcing (Milankovitch pulses, p 216). In the western Pacific they are primarily due to within-species dispersion. In progressively more eastern regions, they are decreasingly due to dispersion and increasingly due to reticulate evolutionary change. Only in the remote northeastern, northwestern and (perhaps) the far eastern Pacific, is isolation likely to have been so long-enduring that purely allopatric origination may have occurred with some species.

This general model is a superficial view of many different processes. It predicts that:

♦ New species are unlikely to 'appear' over a short time interval; rather, very large numbers of new genetic 'species' are likely to form during intervals of weak surface circulation, to subsequently become repackaged during cycles of strong circulation (figure 71, p 217).

♦ 'New' species need have no single place or time of origin. Instead, both place and time will be defined by surface circulation vectors, and the boundaries of both will be arbitrary and ever-changing.

♦ There may be little distinction between central and peripheral origins as species would only evolve over many reticulate events (p 216). In such a reticulate system, there is no fundamental distinction between allopatric and sympatric speciation.

♦ Reticulate patterns create slow rates of evolution because evolutionary modifications are continuously submerged by cycles of hybridisation (p 214).

♦ Rates of speciation and extinction are approximately balanced (p 219).

AN EVOLUTIONARY MODEL:
THE KNOWN TO THE UNKNOWABLE

Most *factual* aspects of this book have been concerned with spatial patterns, for these are largely derived from observations; most *hypothetical* aspects have been concerned with temporal change and systematics, for these cannot be observed directly. The hypotheses that have been presented range, accordingly, from the well founded to the very speculative. When combined, they form a single evolutionary model of reticulate evolution primarily driven by the physical environment. Like all models, the present one is a deliberate simplification of the real world. It has varying relevance to different groups of corals and to different types of studies of corals. Its greatest relevance is to our understanding of biogeographic patterns of diversity and affinity; evolutionary change; and concepts of species. The following is the author's assessment of our current status of understanding of these subjects.

Biogeographic patterns of diversity and affinity Global patterns of taxonomic diversity have been described at family (summarised p 155), generic (summarised p 156) and species (summarised p 160) levels. Global patterns of *taxonomic affinity* have been described at generic level (summarised p 157) and at species level for the central Indo-Pacific only (chapter 10). These distribution patterns all rest on a common taxonomic foundation that is certain to contain errors of omission and detail, but is unlikely to contain errors of quantitative significance, despite the 'what are species' issues raised in this chapter. This is primarily because the taxonomic order on which the distribution patterns are based is something observable in nature, and also because the patterns presented have not been extrapolated beyond this observable taxonomic base (figure 50, p 160 excepted).

These distribution patterns may, in future years, be refined through taxonomic changes and further data collection, but the major advancement to our understanding of them is likely to come from the replacement of presence/absence data with abundance data. The amount of effort required to do this might long be prohibitive, but the results would certainly be highly explanatory for both biogeography and regional ecology.

Future extension of the present generic-level global database (p 155) to species level is well in sight, but this will involve a whole gamut of 'what are species' issues (especially in the central and southeastern Pacific), for which there can never be simple solutions. Inter-regional variation can be studied in great detail using molecular methods, but overall affinities must remain matters of conjecture, because species, however determined, cannot ultimately conform to any common definition (p 3).

Patterns of *systematic diversity* and *systematic affinity,* that is, patterns of biodiversity, are largely unknown and are conceptually unknowable other than by association with patterns of taxonomic diversity and affinity. For corals, there will always be fuzzy interactions between the left and right sides of Fig 74. At present, the directions these interactions are likely to take is unknown, and clearly this subject has only superficially been considered in this book.

Evolutionary mechanisms Corals are an ancient, slowly evolving group. Most extant families have Mesozoic origins (p 108). Most extant genera have a fossil record reaching far back into the Cenozoic, and now have distributions reflecting the ancient circum-global seaways they originally occupied (chapter 8). Species, as far as the fossil record shows, are likewise old and unchanging. They also have wide (but now post-Tethyan and post Central American Seaway) distributions.

In concept, metaspecies may be almost as old as the genera into which they are grouped. Any given metaspecies may have a core 'species' that is both widely dispersed and abundant, and survives cycles of reticulate divergence and hybridisation. This core may be a 'reservoir species' (p 219). 'Sibling species', which may be offshoots of reservoir species, have an arbitrary existence in space and time (figure 71, p 217). The age of metaspecies can be determined by molecular clocks, but that of all lower systematic levels are arbitrary to varying degrees, for both the fossil record and molecular clocks ultimately depend on discontinuities and/or divergences that are subjective divisions within continua.

Reticulate patterns of phylogeny present many conceptual and operational issues for coral taxonomy. Taxonomy creates hierarchical order, whereas reticulate patterns do not necessarily have hierarchies; it creates discontinuities in space that are actually continua; it separates species as genetically independent units, whereas they may have no independence over wide geographic ranges. This book has also hardly scratched the surface of this subject. In all probability it will be advanced through molecular studies of specific clades: some will reveal dichotomous evolution, others will be reticulate.

Concepts of species It is with concepts of species that the model of reticulate evolution has its greatest impact. On the one hand, the model suggests chaos where there may have been an assumption of order. On the other hand, it provides an explanatory basis for the interactions between systematics and taxonomy that, in future years, will hopefully be of practical value. Reticulate evolution, and the concept of species that derives from it, is no more, or less, 'proven' than competing concepts. Whether 'proven' or not, the concept must be taken into account by all who question the nature of species: by biogeographers, taxonomists and geneticists, and all who study the attributes of species. The existence (or not) of reticulate evolution can be demonstrated in corals, just as it has been demonstrated in an increasing variety of plants. However, with or without reticulate evolution, there can be no common solution to 'the species problem': it is now, and forever will be, a philosophical issue.

THE MAIN CONCLUSIONS

♦ Species are only definable as 'the most clearly identifiable discontinuities in continuous variation from the population to the genus'.

♦ In concept, groups of species form metaspecies in geological time, a metaspecies being the lowest taxon level that never hybridises.

♦ With corals, there is a substantial difference between systematic (conceptual) and taxonomic (operational) hierarchies.

♦ The age of multi-purpose species, where the complexity of life is described by a simple hierarchy of nomenclature, has become untenable over recent years. Operational species may belong to different hierarchical levels and have different systematic associations.

♦ At present, operational species may be monophyletic, may be syngameons, or may form syngameons, the components of which are capable of hybridisation. Species may also be single races or even single populations.

♦ In evolutionary time races and species are continually being repackaged to form reticulate patterns of diversity and affinity.

♦ Natural selection creates upper limits of both systematic and species diversity. This limit controls rates of evolution.

♦ Patterns of species diversity in the Indo–west Pacific are mostly the products of bathymetry and connectivity ratchets (p 178). Diversity is exported from the equatorial part of this province in both evolutionary and non-evolutionary time-frames, the evolutionary time-frame becoming increasingly dominant eastward across the Pacific.

APPENDIX

CHARACTERS OF FAMILIES AND GENERA

This appendix is a summary of present and past biogeographic characteristics and systematic order of extant zooxanthellate families and genera. The reader is referred to Veron (1993) for further biogeographic and taxonomic detail, and to chapter 8[1] for references to the Cenozoic fossil record[2].

FAMILY ASTROCOENIIDAE

The systematic position of this small family has been repeatedly changed; the position indicated in figure 25 (p 110), is only weakly supported by the fossil record and awaits further study by molecular techniques. Affinities with the Pocilloporidae seem clear on the basis of morphology of extant genera.

STYLOCOENIELLA NUMBER OF EXTANT SPECIES: At least three. PRESENT DISTRIBUTION: Red Sea to central Pacific. GENERAL ABUNDANCE: Uncommon, cryptic. FOSSIL RECORD: Eocene of the Indo-Pacific, Oligocene of the Caribbean and Tethys.

All three species occur on small, remote Cocos (Keeling) Atoll. Dispersal capability of all species is, presumably, good.

STEPHANOCOENIA NUMBER OF EXTANT SPECIES: One. PRESENT DISTRIBUTION: Western Atlantic. GENERAL ABUNDANCE: Sometimes common. FOSSIL RECORD: ?Cretaceous (Wells 1956), Eocene of the Caribbean, Oligocene of the Caribbean and Tethys.

FAMILY POCILLOPORIDAE

The systematic position of this family in figure 25 (p 110) is dependent on strong morphological similarities with the Astrocoeniidae together with a good fossil record.

The four major extant genera appear to have been abundant and cosmopolitan throughout most of the Cenozoic, becoming progressively restricted in the Tethys during the Miocene and in the Caribbean and Gulf of Mexico in the Plio-Pleistocene. Despite these extinctions, the three major Indo-Pacific genera are all widespread and abundant. All are probably dispersed by both rafting (p 87) and teleplanic planulae (p 87).

POCILLOPORA Number of extant species: Approximately ten. Present distribution: Red Sea and western Indian Ocean to far eastern Pacific. General abundance: Very common, very conspicuous. Fossil record: Eocene of the Caribbean, Oligocene of the Tethys, Miocene of the Pacific.

Pocillopora (probably *P. eydouxi*) occurred in the Caribbean as recently as the Late Pleistocene (Geister 1977). Its extinction there is anomalous, as the genus is more uniformly widespread in the Pacific than any other, with perhaps six species occurring in the far eastern Pacific (p 166), three of which show little taxonomically significant variation across the entire Indo-Pacific. *Pocillopora* can also be abundant in remote localities, including the depauperate reefs of Pacific Panama (Glynn 1976) and Johnston Atoll (Maragos and Jokiel 1986) and in high-latitude non-reef localities of both Australian coasts and the far south Pacific.

Pocillopora damicornis is the most-studied of all corals (Index), especially its genetics and reproductive biology. It is very widespread (figure 63, p 190).

STYLOPHORA Number of extant species: Approximately five. Present distribution: Red Sea and western Indian Ocean to southern Pacific. General abundance: Very common, very conspicuous. Fossil record: Palaeocene of the Pacific, Eocene of the Caribbean and Tethys.

Stylophora is probably the only major genus to have a higher species diversity in the western Indian Ocean and Red Sea than in the central Indo-Pacific.

SERIATOPORA Number of extant species: Approximately five. Present distribution: Red Sea and western Indian Ocean to southern Pacific. General abundance: Very common, conspicuous. Fossil record: Miocene of the Pacific.

Most records and occurrences of the genus are the one species, *S. hystrix* (figure 63, p 190).

MADRACIS Number of extant zooxanthellate species: Approximately four. Present distribution: Western Indian Ocean to eastern Pacific and western to eastern Atlantic. General abundance: Uncommon, mostly cryptic in the Indo-Pacific, much more common in the western Atlantic. FOSSIL RECORD: ?Cretaceous (Wells 1956), Eocene of the Caribbean, Miocene of the Tethys

This genus is the only survivor of the Pocilloporidae in the Atlantic. It has both zooxanthellate and azooxanthellate species in both the Atlantic and Indo-Pacific. Although these species are readily identifiable, they do not form sub-generic groups, and are not separable in the fossil record.

PALAUASTREA NUMBER OF EXTANT SPECIES: One. PRESENT DISTRIBUTION: Central Indo-Pacific. GENERAL ABUNDANCE: Uncommon. FOSSIL RECORD: None.

This species and *Porites cylindrica* are frequently confused *in situ*.

FAMILY ACROPORIDAE

Fossil evidence for the Jurassic origin of the Acroporidae is weak, being dependent on two Mesozoic genera. The extant genera are Cenozoic in origin except for *Astreopora*, which has been separate since the Cretaceous.

MONTIPORA NUMBER OF EXTANT SPECIES: At least eighty. PRESENT DISTRIBUTION: Red Sea and western Indian Ocean to southern Pacific. GENERAL ABUNDANCE: Extremely common, some species inconspicuous. FOSSIL RECORD: ?Eocene of the Pacific, Oligocene of the Tethys.

The often-highlighted taxonomic difficulties of *Montipora* stem from the number of species, and the presence of several species complexes. Otherwise, the genus does not deserve its taxonomic reputation; it has a wide array of useful, if not always conservative, skeletal characters.

Many more species of *Montipora,* especially ramose species, occur in the Philippines than either Japan or (especially) Australia. Many of these are poorly known.

ANACROPORA NUMBER OF EXTANT SPECIES: Approximately six. PRESENT DISTRIBUTION: Western Indian Ocean to western Pacific. GENERAL ABUNDANCE: Uncommon, mostly non-reefal. FOSSIL RECORD: None.

Corallites structures are *Montipora*-like, the colony shape is *Acropora*-like. *Acropora* may have evolved from an *Anacropora*-like ancestor. Because it now occurs in non-reef environments, distribution records are generally unreliable.

ACROPORA NUMBER OF EXTANT SPECIES: At least one hundred and fifty. PRESENT DISTRIBUTION: Cosmopolitan. GENERAL ABUNDANCE: Extremely common, very conspicuous, usually dominant in Indo-Pacific reefs. FOSSIL RECORD: Eocene of the Caribbean, Pacific and Tethys.

Acropora reigns supreme throughout the Indo-Pacific in almost all low-nutrient, high-energy reef environments and probably has done so since the Miocene. This overwhelming evolutionary success appears to be primarily based on three morphological characteristics: small corallites, allowing for fine detail in skeletal development; division of the roles of

Figure 76 The undersurface of an *Acropora* table showing the highly determinate, integrated architecture not found in other corals. There are approximately 800 000 individuals in this colony.

axial and radial corallites, allowing for highly deterministic growth forms; and porous skeletal microstructure, allowing maximum strength for weight. When combined, these characteristics produce morphologies of opportunity: they allow very rapid growth, very determinate growth, a high degree of colony integration and rapid local dispersion through fragmentation. With these characteristics, Indo-Pacific *Acropora* species are able to produce a much wider range of architectures, exploit a wider range of habitats and grow more rapidly than species of any other genus[3].

Perhaps the greatest evolutionary achievements of coral are the 'plate'- and 'table'-forming species (*A. palmata* of the Caribbean, and the *hyacinthus, divaricata* and *loripes* species groups of the Indo-Pacific). These very highly integrated architectures (figure 76) are particularly suited to rapid exploitation of a wide range of environments, including high-energy ones: growth rate, substrate coverage, exposure to sunlight and the sieving of plankton are all maximised, while the quantity of skeletal material required to do so is minimal. These characteristics would be particularly advantageous during times of climatic change, when sea levels, wave turbulence, surface circulations and inorganic nutrients are all in states of flux.

The paucity of *Acropora* in the Caribbean is in conformity with other coral genera, that is, owing to a lack of speciation (p 147). Whether this paucity also applied to the Tethys cannot be determined from the fossil

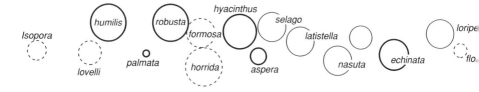

Figure 77 Hypothetical representation of the taxonomic affinities of the sub-generic groupings of *Acropora* species. Affinities within groups indicated by a thick line are relatively clear, those indicated by a broken line are relatively unclear. Except for the sub-genus *Isopora*, these groups are not sufficiently well-defined to justify establishing sub-genera.

record, although there is no suggestion that Tethyan *Acropora* were ever highly diverse or particularly abundant.

Intra-specific latitudinal change reaches an extreme in *Acropora* with the majority of species forming taxonomically meaningful geographic subspecies in most high-latitude regions. Thus, there is a greater degree of intra-specific similarity between the *Acropora* of the tropical Northwest Shelf reefs of Australia and the Great Barrier Reef than there is between these regions and their southern neighbours (p 181). A similar situation occurs in Japan, where common species in most high-latitude localities are taxonomically separable from each other as well as from their tropical counterparts.

Several apparently temporary populations of *Acropora* have been recorded in remote places: these include *A. yongei* at Rottnest Island near Perth, western Australia (Marsh 1994) three species in Hawaii (Grigg 1981; Maragos pers comm) and *A. valida* in the far eastern Pacific (p 85).

Relationships among species of *Acropora* cannot reliably be determined from morphological criteria, thus the genus has only one, small, subgenus *(Isopora)*. To aid identification, Veron and Wallace (1984) divided the genus into fifteen sub-generic groupings (figure 77); these have no taxonomic status and the relative positions of the groups are somewhat arbitrary.

ASTREOPORA NUMBER OF EXTANT SPECIES: Approximately fifteen. PRESENT DISTRIBUTION: Red Sea and western Indian Ocean to southern Pacific. GENERAL ABUNDANCE: Generally common, conspicuous. FOSSIL RECORD: Cretaceous of the Tethys, Eocene of the Caribbean and Tethys. Fossil *Astreopora* are readily confused with *Turbinaria;* this has confused the fossil record.

Astreopora myriophthalma is by far the most common *Astreopora* of the eastern and western Australian coast. The most notable difference between the two faunas is the absence of *A. explanata* in the east.

FAMILY PORITIDAE

In figure 25 (p 110), the Poritidae has been moved from its traditional position within the Fungiina to a suborder of its own because there is little significant correspondence between the microskeletal characters of the Poritidae and those of any other extant family. The four extant genera are morphologically very distinct.

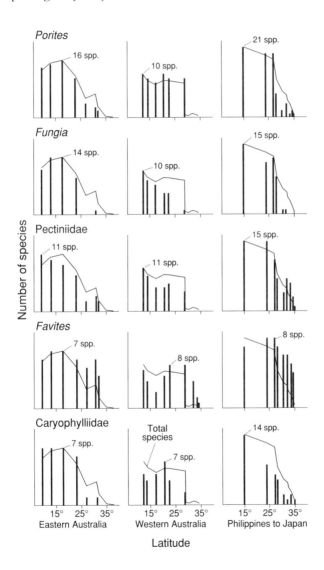

Figure 78 Latitudinal attenuation of total species number in the taxa indicated. Left column = the coast of eastern Australia; centre column = the coast of western Australia; right column = Philippines to Japan. Locations are those indicated in figure 54 (p 172), Vanuatu excluded. Lines are total numbers of all species along respective coastlines.

PORITES NUMBER OF EXTANT SPECIES: Approximately eighty. PRESENT DISTRIBUTION: Cosmopolitan. GENERAL ABUNDANCE: Extremely common, conspicuous at generic level. FOSSIL RECORD: ?Cretaceous (Wells 1956), Eocene of the Caribbean and Tethys. The genus became overwhelmingly dominant in the Miocene Tethys.

More than any other major genus, *Porites* is in need of taxonomic revision, primarily because the majority of massive species do not have distinctive morphologies and are only identifiable *in situ* by minute corallite characteristics. Because of this, and because some species (*P. lobata*, *P. lutea* and *P. australiensis*) are extremely abundant throughout most of the Indo-Pacific, uncommon species or variants of common species may go undetected much more easily than the species of other genera. There are further problems: corallite characters are very variable both *within* colonies (figure 1, p 15) and *between* colonies in different environments, and, as with most coral species, geographic variation goes beyond meaningful species distinctions (p 203). The morphometric taxonomic methods of Budd (p 23) and Budd and Coates (1992) and the electrophoretic methods of Potts et al (in press) are mutually supportive for Caribbean *Porites* (p 28). Electrophoresis especially may provide an insight into the much greater complexity of *Porites* taxonomy and biogeography in the Indo-Pacific; questions of primary interest are the geographic ranges of morphologically meaningful species, the affinities of species in remote regions (the southern-eastern and far eastern Pacific), the structure of species complexes (like *P. compressa* in Hawaii) and the extent of speciation in areas of very high species diversity (especially Indonesia and the Philippines).

Despite the wide geographic range of the genus as a whole, and many of its species, latitudinal attenuation, especially between reef and non-reef environments, occurs more abruptly in *Porites* than in most major genera (figure 78). This occurs not only in species diversity, but also in abundance and colony size: for example the number of species in the southern-most reefs of western Australia is about the same as in tropical reefs, but only one species (*P. lutea*) forms large colonies.

STYLARAEA NUMBER OF EXTANT SPECIES: One. PRESENT DISTRIBUTION: Red Sea and western Indian Ocean to western Pacific. GENERAL ABUNDANCE: Rare, occurs only in shallow, wave-washed biotopes. FOSSIL RECORD: Plio-Pleistocene of Guam (R Randall, pers comm).

Stylaraea punctata is a taxonomic and ecological isolate that, superficially, has as much in common with the primarily Mesozoic Actinacididae as the extant Poritidae. Like *Oulastrea crispata*, it is confined to shallow rocky environments where other corals are seldom found.

GONIOPORA NUMBER OF EXTANT SPECIES: Approximately thirty. PRESENT DISTRIBUTION: Red Sea and western Indian Ocean to southern Pacific. GENERAL ABUNDANCE: Generally common, very conspicuous. FOSSIL

RECORD: Cretaceous (Wells 1956), Eocene of the Caribbean and Tethys.

Goniopora has been a major reef-builder throughout much of the duration of the Cenozoic Tethys. Extant *Goniopora* frequently forms very extensive monospecific or multi-specific stands in inshore environments dominated by terrigenous sediments. Species are easily recognised *in situ* by characters of soft tissues (illustrated, Veron 1986), but these may become unreliable over wide geographic ranges.

The derivation of the poritid pattern of septal fusion from *Goniopora* (Bernard 1905) is one of the few instances in scleractinian taxonomy where one taxon or taxonomic character can be said to be 'primitive' compared with another.

ALVEOPORA NUMBER OF EXTANT SPECIES: Approximately fifteen. PRESENT DISTRIBUTION: Red Sea and western Indian Ocean to southern Pacific. GENERAL ABUNDANCE: Sometimes common, very conspicuous. FOSSIL RECORD: Eocene of the Caribbean and Tethys.

Alveopora is a taxonomically isolated genus with unclear affinities. It is readily confused with *Goniopora,* although these genera are probably not closely related. Morphological differences between the two are demonstrated by all *Goniopora* having twenty-four tentacles per polyp, and all *Alveopora* having twelve.

It is rare to see many *Alveopora* species together in the same biotope, as habitats of individual species are very different, more so than for any other genus. These habitats include protected turbid biotopes (the majority of species), exposed upper reef slopes (eg *A. marionensis*) and high-latitude, non-reef biotopes (eg *A. japonica*).

FAMILY SIDERASTREIDAE

A formerly diverse family in the Caribbean, the Siderastreidae are now represented there only by two species of *Siderastrea. Psammocora* and *Coscinaraea* have clear affinities, with *P. vaughani* being intermediate between them. The other genera within the family are paucispecific, with uncertain relationships.

SIDERASTREA NUMBER OF EXTANT SPECIES: Approximately six. PRESENT DISTRIBUTION: Red Sea and western Indian Ocean to Philippines; far eastern Pacific; western to eastern Atlantic. GENERAL ABUNDANCE: Uncommon in the Indian Ocean, often common in the Caribbean. FOSSIL RECORD: Cretaceous of Tethys; Eocene of the Caribbean.

Recorded in the western Pacific (Philippines) from only a single specimen. Recently found in the Gulf of Panama where it may be an relict endemic (Budd and Guzmán, in press). *Siderastrea radians* has been recorded from both the Atlantic and Indian Ocean. If identification of Indian Ocean occurrences is correct, it is the only species occurring in both the Indo-Pacific and the Atlantic.

PSEUDOSIDERASTREA NUMBER OF EXTANT SPECIES: One. PRESENT DISTRIBUTION: Western Indian Ocean to western Pacific. GENERAL ABUNDANCE: Uncommon, cryptic. FOSSIL RECORD: Pliocene of the Pacific.

PSAMMOCORA NUMBER OF EXTANT SPECIES: Approximately fifteen. PRESENT DISTRIBUTION: Red Sea and western Indian Ocean to far eastern Pacific. GENERAL ABUNDANCE: Generally common, sometimes cryptic. FOSSIL RECORD: Miocene of the Caribbean.

In general, the species of *Psammocora* are distinct. They may show substantial environment-correlated variation, the same variation being repeated in most geographic regions. Except for some colonies of *P. nierstraszi*, coralla from western and eastern Australia are indistinguishable. *Psammocora superficialis* shows no geographic variation throughout the entire Indo-Pacific.

COSCINARAEA NUMBER OF EXTANT SPECIES: Approximately twelve. PRESENT DISTRIBUTION: Red Sea and western Indian Ocean to southern Pacific. GENERAL ABUNDANCE: Generally common, conspicuous. FOSSIL RECORD: ?Eocene of the Caribbean.

Two morphologically similar species (*mcneilli* and *marshae*) are endemic to extra-tropical southern Australia; the remainder have tropical distributions. Two other species (*exesa* and *columna*), also morphologically similar, are common and widely distributed and account for most records of the genus.

ANOMASTRAEA NUMBER OF EXTANT SPECIES: One. PRESENT DISTRIBUTION: Western Indian Ocean only. GENERAL ABUNDANCE: Uncommon, cryptic. FOSSIL RECORD: None.

HORASTREA NUMBER OF EXTANT SPECIES: One. PRESENT DISTRIBUTION: Western Indian Ocean only. GENERAL ABUNDANCE: Generally uncommon. FOSSIL RECORD: None.

FAMILY AGARICIIDAE

This is a Cenozoic Tethyan family, with most genera now extant. The family had a fate in the Atlantic similar to that of the Pocilloporidae, due to the extinction of *Pavona* and *Gardineroseris*, and reduction of *Leptoseris* to a single species. *Agaricia* is now the only speciose Caribbean genus.

AGARICIA NUMBER OF EXTANT SPECIES: Approximately five. PRESENT DISTRIBUTION: Western to eastern Atlantic[4]. GENERAL ABUNDANCE: Generally common, very conspicuous. FOSSIL RECORD: Miocene of the Caribbean and Tethys.

PAVONA NUMBER OF EXTANT SPECIES: Approximately twenty-two. PRESENT DISTRIBUTION: Red Sea and western Indian Ocean to far eastern Pacific. GENERAL ABUNDANCE: Very common, conspicuous. FOSSIL RECORD: ?Eocene of the Caribbean, Oligocene of the Tethys and Pacific.

Pavona is readily divided into 'leafy' and 'non-leafy' groups of species that have ill-defined relationships. The latter are not clearly delineated from *Leptoseris*. Most species of *Pavona* are well-defined, widely distributed, and show relatively little geographic variation in morphology, colour or abundance.

LEPTOSERIS NUMBER OF EXTANT SPECIES: Approximately fourteen. PRESENT DISTRIBUTION: Red Sea and western Indian Ocean to far eastern Pacific and Caribbean and Gulf of Mexico[5]. GENERAL ABUNDANCE: Sometimes common, mostly conspicuous. FOSSIL RECORD: Oligocene of Caribbean, Tethys and Indian Ocean.

Leptoseris species are mostly widespread like *Pavona*, but unlike *Pavona*, they show substantial geographic variation.

GARDINEROSERIS NUMBER OF EXTANT SPECIES: At least two. PRESENT DISTRIBUTION: Red Sea and western Indian Ocean to far eastern Pacific. GENERAL ABUNDANCE: Generally uncommon, sometimes cryptic. FOSSIL RECORD: Miocene of the Caribbean and Pacific.

Gardineroseris planulata is one of the most widespread of Indo-Pacific species. Pliocene fossils are indistinguishable from the extant (Veron and Kelley 1988) and the same species is recorded by Budd et al (in press) in the Pliocene of the Caribbean.

COELOSERIS NUMBER OF EXTANT SPECIES: One. PRESENT DISTRIBUTION: Central Indian Ocean to central Pacific. GENERAL ABUNDANCE: Generally uncommon, sometimes cryptic. FOSSIL RECORD: ?Miocene of the Pacific.

PACHYSERIS NUMBER OF EXTANT SPECIES: Approximately four. PRESENT DISTRIBUTION: Red Sea and western Indian Ocean to western Pacific. GENERAL ABUNDANCE: Very common, very conspicuous. FOSSIL RECORD: ?Eocene of the Pacific, Miocene of the Tethys.

FAMILY FUNGIIDAE

In several important respects, the taxonomy, biogeography and palaeontology of the Fungiidae all stand apart from those of the other major families. Most species are solitary and, like the solitary azooxanthellates, lack the growth form variation that complicates the taxonomy of most colonial corals. Probably because of this, the family has been taxonomically revised more than any other (Gardiner 1909; Wells 1966; Veron and Pichon 1979; Hoeksema 1989). These 'revisions' have involved much name changing (p 19), but they have resulted in the most complete taxonomic compilation of all the major families.

Biogeographically, the Fungiidae, alone of the major families, is restricted to the Indo-Pacific. Palaeontologically, it has the shortest and least-known fossil record of any major family, despite the fact that the genera, including the principal genus, *Fungia,* form the largest and heaviest skeletons of all coral polyps and therefore have the greatest preservation potential.

Wells (1956, 1966) suggested a Tethyan Cretaceous origin of the Fungiidae from the Synastreidae, but this is by analogy across a family-level gap. Whether this is so or not, there has been no confirmation of *Cycloseris* in Europe: all confirmed fossil Fungiidae are Indo-Pacific. A second aspect of the fossil record proposed by Wells (1966) is that all colonial and polystomatous genera arose from *Fungia* during the Miocene and post-Miocene. This observation may simply reflect a weakness in the Indo-Pacific fossil record; it can readily be tested using molecular means. Most Miocene records used by Wells are from studies in the Indonesian region earlier this century that are not reliably dated.

CYCLOSERIS NUMBER OF EXTANT SPECIES: Approximately sixteen. PRESENT DISTRIBUTION: Red Sea and western Indian Ocean to far eastern Pacific. GENERAL ABUNDANCE: Generally uncommon, non-reefal. FOSSIL RECORD: ?Cretaceous of the Indian Ocean (see above).

Usually found only in non-reef (usually inter-reef) biotopes, and thus distribution records are likely to be incomplete. The highly distinctive *Cycloseris cyclolites* is the only commonly encountered central Indo-Pacific species. As with *Fungia, Cycloseris* can form temporary populations in high latitudes of Japan (p 85) and the Galapagos.

DIASERIS NUMBER OF EXTANT SPECIES: At least three. PRESENT DISTRIBUTION: Red Sea and western Indian Ocean to far eastern Pacific. GENERAL ABUNDANCE: Generally uncommon, non-reefal. FOSSIL RECORD: None.

Like *Cycloseris, Diaseris* is usually found only in non-reef (usually inter-reef or sea-grass) biotopes and thus distribution records are likely to be incomplete. *Diaseris fragilis* from the far eastern Pacific are identical to those from the western Pacific.

The asexual reproductive capacity of *Diaseris fragilis* by autotomy is much greater than that of any other coral that employs autotomy, and it can occur so frequently that individuals become very small (less than 10 mm diameter), forming a '*Diaseris* gravel' that covers the substrate in a living layer.

FUNGIA NUMBER OF EXTANT SPECIES: Approximately thirty-three. PRESENT DISTRIBUTION: Red Sea and western Indian Ocean to southern Pacific. GENERAL ABUNDANCE: Very common, very conspicuous. FOSSIL RECORD: Miocene of the Pacific.

Species generally show little geographically or environmentally correlated variation, partly because they are not colonial, but also because they seldom occur on exposed reef fronts or in high-latitude, non-reef localities, both of which are common environmental extremes for other corals.

Of all major genera, this is the most restricted to tropical waters (figure 78, p 245) and may occur in very large concentrations in equatorial regions. *Fungia scutaria* only occurs in higher latitudes of eastern Australia (Elizabeth and Middleton Reefs) and Japan (Tanegashima). This species also has the widest longitudinal distribution (to Hawaii and Pitcairn Islands) (p 190).

Long-term survival (probably temperature tolerance) rather than dispersal ability may be limiting distributions: a single specimen of *Fungia repanda* has been recorded from the Houtman Abrolhos Islands and temporary populations of *Fungia* have occurred in mainland Japan (p 85).

HELIOFUNGIA NUMBER OF EXTANT SPECIES: One. PRESENT DISTRIBUTION: Eastern Indian Ocean to western Pacific. GENERAL ABUNDANCE: Generally common, very conspicuous. FOSSIL RECORD: ?Miocene of the Pacific.

Heliofungia has a polyp morphology very unlike that of *Fungia*. This is correlated with an ecological preference for turbid environments, where the species is, presumably, a detritus feeder. Latitudinal restrictions are, however, similar to *Fungia* and are probably temperature correlated.

CTENACTIS NUMBER OF EXTANT SPECIES: Three. PRESENT DISTRIBUTION: Red Sea to western Pacific. GENERAL ABUNDANCE: Generally common, very conspicuous. FOSSIL RECORD: ?Miocene of the Pacific.

Species are ecologically and biogeographically similar to *Fungia*, which the genus as a whole closely resembles.

HERPOLITHA NUMBER OF EXTANT SPECIES: Two. PRESENT DISTRIBUTION: Red Sea and western Indian Ocean to Caribbean and Gulf of Mexico. GENERAL ABUNDANCE: Generally common, very conspicuous. FOSSIL RECORD: Pliocene of the Pacific.

POLYPHYLLIA NUMBER OF EXTANT SPECIES: Three. PRESENT DISTRIBUTION: Western Indian Ocean to southern Pacific. GENERAL ABUNDANCE: Sometimes common, very conspicuous. FOSSIL RECORD: None.

Polyphyllia novaehiberniae has an unusual distribution in that it occurs in the western Pacific but not in the central Indo-Pacific. This species also has the most extreme record of a disjunct distribution, having also been reported from Kenya[6].

SANDALOLITHA NUMBER OF EXTANT SPECIES: Two. PRESENT DISTRIBUTION: Central Indian Ocean to southern Pacific. GENERAL ABUNDANCE: Sometimes common, very conspicuous. FOSSIL RECORD: Pliocene of the Pacific.

HALOMITRA NUMBER OF EXTANT SPECIES: One. PRESENT DISTRIBUTION: Western Indian Ocean to southern Pacific. GENERAL ABUNDANCE: Generally uncommon, very conspicuous. FOSSIL RECORD: Miocene of the Tethys and Pacific.

ZOOPILUS NUMBER OF EXTANT SPECIES: One. PRESENT DISTRIBUTION: Central-west Pacific. GENERAL ABUNDANCE: Uncommon, very conspicuous. FOSSIL RECORD: None.

This is the only genus of Fungiidae not to occur in Australia, although it is common in southern Papua New Guinea immediately to the north of the Great Barrier Reef.

LITHOPHYLLON NUMBER OF EXTANT SPECIES: Approximately four. PRESENT DISTRIBUTION: Eastern Indian Ocean to Caribbean and Gulf of Mexico. GENERAL ABUNDANCE: Generally uncommon, conspicuous. FOSSIL RECORD: ?Oligocene of the Pacific (Wells, 1966).

PODABACIA NUMBER OF EXTANT SPECIES: Two. PRESENT DISTRIBUTION: Red Sea and western Indian Ocean to southern Pacific. GENERAL ABUNDANCE: Sometimes common, conspicuous. FOSSIL RECORD: Pliocene of the Pacific.

CANTHARELLUS NUMBER OF EXTANT SPECIES: Two. PRESENT DISTRIBUTION: Red Sea to western Pacific. GENERAL ABUNDANCE: Rare, cryptic. FOSSIL RECORD: ?Miocene of the Indo-Pacific.

This genus has doubtful validity (as being distinct from *Cycloseris),* as the separation is dependent on the single character of polyps remaining attached to the substrate throughout life. An undescribed species of Japanese *Fungia* does likewise (M Nishihira pers comm).

FAMILY OCULINIDAE

This is a primarily Cenozoic family. It has a poor fossil record and doubtful affinities with the Rhipidogyridae, a Cretaceous family of doubtful validity (p 110).

OCULINA NUMBER OF EXTANT ZOOXANTHELLATE SPECIES: Approximately five. PRESENT DISTRIBUTION: Caribbean, Gulf of Mexico and Bermuda. GENERAL ABUNDANCE: Generally uncommon. FOSSIL RECORD: ?Cretaceous (Wells 1956), Oligocene of the Tethys, Miocene of the Tethys.

Like *Astrangia*[7], *Cladocora* and *Madracis* (p 114), the genus contains zooxanthellate and azooxanthellate species. The two groups do not have morphologically distinctive characters.

SCHIZOCULINA NUMBER OF EXTANT SPECIES: One. PRESENT DISTRIBUTION: Eastern Atlantic only. GENERAL ABUNDANCE: Uncommon. FOSSIL RECORD: None.

The distribution of *Schizoculina fissipara,* endemic to the east African coast, is unlike that of any other coral.

GALAXEA NUMBER OF EXTANT SPECIES: Approximately five. PRESENT DISTRIBUTION: Red Sea and western Indian Ocean to southern Pacific. GENERAL ABUNDANCE: Very common, very conspicuous. FOSSIL RECORD: ?Miocene of the Pacific.

Galaxea fascicularis forms some of the largest of all coral colonies.

ACRHELIA NUMBER OF EXTANT SPECIES: One. PRESENT DISTRIBUTION: Eastern

Indian Ocean to southern Pacific. GENERAL ABUNDANCE: Generally uncommon, conspicuous. FOSSIL RECORD: None.

SIMPLASTREA NUMBER OF EXTANT SPECIES: One. PRESENT DISTRIBUTION: Indonesia only. GENERAL ABUNDANCE: Rare. FOSSIL RECORD: None.

The genus is known only from the holotype of *S. vesicularis,* an eroded corallum which, nevertheless, cannot be placed in another genus.

PARASIMPLASTREA NUMBER OF EXTANT SPECIES: One. PRESENT DISTRIBUTION: Oman only. GENERAL ABUNDANCE: Rare. FOSSIL RECORD: Miocene of the Pacific, common in the Pliocene of the central Indo-Pacific.

Sheppard's (1985) record of this genus in Oman is one of the two instances[8] where a genus formerly believed extinct has been found alive[9]. However, the systematic position of this species, especially in relation to the faviid species *Leptastrea beewickensis,* warrants further study.

FAMILY PECTINIIDAE

All genera of the family are extant. The fossil record is poor; affinities with the Mussidae, which seem clear, are primarily based on skeletal morphology.

There is stronger latitudinal attenuation of combined species of the family down the eastern and western coasts of Australia than in Japan (figure 78, p 245) due to differences in species compositions. Along all three coastlines, attenuation becomes very pronounced with abundance data.

ECHINOPHYLLIA NUMBER OF EXTANT SPECIES: Approximately eight. PRESENT DISTRIBUTION: Red Sea and western Indian Ocean to southern Pacific. GENERAL ABUNDANCE: Very common, conspicuous. FOSSIL RECORD: Miocene of the Indo-Pacific.

OXYPORA NUMBER OF EXTANT SPECIES: At least three. PRESENT DISTRIBUTION: Western Indian Ocean to southern Pacific. GENERAL ABUNDANCE: Generally common, conspicuous. FOSSIL RECORD: Pliocene of the Pacific.

MYCEDIUM NUMBER OF EXTANT SPECIES: At least two. PRESENT DISTRIBUTION: Red Sea and western Indian Ocean to southern Pacific. GENERAL ABUNDANCE: Generally common, conspicuous. FOSSIL RECORD: Miocene of the Tethys.

PHYSOPHYLLIA NUMBER OF EXTANT SPECIES: One. PRESENT DISTRIBUTION: Western Pacific. GENERAL ABUNDANCE: Rarely common, conspicuous. FOSSIL RECORD: Pleistocene of the Pacific. A poorly defined genus closely allied to Pectinia.

PECTINIA NUMBER OF EXTANT SPECIES: Approximately seven. PRESENT DISTRIBUTION: Western Indian Ocean to southern Pacific. GENERAL ABUNDANCE: Generally common, very conspicuous. FOSSIL RECORD: ?Pliocene of the Pacific.

FAMILY MEANDRINIDAE

This is a major family of the Cretaceous, with extant genera being remnants of Tethyan (not Cretaceous/Tertiary (p 119)) extinctions. The family is represented in the Indo-Pacific by one species *Ctenella chagius*, from Chagos in the central Indian Ocean (Dinesen 1977), which thus has the most anomalous distribution of all corals.

MEANDRINA NUMBER OF EXTANT SPECIES: At least two. PRESENT DISTRIBUTION: Caribbean, Gulf of Mexico and Bermuda. GENERAL ABUNDANCE: Generally common, conspicuous. FOSSIL RECORD: ?Eocene of the Caribbean, Oligocene of the Tethys.

DICHOCOENIA NUMBER OF EXTANT SPECIES: At least two. PRESENT DISTRIBUTION: Caribbean, Gulf of Mexico and Bermuda. GENERAL ABUNDANCE: Sometimes common, conspicuous. FOSSIL RECORD: ?Cretaceous (Wells 1956), Eocene of the Caribbean, Miocene of the Tethys.

DENDROGYRA NUMBER OF EXTANT SPECIES: One. PRESENT DISTRIBUTION: Caribbean and Gulf of Mexico. GENERAL ABUNDANCE: Uncommon, very conspicuous. FOSSIL RECORD: Miocene of the Tethys.

GOREAUGYRA NUMBER OF EXTANT SPECIES: One. PRESENT DISTRIBUTION: Caribbean (Bahamas) only. GENERAL ABUNDANCE: Rare. FOSSIL RECORD: None.

Type specimens of *G. memoralis,* the only specimens of the genus, appear superficially distinct from *Dendrogyra* and *Meandrina,* but the status of the genus is doubtful.

CTENELLA NUMBER OF EXTANT SPECIES: One. PRESENT DISTRIBUTION: Central Indian Ocean. GENERAL ABUNDANCE: Uncommon, conspicuous. FOSSIL RECORD: None.

Ctenella chagius, confined to the central Indian Ocean, is the only Indo-Pacific species of Meandrinidae which is otherwise confined to the Caribbean. Even at family level, extant distributions are not reliable indicators of place of origination.

FAMILY MUSSIDAE

This is a Cenozoic Tethyan family, with most genera now extant. It has clear morphological affinities with the Meandrinidae; the fossil record of this affinity is weak.

BLASTOMUSSA NUMBER OF EXTANT SPECIES: Three. PRESENT DISTRIBUTION: Red Sea and western Indian Ocean to western Pacific. GENERAL ABUNDANCE: Generally uncommon, sometimes inconspicuous. FOSSIL RECORD: Pleistocene of the Pacific.

Blastomussa wellsi usually has phaceloid corallites, but in most high-latitude regions colonies become cerioid. This is environment correlated and can be traced from one extreme to the other over a wide latitudinal range.

CYNARINA NUMBER OF EXTANT SPECIES: One. PRESENT DISTRIBUTION: Red Sea and western Indian Ocean to western Pacific. GENERAL ABUNDANCE: Generally uncommon, very conspicuous. FOSSIL RECORD: ?Oligocene (Wells 1956), Pliocene of the Pacific.

INDOPHYLLIA NUMBER OF EXTANT SPECIES: One. PRESENT DISTRIBUTION: Indonesia only. GENERAL ABUNDANCE: Rare, inconspicuous. FOSSIL RECORD: Oligocene (Wells 1956).

This genus, formerly thought long extinct, was rediscovered alive in Indonesia (Borel Best and Hoeksema 1987).

SCOLYMIA NUMBER OF EXTANT SPECIES: Approximately four. PRESENT DISTRIBUTION: Red Sea and western Indian Ocean to southern Pacific; Caribbean and Gulf of Mexico. GENERAL ABUNDANCE: Generally uncommon, conspicuous. FOSSIL RECORD: Oligocene of the Tethys, Miocene of the Caribbean.

Scolymia is readily confused with *Mussa,* the only difference between these genera in the Atlantic being whether they are free living or not; and this appears to be confused with extant corals (Fenner 1993). *Scolymia lacera* of the Indo-Pacific is relatively distinct.

AUSTRALOMUSSA NUMBER OF EXTANT SPECIES: One. PRESENT DISTRIBUTION: Central Indo-Pacific. GENERAL ABUNDANCE: Sometimes common, very conspicuous. FOSSIL RECORD: Pliocene of the Pacific.

ACANTHASTREA NUMBER OF EXTANT SPECIES: Approximately six. PRESENT DISTRIBUTION: Red Sea and western Indian Ocean to southern Pacific. GENERAL ABUNDANCE: Generally uncommon, *Favites*-like. FOSSIL RECORD: ?Miocene of the Pacific.

Several species (*A. hillae, A. bowerbanki, A. lordhowensis* and possibly *A. amakusensis*) are much more common in high-latitude non-reef localities than in the tropics. *Acanthastrea echinata* is the only widespread common species of the genus.

MUSSISMILIA NUMBER OF EXTANT SPECIES: Approximately three. PRESENT DISTRIBUTION: Brazil only. GENERAL ABUNDANCE: Generally uncommon, conspicuous. FOSSIL RECORD: Miocene of the Tethys.

Recorded from the Tethys (southern France and Spain) of the Early and Middle Miocene (Chevalier 1961; Oosterbaan 1990), now endemic to Brazil. The genus is in need of revision; two of the three species qualify for inclusion in *Acanthastrea* (pp 168, 278).

LOBOPHYLLIA NUMBER OF EXTANT SPECIES: Approximately nine. PRESENT DISTRIBUTION: Red Sea and western Indian Ocean to southern Pacific. GENERAL ABUNDANCE: Very common, very conspicuous. FOSSIL RECORD: Miocene of the Tethys.

Some of the more uncommon species of this genus are well-defined, others, including the very common *L. hemprichii* appear to be species complexes which show little geographic variation (p 22).

SYMPHYLLIA NUMBER OF EXTANT SPECIES: Approximately six. PRESENT DISTRIBUTION: Red Sea and western Indian Ocean to southern Pacific. GENERAL ABUNDANCE: Generally common, very conspicuous. FOSSIL RECORD: Pliocene of the Pacific.

MUSSA NUMBER OF EXTANT SPECIES: One. PRESENT DISTRIBUTION: Caribbean and Gulf of Mexico. GENERAL ABUNDANCE: Uncommon, very conspicuous, but confused with *Scolymia*. FOSSIL RECORD: Miocene of the Caribbean.

   The distinction between *Mussa* and Atlantic *Scolymia* is somewhat arbitrary (see above).

ISOPHYLLIA NUMBER OF EXTANT SPECIES: Probably two. PRESENT DISTRIBUTION: Caribbean, Gulf of Mexico and Bermuda. GENERAL ABUNDANCE: Generally common, conspicuous. FOSSIL RECORD: Miocene of the Caribbean.

ISOPHYLLASTRAEA NUMBER OF EXTANT SPECIES: One. PRESENT DISTRIBUTION: Caribbean and Gulf of Mexico. GENERAL ABUNDANCE: Generally uncommon, conspicuous. FOSSIL RECORD: Miocene of the Caribbean and Tethys.

MYCETOPHYLLIA NUMBER OF EXTANT SPECIES: Approximately five. PRESENT DISTRIBUTION: Caribbean and Gulf of Mexico. GENERAL ABUNDANCE: Generally common, conspicuous. FOSSIL RECORD: Oligocene of the Tethys, Miocene of the Caribbean.

FAMILY MERULINIDAE

This is a Cenozoic Tethyan family, with most genera now extant. Its affinities with the Mussidae are clear on morphological grounds.

HYDNOPHORA NUMBER OF EXTANT SPECIES: Approximately seven. PRESENT DISTRIBUTION: Red Sea and western Indian Ocean to southern Pacific. GENERAL ABUNDANCE: Generally common, very conspicuous. FOSSIL RECORD: ?Cretaceous (Wells 1956; Beauvais 1984); Eocene of the Tethys, Oligocene of the Caribbean.

PARACLAVARINA NUMBER OF EXTANT SPECIES: One. PRESENT DISTRIBUTION: Central Indo-Pacific. GENERAL ABUNDANCE: Generally uncommon, conspicuous. FOSSIL RECORD: None.

MERULINA NUMBER OF EXTANT SPECIES: Three. PRESENT DISTRIBUTION: Red Sea and western Indian Ocean to southern Pacific. GENERAL ABUNDANCE: Sometimes common, conspicuous. FOSSIL RECORD: Pliocene of the Pacific.

BONINASTREA NUMBER OF EXTANT SPECIES: One. PRESENT DISTRIBUTION: Western Pacific. GENERAL ABUNDANCE: Rare. FOSSIL RECORD: None.

   Currently known from only two disjunct localities, in the Ogasawara (Bonin) Islands of Japan, and in Indonesia (p 279).

SCAPOPHYLLIA NUMBER OF EXTANT SPECIES: One. PRESENT DISTRIBUTION: Eastern Indian Ocean to southern Pacific. GENERAL ABUNDANCE: Generally uncommon, conspicuous. FOSSIL RECORD: Miocene of the Tethys (Chevalier 1961).

FAMILY FAVIIDAE

The Faviidae contains more genera (figure 79) than any other scleractinian family (figure 25, p 110) and is second only to the Acroporidae in number of extant species, as well as overall abundance throughout the Indo-Pacific and Caribbean. With few exceptions the genera are well-defined and widely distributed. Two genera (*Favia* and *Montastrea*) are common to both the Indo-Pacific and Atlantic; two (*Astreosmilia* and *Erythrastrea*) are restricted to the western Indian Ocean.

Because of their solid construction and wide geographic distribution, most faviid genera are readily preserved as fossils and have a good fossil record. The Faviidae is the only family to have been a dominant reef-builder in both the Mesozoic and Cenozoic eras. It fared better than any other family in the Cretaceous/Tertiary extinctions (p 119).

Most species are also widely distributed, both longitudinally and latitudinally. They usually exhibit less inter-regional variation within the central Great Barrier Reef than other major groups of corals and this, combined with the rarity of endemic species, gives a relatively uniform Indo–west Pacific fauna.

Some species of faviids are restricted to inter-tidal habitats and upper reef slopes, but most occur over a wide range of environments. These species have similar, correspondingly wide, range of skeletal variation. Coralla from high energy environments exposed to strong sunlight have heavily calcified skeletal structures and corallites. Those from deep or turbid water, with poorly illuminated environments, are always lightly calcified and have relatively small corallites separated by a blistery coenosteum. This similarity in response to environmental gradients frequently results in coralla of different species from the same environment looking superficially more alike than coralla of the same species from very different environments (p 18)[10].

Intra-specific geographic variation is very great between high-latitude and tropical locations of the same region. Coralla from high-latitude regions are usually heavily calcified. Coralla from different tropical regions frequently show minimal geographic variation. Many species show major variation in the relative abundance in the tropics.

ASTREOSMILIA NUMBER OF EXTANT SPECIES: One. PRESENT DISTRIBUTION: Western and central Indian Ocean. GENERAL ABUNDANCE: Uncommon, conspicuous. FOSSIL RECORD: None.

CAULASTREA NUMBER OF EXTANT SPECIES: Approximately four. PRESENT

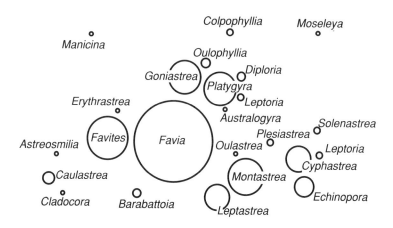

Figure 79 Hypothetical taxonomic affinities between the genera of the Faviidae. The diameter of the circles is proportional to the number of species.

DISTRIBUTION: Red Sea and western Indian Ocean to southern Pacific. GENERAL ABUNDANCE: Generally common, very conspicuous. FOSSIL RECORD: Eocene of the Tethys, Oligocene of the Caribbean.

Caulastrea tumida is most abundant in turbid water and occurs commonly in high-latitude non-reef environments of Japan, while the other species are typically found in clear-water reef environments.

ERYTHRASTREA NUMBER OF EXTANT SPECIES: Probably one. PRESENT DISTRIBUTION: Red Sea. GENERAL ABUNDANCE: Rare. FOSSIL RECORD: None.

MANICINA NUMBER OF EXTANT SPECIES: One. PRESENT DISTRIBUTION: Caribbean and Gulf of Mexico. GENERAL ABUNDANCE: Generally uncommon, conspicuous, often non-reefal. FOSSIL RECORD: Oligocene of the Caribbean.

FAVIA NUMBER OF EXTANT SPECIES: At least thirty. PRESENT DISTRIBUTION: Cosmopolitan. GENERAL ABUNDANCE: Extremely common, conspicuous. FOSSIL RECORD: ?Cretaceous (Wells 1956), Eocene of the Caribbean, Tethys and Pacific.

The number of species of *Favia* has been generally understated in the taxonomic literature. It is one of the most widely and uniformly distributed of all coral genera, in both the Indo-Pacific and Atlantic. Individual species are also very widely distributed in the Indo–west Pacific, and a high proportion have distribution limits extending beyond the latitudinal limits of reefs. *Favia fragum*, the principal species of the genus in the Atlantic, is very widespread; it differs substantially from any Indo-Pacific species.

BARABATTOIA NUMBER OF EXTANT SPECIES: Approximately three. PRESENT DISTRIBUTION: Eastern Indian Ocean to southern Pacific. GENERAL ABUNDANCE: Sometimes common, readily confused with *Favia*. FOSSIL RECORD: None.

This is, perhaps, a genus of convenience to accommodate a few species that appear to have affinities with each other but which are normally outside the boundaries of *Favia*.

FAVITES NUMBER OF EXTANT SPECIES: Approximately fifteen. PRESENT DISTRIBUTION: Red Sea and western Indian Ocean to southern Pacific. GENERAL ABUNDANCE: Very common, conspicuous. FOSSIL RECORD: Eocene of the Caribbean, Oligocene of the Tethys.

As with *Favia*, there are likely to be more *Favites* species on most Indo–west Pacific reefs than have so far been studied.

The distribution range of Indo-Pacific *Favites* is similar to that of *Favia*. *Favites* is particularly common in higher latitudes, occurring abundantly on the southwest and southeast coasts of Australia and mainland Japan, in all cases well beyond the latitudinal ranges of reefs (figure 78, p 245). Most *Favites* species are widely and uniformly spread within the central Indo-Pacific, with many minor regional differences in colour, skeletal detail and abundance.

Coralla of most species from high latitudes are heavily calcified with thick septa and elongate (*Acanthastrea*-like[11]) septal dentations. Such coralla are readily distinguished from, but intergrade with, coralla from tropical locations.

GONIASTREA NUMBER OF EXTANT SPECIES: Approximately twelve. PRESENT DISTRIBUTION: Red Sea and western Indian Ocean to southern Pacific. GENERAL ABUNDANCE: Very common, generally conspicuous. FOSSIL RECORD: Eocene of the Caribbean, Oligocene of the Tethys and ?Pacific.

For most *Goniastrea* species, there are significant differences in colour and/or skeletal detail, between colonies from temperate and tropical central Indo-Pacific locations. *Goniastrea* species are often the dominant corals of inter-tidal mudflats, rock platforms and some outer reef flats. The genus includes some of the most tolerant of all coral species to aerial exposure, the same set of species occurring in inter-tidal environments throughout much of the Indo–west Pacific.

Most species are well defined and widely distributed, showing little taxonomically significant geographic variation. Environment-correlated variation may reach extremes in some inter-tidal habitats where, for example, normally cerioid species may develop colonies with meandroid upper surfaces.

PLATYGYRA NUMBER OF EXTANT SPECIES: Approximately twelve. PRESENT DISTRIBUTION: Red Sea and western Indian Ocean to southern Pacific. GENERAL ABUNDANCE: Extremely common, conspicuous but may be confused with *Goniastrea*. FOSSIL RECORD: ?Eocene of the Pacific, Oligocene of the Caribbean and Tethys.

The eight species of Indo–west Pacific *Platygyra* recognised in this study all have similar skeletal characters. All show similar skeletal modifications along environmental gradients and some, especially *P. daedalea* and *P. lamellina,* may be difficult to distinguish unless they are collected from the same biotope.

AUSTRALOGYRA NUMBER OF EXTANT SPECIES: One. PRESENT DISTRIBUTION: Central Indo-Pacific. GENERAL ABUNDANCE: Generally uncommon, very conspicuous. FOSSIL RECORD: None.

LEPTORIA NUMBER OF EXTANT SPECIES: Two. PRESENT DISTRIBUTION: Red Sea and western Indian Ocean to southern Pacific. GENERAL ABUNDANCE: Sometimes common, conspicuous. FOSSIL RECORD: ?Cretaceous (Wells 1956), Eocene of the Caribbean and Pacific, Oligocene of the Tethys.

OULOPHYLLIA NUMBER OF EXTANT SPECIES: Approximately three. PRESENT DISTRIBUTION: Red Sea and western Indian Ocean to western Pacific. GENERAL ABUNDANCE: Sometimes common, conspicuous. FOSSIL RECORD: ?Oligocene of the Tethys, Pleistocene of the Pacific.

COLPOPHYLLIA NUMBER OF EXTANT SPECIES: Two. PRESENT DISTRIBUTION: Caribbean and Gulf of Mexico. GENERAL ABUNDANCE: Very common, very conspicuous. FOSSIL RECORD: Eocene of the Caribbean and Tethys.

DIPLORIA NUMBER OF EXTANT SPECIES: Three. PRESENT DISTRIBUTION: Caribbean, Gulf of Mexico and Bermuda. GENERAL ABUNDANCE: Very common, conspicuous. FOSSIL RECORD: ?Cretaceous (Wells 1956), Eocene of the Tethys, Oligocene of the Caribbean.

MONTASTREA NUMBER OF EXTANT SPECIES: Approximately thirteen. PRESENT DISTRIBUTION: Cosmopolitan. GENERAL ABUNDANCE: Generally common, conspicuous. FOSSIL RECORD: ?Late Jurassic Wells (1956), ?Cretaceous of the Tethys (Budd and Coates 1992 depending on taxonomic issues, p 149), Eocene of the Caribbean.

Montastrea is a poorly defined genus containing mostly distinctive species within a given region, but over wider geographic ranges within the central Indo-Pacific, it has several species that form distinctive geographic subspecies of doubtful taxonomic affinity. It is the chief frame builder of the Caribbean.

CLADOCORA NUMBER OF EXTANT SPECIES: One. PRESENT DISTRIBUTION: Western to eastern Atlantic. GENERAL ABUNDANCE: Generally uncommon, conspicuous. FOSSIL RECORD: ?Cretaceous (Wells 1956), Eocene of the Caribbean and Tethys.

OULASTREA NUMBER OF EXTANT SPECIES: One. PRESENT DISTRIBUTION:

Central Indo-Pacific. GENERAL ABUNDANCE: Uncommon, occurs in non-reef biotopes. FOSSIL RECORD: None.

Oulastrea crispata occurs in the Sea of Japan where water temperatures go to near freezing. It reportedly still has zooxanthellae in these conditions (p 91).

PLESIASTREA NUMBER OF EXTANT SPECIES: At least two. PRESENT DISTRIBUTION: Red Sea and western Indian Ocean to far eastern Pacific. GENERAL ABUNDANCE: Sometimes common. FOSSIL RECORD: Miocene of the Tethys.

Plesiastrea versipora is the only tropical species to occur along the full southern coast of Australia, where it forms a distinct geographic subspecies (formerly considered a separate species, *P. urvillei*).

DIPLOASTREA NUMBER OF EXTANT SPECIES: One. PRESENT DISTRIBUTION: Red Sea and western Indian Ocean to southern Pacific. GENERAL ABUNDANCE: Generally common, very conspicuous. FOSSIL RECORD: ?Cretaceous (Wells 1956), Eocene of the Caribbean and ?Indian Ocean, Oligocene of the Tethys.

Diploastrea heliopora has very little morphological variation, either environment correlated or geographic. It is also conservative in the fossil record, the one common 'chronospecies' being at least Early Miocene in age. Affinities in the Faviidae are obscure: it is as close to *Montastrea cavernosa* as any other species.

LEPTASTREA NUMBER OF EXTANT SPECIES: Approximately eight. PRESENT DISTRIBUTION: Red Sea and western Indian Ocean to southern Pacific. GENERAL ABUNDANCE: Generally common, conspicuous. FOSSIL RECORD: Oligocene of the Indo-Pacific, Miocene of the Tethys.

CYPHASTREA NUMBER OF EXTANT SPECIES: Approximately nine. PRESENT DISTRIBUTION: Red Sea and western Indian Ocean to southern Pacific. GENERAL ABUNDANCE: Very common, conspicuous. FOSSIL RECORD: ?Oligocene (Wells 1956), Miocene of the Tethys.

Cyphastrea serailia is by far the most common and widespread species of the genus yet is poorly defined and probably masks other, less common species (p 22).

SOLENASTREA NUMBER OF EXTANT SPECIES: Two. PRESENT DISTRIBUTION: Caribbean and Gulf of Mexico. GENERAL ABUNDANCE: Sometimes common, generally inconspicuous. FOSSIL RECORD: Oligocene of the Caribbean, Miocene of the Tethys.

ECHINOPORA NUMBER OF EXTANT SPECIES: Approximately seven. PRESENT DISTRIBUTION: Red Sea and western Indian Ocean to southern Pacific. GENERAL ABUNDANCE: Very common, conspicuous. FOSSIL RECORD: Miocene of the Pacific.

Echinopora mammiformis is one of a number of species that can have two completely different growth forms (in this case, flat plates and thin branches[12]), which have at times been placed in different genera (figure 3, p 16).

MOSELEYA NUMBER OF EXTANT SPECIES: One. PRESENT DISTRIBUTION: Central Indo-Pacific. GENERAL ABUNDANCE: Generally uncommon, very conspicuous. FOSSIL RECORD: None.

A well-defined monospecific genus that shows affinities with both the Faviidae and the Trachyphylliidae.

FAMILY TRACHYPHYLLIIDAE

TRACHYPHYLLIA NUMBER OF EXTANT SPECIES: One. PRESENT DISTRIBUTION: Red Sea and western Indian Ocean to southern Pacific. GENERAL ABUNDANCE: Uncommon, very conspicuous. FOSSIL RECORD: Eocene Tethys and Indian Ocean, Oligocene of the Caribbean.

Only one extant genus is included in this family, which is closely related to the Faviidae in general and *Moseleya* in particular. *Trachyphyllia* is repeatedly confused with *Antillia* in the fossil record.

FAMILY CARYOPHYLLIIDAE

Along with the Faviidae, the Caryophylliidae thrived in both the Cretaceous and the Cenozoic. It has more genera than any other family, but most of these are azooxanthellate. The six zooxanthellate Indo-Pacific genera and *Eusmilia* of the Caribbean and Gulf of Mexico are readily distinguished from their azooxanthellate relatives. Most species display relatively little environment-correlated or geographic variation.

EUPHYLLIA NUMBER OF EXTANT SPECIES: Nine. PRESENT DISTRIBUTION: Red Sea and western Indian Ocean to southern Pacific. GENERAL ABUNDANCE: Generally common, very conspicuous. FOSSIL RECORD: ?Eocene of the Indian Ocean, Oligocene of the Caribbean and Tethys.

Two species groups of *Euphyllia* cannot be identified from skeletons alone. The first, originally described as *Euphyllia fimbriata,* contains two species: *E. divisa* and *E. ancora.* The second contains five species: *E. glabrescens, E. paradivisa, E. paraancora* and *E. paraglabrescens.* The last species is known only from one site, at Tanegashima, Japan (p 232).

CATALAPHYLLIA NUMBER OF EXTANT SPECIES: Probably one. PRESENT DISTRIBUTION: Central Indo-Pacific. GENERAL ABUNDANCE: Generally uncommon, very conspicuous. FOSSIL RECORD: None.

CataXaphyllia is generally uncommon throughout the recorded distribution range. It occurs in localised areas of Honshu, Japan but has not yet been found elsewhere in Japan (figure 65, p 193). These colonies have the same general appearance as those from the Philippines and Australia, with the same grey-green tentacles with pink tips. There are, however, minor colour and ecological differences between Japanese and Australian colonies that may amount to distinct geographic subspecies.

PLEROGYRA Number of extant species: Three. Present distribution: Red Sea and western Indian Ocean to southern Pacific. General abundance: Generally uncommon, very conspicuous. Fossil record: None.

A single species, *P. sinuosa,* occurs throughout most of the geographic range of this genus. Like most other Caryophylliidae, *Plerogyra* species are most commonly found in turbid water, but their occurrence is often unpredictable.

PHYSOGYRA Number of extant species: Three. Present distribution: Red Sea and western Indian Ocean to southern Pacific. General abundance: Sometimes common, very conspicuous. Fossil record: ?None.

GYROSMILIA Number of extant species: One. Present distribution: Western Indian Ocean. General abundance: Uncommon. Fossil record: None.

EUSMILIA Number of extant species: One. Present distribution: Caribbean and Gulf of Mexico. General abundance: Sometimes common, very conspicuous Fossil record: ?Oligocene of the Tethys, Miocene of the Caribbean.

MONTIGYRA Number of extant species: One. Present distribution: NW Australia only. General abundance: Rare. Fossil record: None.

This genus is known from a single specimen. Present indications are that it could now be extinct.

HETEROCYATHUS Number of extant species: Probably one zooxanthellate species. Present distribution: Western Indian Ocean to central Pacific. General abundance: Restricted to inter-reef habitats. Fossil record: Pliocene of the Pacific.

Family Dendrophylliidae

This is an essentially azooxanthellate family with distinctly different Cretaceous and Cenozoic genera. The zooxanthellate genera *Turbinaria* and *Duncanopsammia* are very dissimilar; the latter is morphologically much closer to azooxanthellate relatives (below).

TURBINARIA Number of extant species: Approximately fifteen. Present distribution: Red Sea and western Indian Ocean to southern Pacific. General abundance: Very common, very conspicuous. Fossil record: Oligocene of the Caribbean, Miocene of the Tethys.

On both western and eastern Australian coasts, and to a lesser extent in Japan, *Turbinaria* species have distinctive distribution patterns, most species being more abundant, and forming larger colonies in non-reef habitats of high-latitude locations. In general, there are greater similarities between coralla from high-latitude locations on the eastern and western Australian coasts than there are between high- and low-latitude locations on the same coast. Most *Turbinaria* species exhibit very great environment-correlated variation and lack conservative skeletal characters, making some species especially difficult to separate.

DUNCANOPSAMMIA NUMBER OF EXTANT SPECIES: One. PRESENT DISTRIBUTION: Central Indo-Pacific. GENERAL ABUNDANCE: Uncommon, very conspicuous. FOSSIL RECORD: ?Miocene of the Pacific.

Duncanopsammia, more than any other genus, blends zooxanthellate with non-zooxanthellate characters. Based on corallite morphology alone, *D. axifuga* would fit comfortably in the azooxanthellate genus *Balanophyllia*, indicating a Cenozoic origin from azooxanthellate ancestors independently of other Dendrophylliidae.

HETEROPSAMMIA NUMBER OF EXTANT SPECIES: At least two. PRESENT DISTRIBUTION: Red Sea and western Indian Ocean to western Pacific. GENERAL ABUNDANCE: Uncommon, occurs on inter-reef sand flats. FOSSIL RECORD: Oligocene of the Tethys.

NUMBERS AND MORPHOLOGICAL BOUNDARIES OF TAXA

> Here must be entered a caveat to the effect that the validity of any discussion of the numbers of species on Indo-Pacific reefs is seriously limited at present by two factors, 1) the species problem in corals in general, and 2) the incompleteness of reef exploration. (J Wells 1969)

The number of genera in families (figure 80) is strongly skewed, with only three families having fewer than seven genera. The Faviidae have by far the greatest number of zooxanthellate genera (n=25); the Caryophylliidae have the greatest number of azooxanthellate genera (n=46; S Cairns pers comm). These numbers are only equalled in the Cenozoic by the number of genera of Miocene faviids, but similar numbers are found in many Mesozoic families (figure 25, p 110).

At family level, most taxonomic boundaries are well-defined. The relationship between the Trachyphylliidae and the Faviidae is a doubtful area only because the former family has only one extant genus (*Trachyphyllia*), which has affinities with only one other species (*Moseleya latistellata*, in the Faviidae).

Estimates of numbers of genera (zooxanthellate and azooxanthellate) in all families extant and extinct, are given in chapter 8 (figure 36, p 134). Most genera have five species or fewer (figure 80), *Acropora* having, by far, the greatest number.

There are many taxonomic issues created by species that have intermediate taxonomic positions. Common examples are: in the Agariciidae, some leafy *Pavona* species may be *Leptoseris*-like; both *Leptoseris mycetoseroides* and *L. yabei* are *Pavona*-like; in the Fungiidae, some *Cycloseris* species are *Fungia*-like; in the Pectiniidae, *Physophyllia ayleni* is *Pectinia*-like; in the Mussidae, *Lobophyllia hattai* is *Symphyllia*-like, *Symphyllia valenciennesii* is *Lobophyllia*-like, two *Mussismilia* are *Acanthastrea*-like; in

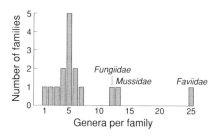

Figure 80 Number of genera in the families of zooxanthellate corals.

the Faviidae, *Barabattoia amicorum* is *Favia*-like, several *Favites* are *Favia*-like, *Oulophyllia bennettae* is *Favites*-like, *Australogyra zelli* is *Platygyra*-like and *Leptastrea agassizi* is *Cyphastrea*-like.

The placing of a species in one genus or another (including some of the above cases) often involves a substantial degree of uncertainty, and thus the species may later be re-assigned to another genus (p 49). Species of uncertain genera such as these have an insignificant quantitative affect on the value of most genera as taxonomic units, but they do reflect the unsatisfactory qualitative status of many. These genera unavoidably become units of identification rather than phylogeny.

A second qualitative problem with genera as phylogenetic units is that many contain species with such wide environment-correlated skeletal variation that generic boundaries are crossed (figure 2, p 15). Common examples are as follows: *Seriatopora caliendrum* can become *Stylophora*-like; *Caulastrea tumida* can become *Favia*-like; *Favia rotundata* can become *Favites*-like; *Plerogyra turbida* can become *Physogyra*-like, *Gardineroseris planulata* can become *Pavona*-like.

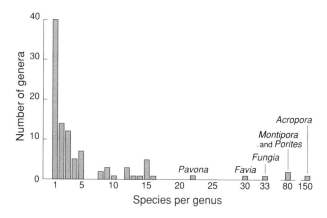

Figure 81 Number of species in the genera of zooxanthellate corals. The number of species in *Favia* and (especially) *Porites* is much less certain than for other genera.

GLOSSARY

'When I use a word', Humpty Dumpty said, in a rather scornful tone, 'it means just what I choose it to mean — neither more nor less'. 'The question is', said Alice, 'whether you can make words mean so many different things'.
(Lewis Carroll's *Through the Looking Glass*)

Inevitably, all authors, the present one included, do what Humpty Dumpty did. The following explains terms specifically as they are used in this book.

Acolonial corals Solitary individuals that do not form colonies.

Agglomeration The process whereby individual points or objects are accumulated or clustered into a single group.

Algorithm A part of a computer program that provides the solution of a specific numeric problem.

Allele A particular form of a gene at a particular locus.

Allopatric populations Populations that are contiguous but separated by space across which migration occurs only at very low frequency (figure 12, p 43).

Allopatric speciation The splitting of a widespread population into two or more isolates by a geological or ecological isolating barrier ('dumbbell' speciation or speciation by subdivision) and subsequent differentiation into a new taxon, or the dispersal of a few propagules across a pre-existing barrier and subsequent differentiation into a new taxon. See 'concepts of species — terminology issues'.

Allopatric species Species that have arisen by allopatric speciation.

Allozymes An allele of an enzyme.

Amphitropical distributions Where conspecific disjunct distributions occur either side of the equator.

Anagenesis Progressive evolution towards higher levels of organisation or specialisation, now generally used to describe directional evolution of a feature over an arbitrary short length of a lineage.

Aposymbiotic corals Corals that can live indefinitely with or without zooxanthellae (p 97).

Aragonitic skeletons Skeletons primarily composed of the aragonite form of calcium carbonate. All Scleractinia have aragonitic skeletons (cf calcitic skeletons). Aragonite turns into calcite through the process of diagenesis and in so doing usually loses much skeletal detail.

Askeletal corals and larvae Corals and larvae that do not have skeletons.

Attribute One of a set of descriptive terms, a character.

Azooxanthellate corals and larvae Corals and larvae that do not have zooxanthellae.

Binary data Data in the form of two alternatives (+/-, 0/1, or presence/absence).

Biodiversity The term has recently acquired many meanings, but can be considered synonymous with 'systematic diversity'. Biodiversity thus has the same relationship to taxonomic diversity as systematics has to taxonomy. Patterns of taxonomic diversity (chapter 9, p 155) are indicative only of patterns of biodiversity.

Biogeography The study of the geographic distribution of life and the reasons for it. In practice, biogeography is divisible into observations of distributions and explanations of those observations.

Biogeographically significant space and time See 'descriptors of space and time' (p 269).

Biological species concept Species as are defined as genetically similar populations capable of interbreeding and which, through genetically determined isolation mechanisms, evolve in a way isolated or distinct from other populations (Mayr, 1963; Grant, 1981).

Biotope A geographic area that is under the influence of environmental parameters, the dominant characteristics of which are homogeneous. Biotopes are generally the smallest ecological units that can be delimited by convenient boundaries and which are characterised by their biota.

Black-box A functional unit or process of any nature, the external parameters of which are known, but the internal parameters are not known, eg the external environmental factors controlling the composition of a community may be known, but the internal process which governs that control may be an unknown black box.

Boreal Pertaining to cool or cold temperate regions of the northern hemisphere.

Calcitic skeletons Skeletons primarily composed of the calcite form of calcium carbonate. All Rugosa and molluscs have calcite skeletons (cf aragonitic skeletons).

Calice The upper surface of a corallite to which the soft parts of an individual polyp are attached.

Carbonate compensation depth The depth at which the rate of dissolution of calcium carbonate equals the rate of supply. This lies within the deep zone of undersaturation of various carbonates and below the saturated and supersaturated higher layers in all oceans.

Cenozoic The last era in geological time (figure 24, p 109).

Central American Seaway A former seaway between north and south America, now closed by the Isthmus of Panama.

Cerioid corals Massive corals that have corallites sharing common walls (eg the upper surface of the corallum illustrated by figure 2 (p 15).

Chronospecies A conceptual species that exists through evolutionary time.

Clade A phylogeny inferred to be monophyletic, a group of taxa sharing a closer common ancestry with one another than with members of any other clade.

Cladistics The study of clade relationships using a numerical method of grouping taxa by their shared derived characters.

Cladogram A diagram, in the form of a tree, grouping taxa by their shared derived characters by cladistic methods (p 24).

Cline Gradual sequential geographic change in morphological characters or genetic composition within species.

Clones Asexually produced replicates of colonies.

Coenosteum Thin horizontal skeletal plates between corallites.

Colonial corals Corals composed of many individuals. There is no clear distinction (eg in fungiids) between single individuals with many mouths and colonies of individuals with single mouths.

Community A group of organisms of different species that co-occur in the same habitat or area and interact through trophic and spatial relationships. Communities are typically characterised by reference to one or more dominant species.

Concepts of species — terminology issues Common terms and concepts are used differently by different authors. This is largely because our understanding of evolutionary processes is rapidly evolving, and doing so in a wide range of different disciplines that share the same terms and similar or related concepts. It is also because common terms used in the context of one theory or process often take on a different meaning when used in the context of another, and again change when theory, process, concept and observation are intermixed, and again when meeting the needs of different taxa or different research fields. These changes are often subtle and tend to confuse and frustrate, even when they are well explained.

Standard terms within the literature of evolution — 'allopatric', 'dichopatric', 'parapatric', 'peripatric' and 'sympatric' (figure 12, p 43) are all geographically (area) based. Such is the close link between *patterns* resulting from evolution and the *process* of evolution itself that these terms continue to be used in reference to both, without causing undue confusion (discussed by Cracraft 1987; Chandler and Gromko 1989).

Conservative characters Characters of species that show relatively little environment-correlated and/or geographic variation.

Congeneric Belonging to the same genus.

Conspecific Belonging to the same species.

Contiguous distributions Sequential distributions so that one area is immediately adjacent to the next.

Co-occurring colonies (co-occurrence) Colonies that occur together in the same biotope. Such colonies have minimal environment-correlated variation and thus variation between colonies is indicative of genetic differences.

Corallite The skeleton of an individual coral polyp.

Corallum (plural coralla) The skeleton of a coral colony.

Coral For the specific purposes of this book, the term 'coral' is short for zooxanthellate scleractinian corals. It is sometimes used in a wider sense; if so, the intended meaning should be clear from the context. Sometimes a qualifying term is added (eg non-scleractinian, rugose, azooxanthellate). 'Corals' is also frequently used in narrowed sense with the addition of a qualifier (eg colonial corals, free-living corals, askeletal corals); if so, the term 'coral' retains the above meaning unless this is also qualified (eg 'colonial azooxanthellate corals').

Cretaceous A period in geological time (figure 24, p 109).

Cryptic species Species that are difficult to distinguish *in situ*.

Cyanobacteria Photosynthetic blue-green algae, intermediate between bacteria and higher plants.

Dendrogram A tree-like hierarchical classification with a single root and branches representing levels of dissimilarities of objects. In this book, the objects are localities and dendrograms measure dissimilarities in coral species compositions.

Derived characters Non-ancestral characters.

Diagenesis In coral, the processes by which loose aggregations of aragonitic calcium carbonate become transformed into calcite and thence dolomite. This transformation is greatly enhanced by water movement through porous structures.

Dichopatric populations Populations that are widely separated by space across which there is no migration (figure 12, p 43).

Descriptors of space and time The word-concepts *ecologically-significant space* and *biogeographically-significant space* are used throughout this book. The intended meaning of the former is the 'space needed for ecologically-significant variation to occur'. The absolute measure of this space will depend on the particular situation; it may be an individual biotope. The intended meaning of the latter is 'space needed for biogeographically-significant variation to take place'; implying a continental coastline or larger than regional space (p 270).
The word-concepts *ecologically-significant time* and *evolutionarily significant time* are also frequently used. The intended meaning of the former is the 'time needed for ecologically-significant change to take place', and of the latter, the 'time needed for genetically-significant evolutionary change to take place'. The absolute measure of these will depend entirely on the particular situation.

Dichopatric speciation The splitting of a widespread population into two or more isolates by a geological or ecological isolating barrier ('dumbbell' speciation or speciation by subdivision) and subsequent differentiation into a new taxon.). See 'concepts of species — terminology issues'.

Dichotomous evolution Evolution by the division of phylogenies.

Disjunct distributions Non-continuous distributions between widely separated populations.

Dispersal The process of movement of propagules resulting in dispersion. Synonymous with 'migration' except that the latter implies an undertaking specific in time or place.

Dispersion Synonymous with 'spatial pattern' in ecology and biogeography indicating an achieved state, but often used synonymously with 'dispersal' (a process) in ordinary English. Dispersion range is sometimes used (not in this book) to denote distribution range plus additional area that can be potentially reached by propagules.

Displaced terrains All crustal plates have been 'displaced' from original positions as continents 'drift'; displaced terrains, however, have 'drifted' relatively great distances. Some very large land masses are composed of aggregations of terrains that have very different geographic origins.

Ecologically-significant space and time See 'Descriptors in space and time'.

Ecomorphs Morphological variants of species that may have an environmental and/or genetic origin. Most colonial coral species are divisible into many ecomorphs, each of which can be associated with a particular type of habitat or geographic location (p 229).

El Niño Southern Oscillation (ENSO) The occasional appearance of large masses of warm water (towards year ends), off the coasts of Ecuador and northern Peru, where the water normally is cold. These appearances are linked to atmospheric changes around the world, leading to major climatic and oceanographic anomalies.

Electrophoresis The migration of proteins under the influence of an electrical field. Data about isozymes and allozymes thus separated are generally considered to be closely linked to systematic affinities.

Endemic Restriction of a species to a specified region.

Endosymbiotic algae Symbiotic algae living within the cells of the host animal.

Eocene An epoch in geological time (figure 24, p 109).

Epicontinental sea Seas that flood major continental areas at high stands.

Eustatic Worldwide changes of sea level. These are primarily due to the withdrawal and release of water by the growth and decay of polar ice caps and tectonic movements of the sea floor and landmasses (tectono-eustacy).

Explanate Plate-like or foliaceous.

Extant Now-living (cf extinct).

Extinct No longer living (cf extant), also used in the context of a restricted area where there has been 'regional extinctions'.

Foraminifera Protozoa of the Order Foraminiferida which are abundant in the plankton and benthos of all oceans and which typically develop tests of microscopic size up to 50 mm diameter. Foraminifera are extensively used in dating of geological strata

Founder effect The principle that founders of a new population carry only a fraction of the genetic diversity of the parent population.

Free-living coral Corals that are not attached to the substrate.

Geminate species Similar species, that are the product of relatively recent speciation, occupying adjacent areas (see 'sibling species').

Gene flow The sum of successful dispersal and breeding of individuals originating in one population but arriving in another.

Genetic distance Any of several measures of the degree of genetic difference between populations, based on differences in allele frequencies.

Genetic drift Random changes in the frequencies of two or more alleles or genotypes within a population.

Genetic plasticity Where morphological variation is largely determined by physical environment, not genotype.

Genotype The set of genes possessed by an individual organism.

Geographic subspecies A taxonomically definable subgroup of a species with a restricted geographic range. Only in rare instances do geographic subspecies of corals have taxonomically meaningful names, as frequently found in plants.

Geographic terminology Throughout this book, geographic areas of different sizes are distinguished by use of the term *site* to indicate an area of a few square kilometres, *location* to indicate a major group of reefs, *region* to indicate a continental coastline or the like, *province* to indicate major geographic subdivisions of the world and *realm* to indicate an entire ocean basin or area of similar size. The terms *biotope*, *population* and *community* are used in an ecological or genetic context only, without geographic connotation.

Geological intervals See figure 24, p 109.

Glacio-eustacy See 'eustatic'.

Gonochoric Individuals that have separate sexes (cf hermaphrodite).

Guild A group of species having similar ecological resource requirements and therefore having similar roles in the community.

Habitat A vague word indicating the particular type of environment occupied by an organism.

Hardy-Weinberg equilibrium In a large, random-mating population, the frequencies of homozygotes and heterozygotes with respect to allelic variation in a single gene should remain constant (conform algebraically to the Hardy-Weinberg equilibrium) from generation to generation unless outside forces (selection, mutation and migration) act to change it.

Hermaphrodite Individuals that are both male and female (cf gonochoric).

Hermatypic Literally 'reef building' but commonly used as a descriptor for marine invertebrates that have photosynthetic plants living symbiotically within their tissues. Because the word is a misnomer, several terms, including 'reef-building', 'symbiotic' and 'zooxanthellate', are used synonymously. Of these, the former two are ambiguous and the latter is restricted to extant corals and other taxa with zooxanthellae (see 'zooxanthellate corals').

Historical biogeography The past history of present distributions. In the context of this book, biogeography and historical biogeography are synonymous.

Holocene The last epoch in geological time (figure 24, p 109).

Holotype The specimen on which a named species is based.

Homoplasy Structural resemblance due to parallel, convergent, or reversed evolution rather than common ancestry.

Hybrid An individual formed by hybridisation between unlike forms, usually species.

Hydro-isostasy Deformation of continental shelves due to the weight of overlying water.

Introgression The permanent incorporation of genes from one set of differentiated populations into another.

Isopangeneric contours Contours of equal generic diversity.

Isotherm Of equal temperature.

Isozyme One of several forms of an enzyme produced by different, nonallelic, loci in an individual organism's genome. (A subset of an allozyme.)

Jurassic A period in geological time (figure 24, p 109).

Karyotype The structural characteristics of the chromosomes.

Lithosphere The solid part of the earth including the crust and upper mantle.

Location See 'geographic terminology'.

Ma Million years ago.

Macroevolution A vague term for observable evolutionary change among species and higher taxon levels.

Marsden square A unit of area commonly used in oceanography, being $1°$ latitude $\times$ $1°$ longitude.

Mass extinction An extinction that is characterised by loss of many taxa in a geologically brief time period.

Meandroid Massive corals that have corallite mouths aligned in valleys such that there are no individual polyps.

Mesozoic An era in geological time (figure 24, p 109).

Metaspecies A conceptual taxon that is genetically isolated in evolutionary time, but which consists of species that are capable of hybridising (p 228).

Microenvironment The immediate environment of an individual coral colony.

Microevolution A vague term for slight evolutionary change within species.

Microhabitat A vague word indicating the particular type of habitat occupied by a coral colony.

Migration Large-scale movement of a population. Synonymous with dispersion except implying an activity specific in time or space.

Milankovitch cycles Cycles of orbital forcing (pp 122, 128).

Minimum spanning tree A network algorithm that is specified by forming a complete linkage (joining all points or objects) where the total length of the connections is minimal and where no loops or circuits occur.

Miocene An epoch in geological time (figure 24, p 109).

Mitochondrial DNA DNA contained in mitochondria and therefore additional to nuclear DNA. Mitochondrial DNA is maternally inherited and especially useful for molecular systematics because it has a high mutation rate.

Molecular clock The hypothesis that the rate of evolutionary change in DNA is constant over geologically long periods of time and thus appropriate DNA techniques can determine points in time of phylogenetic divergence.

Monophyletic Derived from the same ancestral taxon.

Morphometric Using measurement of morphological (usually skeletal) characters.

Multivariate Using more than a single attribute (variable).

Mutation A vague term for processes that cause a change in a nucleotide sequence in an organism.

My Million years.

Natural selection The non-random and differential reproduction of different genotypes acting to preserve favourable variants and to eliminate less favourable variants; viewed as the creative force that directs the course of evolution by preserving those variants or traits best adapted in the face of natural competition (Lincoln et al 1982).

Neogene A period in geological time (figure 24, p 109).

Neontology The study of living organisms (cf palaeontology).

Neoplasm Cancerous growths commonly found on corals.

Nomenclatorial priority Priority of name, usually based on the oldest name.

Nominal species Species that exist in name only. These are usually synonymised with operational species.

Nucleotides A subunit of DNA or RNA molecules. The sequence of nucleotide codes for the structure of proteins synthesised in cells.

Oligocene An epoch in geological time (figure 24, p 109).

Operational species Described species that are actually used in taxonomy or in other fields (cf nominal species).

Ordination A general term for techniques that attempt to condense information associated with a set of attributes to a limited number of new attributes.

Palaeocene An epoch in geological time (figure 24, p 109).

Palaeogene A period in geological time (figure 24, p 109).

Palaeontology The study of past life (cf 'neontology').

Panmictic Continuous populations or races within which interbreeding is random.

Panbiogeography A method of determining biogeographic patterns (p 39).

Parapatric populations Populations occupying separate but adjoining areas, such that only a small fraction of individuals in each encounters the other (figure 12, p 43).

Parapatric speciation Speciation between parapatric populations, initiated by both spatial segregation and spatial differentiation, leading to the evolution of isolating mechanisms.). See 'concepts of species — terminology issues'.

Parsimony The use of the shortest number of evolutionary steps as a criterion for constructing a cladogram.

Paucispecific Having a low number of species.

PCR See 'polymerised chain reaction'.

Peripatric populations Populations of very different sizes occupying separate but adjoining areas, such that only a small fraction of individuals in each encounters the other (figure 12, p 43).

Peripatric speciation the dispersal of a few propagules across a pre-existing barrier (founder effect) and subsequent differentiation into a new taxon.

Phaceloid corals Corals that have corallites adjoined only towards their base.

Phenetic species concept Species are defined by morphological characters that are considered sufficient to warrant species status.

Phenotype The sum total of observable structural and functional properties of an organism; the product of the interaction between the genotype and the environment.

Philopatry A tendency to remain in the native locality because of limited capacity to disperse.

Photoadaptation Adaptation of metabolic processes to changing light levels.

Phototrophic corals Corals that obtain metabolic energy from sunlight by photosynthesis of their symbiotic zooxanthellae.

Phylogeny The evolutionary history of a group or lineage.

Phylogenesis The evolutionary history of a taxon.

Phylogenetics The description of evolutionary relationships using cladistic methods.

Phylogeography The determination of geographic distribution of phylogenies.

Planulae Larvae of coral.

Pleistocene An epoch in geological time (figure 24, p 109).

Pliocene An epoch in geological time (figure 24, p 109).

Plocoid coral Massive corals that have corallites with separate walls (cf cerioid corals).

Polymerase chain reaction (PCR) A technique where specific segments of DNA are amplified using specific primers in repeated rounds of replication of the PCR. Only minute amounts of original material are needed and the product can be used for RFLP analysis, sequencing, or other procedures.

Polymorphism The term has two meanings. Throughout this book it is used in reference to morphology. Genetic polymorphisms (as in RFLPs) have no implications for morphological polymorphism.

Polyp An individual coral including soft tissues and skeleton.

Polyphyly Derived from more than one ancestral taxon.

Polyploid Possessing more than two entire chromosome complements.

Population A group of conspecific organisms that exhibit reproductive continuity. It is generally presumed that ecological and reproductive interactions are more frequent among members within a population than with members of other populations.

Primitive characters Characters found in many taxa that have been retained from a common ancestor.

Province See 'geographic terminology'.

Pseudo-extinction Apparent extinction of a clade, but where a significant proportion of the genetic composition of the clade has been retained in another clade.

Pseudo-speciation Apparent origination of a clade, but where a significant proportion of the genetic composition of the clade occurs in another clade.

Quaternary A period in geological time (figure 24, p 109.

Race A vague term for a group of populations within a species that differ in the composition of their gene pools, and usually in their genetically determined phenotypic characters, from other conspecific populations.

Radioimmunoassay An immunological assay that recognises major antigenic sites and can be used to study very small quantities of proteins, including those from fossils.

Rafting The transport of biota on floating objects.

Realm See 'geographic terminology'.

Red Queen Hypothesis That each evolutionary advance by any one species represents a deterioration of the environment of other species so that each species must evolve as fast as it can merely to survive.

Reef flat The flat intertidal part of reefs that are exposed to wave action.

Reef slope The sloping part of reefs below the reef flat.

Region See 'geographic terminology'.

Reinforcement The evolution of prezygotic mating barriers in hybrid zones.

Relict species Species that occur in an isolated region after the species has become extinct over its former, larger, range.

Restriction endonuclease Enzymes that cleave double-stranded DNA at a constant position within a specific recognition sequence, typically 4–6 base pairs long.

Restriction fragment length polymorphisms A genetic polymorphism in an individual, population or species, defined by restriction fragments of a distinctive length caused by gain or loss of a restriction site or insertion or deletion of a length of DNA between restriction sites.

Reticulate evolution Evolution dominated by sequential division and hybridisation of clades (p 216).

RFLP See 'restriction fragment length polymorphisms'.

Rudists Mesozoic bivalves that dominated reefs throughout much of the Cretaceous and became extinct at the close of the Cretaceous.

Rugose corals A major group of non-scleractinian corals that became extinct at the close of the Palaeozoic era.

Saturation The term has different meanings in genetics and ecology. In this book it is used in a genetic context to mean 'saturated with respect to biodiversity', not ecologically to mean 'saturated with respect to space'.

Scleractinian corals Most 'hard' corals are Scleractinia (other orders may, however, have superficially similar colonial forms).

Sibling species Similar species that are the product of relatively recent speciation that may or may not be allopatric (see 'geminate species').

Site See 'geographic terminology'.

Skeletogenesis The evolution of skeletons.

Soft coral General term for askeletal Anthozoa.

Species See p 3 and chapter 13.

Stasipatric speciation The formation of a new species as a result of chromosomal rearrangements giving homozygotes that are adaptively superior in a particular part of the geographic range of the ancestral species.

Stepping-stones Points along the dispersion path of corals such that dispersal can occur from one point to the next.

Stromatolites Mounds of limestone formed by the growth of blue-green algae. Common in the Precambrian and still extant.

Subspecies See 'geographic subspecies'.

Surface circulation vicariance See p 212.

Symbiosis The close association of two organisms where there is substantial mutual benefit.

Sympatric populations Populations that encounter one another with 'moderate' frequency. Such populations may be ecologically segregated, or may breed in different seasons, or have genetic isolating mechanisms (p 65).

Sympatric speciation Speciation between sympatric populations. See 'concepts of species — terminology issues'.

Syngameon A complex of species that can interbreed. Syngameons in plants have the characteristics of well-isolated 'biological' species at their outer boundary, but differ in having a more complex internal structure.

Synonymy The list of names considered by a taxonomist to apply to a given taxon other than the name by which the taxon should be known.

Systematics The study of evolutionary and genetic relationships of organisms.

Taxon A taxonomic unit. Taxa are arranged in hierarchies of taxonomic levels.

Taxonomy The naming and classification of organisms.

Tectono-eustacy See 'eustatic'.

Thermohaline Cold, with a high salt content.

Teleplanic larvae Pelagic larvae that have a protracted planktonic existence.

Tertiary A period in geological time (figure 24, p 109).

Tethys Sea The ancient tropical sea that once connected the Indian and Atlantic Oceans (figures 26 to 30, p 111 to 122).

Transilience Fixation of a small part of the founder genome by the overcoming of isolating barriers.

Triassic A period in geological time (figure 24, p 109).

Type locality The locality from which a type specimen was collected.

Type specimen The specimen on which a nominal (named) species is based.

Upwelling The upward movement of cold, deep, usually nutrient-rich water up continental slopes.

Van Valen's law The probability of extinction within any group remains constant through time.

Vicariance The process that occurs when a formerly continuous population is divided by a barrier and evolves into two or more species (figure 10, p 40).

Vicariance biogeography Historical biogeography that assumes that geographical distributions of organisms mostly result from vicariance processes (p 40).

Vicariant species Ecologically similar, but geographically separated species, or ecologically different species occupying the same area, or closely related species occupying adjacent areas separated by a barrier. Vicariant species arise by the historical process of vicariance.

Viviparous Producing live offspring from within the body of the parent.

Zoochlorellae Small green algae or flagellate Protozoa that live symbiotically in freshwater protozoans and invertebrates.

Zooxanthellate corals and larvae Corals and larvae that have photosynthetic endosymbiotic algae. In this book, 'zooxanthellate' is used as a non-specific term that is applicable to all marine invertebrates, extant or extinct (see 'hermatypic').

ENDNOTES

Chapter 2

1 Henry Bernard, the most prolific coral taxonomist of his time, expressed 'no regrets' at never having seen a living coral. He abandoned taxonomy and became a minister of religion.

2 Most species are composites of a spectrum of environment-correlated and genetically determined morphological variants. These 'ecomorphs' are usually readily recognisable *in situ* and become as familiar to the coral biologist as their terrestrial plant equivalents are to botanists. Figure 4 illustrates six ecomorphs of *Pocillopora damicornis*.

3 Times of genetic divergence may be estimated if one assumes that a 'molecular clock' ticks at a particular rate for a particular segment of DNA in a particular taxon. These rates may vary episodically as well as among genes and among taxa. Patterns of branching (reflecting systematic relationships) and branch lengths (reflecting rates of divergence) are not necessarily closely related.

4 In the present context, the focus of interest is on the separation of species. The composition and behaviour of populations within a given reef area is relevant to this, but marginally so. Likewise, higher taxonomic affinities, such as are determined from comparisons of highly conserved sequences, are more interesting in the context of higher-level systematics and the fossil record than taxonomy. What is of interest in the present context is the genetic distance between gene pools, especially if these are associated with observable non-genetic attributes that allow them to be used in operational taxonomy.

5 Conflicts between taxonomic methods have already been the subject of study; others are waiting to be studied (in the light, the author hopes, of the concluding chapters of this book). Most comparisons of taxonomic methods to date are based on *identifications* of different source material. It is essential that different taxonomic methods, if they are compared, be applied to the same source material so that supposed comparisons are not interpreted through identifications that can never be more than matters of opinion.

Chapter 3

1   The Darwinian term 'natural selection' has been given many different meanings by different authors. The author follows that of Lincoln et al (1982) (p 271).

2   This is not necessarily so: the time-frame of continental drift matches that of higher taxa divergence; species may well have evolved in centres that no longer exist (eg the Tethys Sea).

3   Larval dispersion paths due to surface circulation patterns are, in principle, comparable to Croizat's tracts.

4   Classical vicariance is a term created in this book to distinguish vicariant speciation from the process of surface circulation vicariance, which is a balanced mixture of speciation and hybridisation (chapter 12, p 211).

5   Generic contours of diversity are sometimes called isopangeneric contours.

6   The distribution range of a species must include its type locality, the place where the species was originally described.

7   Electronic copies of the author's primary distribution data have long been available for other users; updated versions, which will include reference data, should be available soon after publication of this book. The differing status of these data sets should be noted before they are used in numerical analyses (p 8).

8   In biogeography, most observations appear *originally* to have had special relevance to specific situations (eg Darwin's finches in the Galapagos Islands). Reduction of these observations into general theory (eg Darwin's biogeography subsequent to the finches) stimulates thought, but it often diminishes the applicability of the original observation.

Chapter 4

1   Exceptions are in some experimental situations, eg as with *Drosophila*.

2   All changes to DNA are due to mutations. These can be fixed in populations by genetic drift or natural selection; the same two mechanisms can also prevent the fixation of a mutation.

3   Habitat and niche diversity: the former is primarily environment-determined, the latter determined by a combination of habitat and species characteristics.

4   The term 'saturation' can have different meanings in genetics and ecology: it is used here to mean 'saturated with respect to biodiversity', not ecologically to mean 'saturated with respect to space'.

Chapter 5

1   Polyploidy would be strongly selected for in reticulate evolution where species have not been separated by natural selection.

2   They would, however, be unlikely to closely resemble naturally occurring equivalents as aquarium-raised corals usually develop abnormal morphological characteristics.

3   Subsequent decrease in species diversity at Tateyama is due to land reclamation.

Chapter 7

1   This database is not reproduced in this book other than in summary diagrams (figure 25 and the histograms of chapter 8) because of its length and inadequacies. It is used here, in the absence of any alternative, to elucidate trends and principals, not taxonomic detail.

2   If the first 'scleractinians' were the first anemones or corallimorphs to secrete skeletons, then the earliest record may be as old as the Ordovician (500 to 440 million years ago). A corallum extraordinarily like *Fungiacyathus* has recently been found in the Ordovician of Scotland (Scrutton and Clarkson 1991). Unfortunately the composition of the skeleton cannot be determined (C Scrutton pers comm).

3   There is no reason to suppose that Mesozoic corals did have 'zooxanthellae' as their symbionts unless all taxonomic meaning is dropped from the term 'zooxanthellae' (p 97).

4   This could well be a function of preservation, especially where there have been active subduction zones.

5   Milankovitch cycles are generated, respectively, by precession variation of equinoxes (19 thousand years and 23 thousand years); oscillation of the earth's axis (41 thousand years); and eccentricities of the earth's orbit (100 thousand years) (Imbrie and Imbrie 1979; Bradley 1985). These cycles are believed to have paced Quaternary glaciations through their control of seasonal and latitudinal distribution of incident solar radiation (Bartlein and Prentice 1989; Imbrie et al 1992). In one form or another, they have existed throughout geological time (Fischer 1986), although their presence becomes increasingly difficult to detect in earlier geological intervals (eg Park and Herbert 1987).

6   Atmospheric carbon dioxide concentrations were about 200 ppm during glacial advances and about 280 ppm during inter-glacial intervals (Delmas 1992). The significance of these differences is central to interactions among the components of the global carbon cycle. Reefs are a component of this cycle, playing a role in the many ways in which carbon is stored (organically and inorganically), translocated, and released, a large part of which is through changing physical and biological properties of the oceans.

Chapter 8

1   There are unresolved taxonomic issues with Miocene *Acanthastrea*, especially in relation to now-living *Mussismilia* of Brazil.

2   Summary diagrams involve different admixtures of total *numbers* of genera in the categories indicated and genera *identifications*, that is, admixtures of simple statistics and taxonomic decisions. Details of the latter are outside the scope of this book.

3   This varies regionally (total number of genera decrease from the Eocene to Recent in the Caribbean) and is affected by taxonomic decisions (footnote above).

4   Several studies, notably the combined work of G Stanley and L Beauvais, which give details of longevities of genera within specific regions and/or specific periods within the Triassic and Jurassic, are probably not open to this criticism.

5   There are varying degrees of certainty associated with identifications, depending on state of preservation and differences between Era bed fossils and now-living equivalents; available details are in Veron and Kelley (1988).

6   The importance of these fossils has long been recognised by the Japanese: part of the deposit is now enclosed in a monument.

7   There have since been widespread subsequent extinctions due to land reclamation.

8   A term taken from Lewis Carroll's *Through the Looking Glass*: The Red Queen said to Alice, 'Now here, you see, it takes all the running you can do, to keep in the same place'.

Chapter 9

1   Different fusion strategies do not substantially influence this pattern.

2   The diversity contours of figures 46, 48 and 50 are generated from *distribution ranges* of taxa, not *records,* a procedure made possible by GIS computer programs. Ranges are used in preference to records as less error is involved in assuming that taxa occur throughout their range than assuming distribution records are complete (discussed by Belasky 1992). This should be remembered when comparing figure 48 with its predecessors (figure 13, p 46), which are all based on records, not ranges. The absence of the Chagos stricture throughout figure 13 is an artefact of hand contour drawing. There are two principal sources of error in figure 50 as these contours were not generated from actual species distribution ranges. These sources of error are: first, the assumption that species within a genus have the same pattern of relative diversity as genera have. This source of error can only be offset by comprehensive inter-regional studies; second, the assumption, in calibrating contours (legend of figure 50), that the species diversity of remote regions can be 'ground-truthed' from

observations in the central Indo-Pacific. This source of error appears to be minor as ground-truthing based *independently* on species data from the Caribbean/Gulf of Mexico and each of the three central Indo-Pacific continental coastlines (figure 54, p 172), produces only minor differences in the pattern of contours. The addition of a 450 species contour in figure 50 is a numerical correction for the absence of a 75 genera contour in figure 48.

3   There are orders-of-magnitude differences in the quantity and quality of data required for inter-regional taxonomy, intra-regional taxonomy, and intra-regional estimates of species diversity. Taxonomic uncertainties, however, need not necessarily result in poor intra-regional estimates of species diversity.

4   These studies and observations necessarily override consideration of geographic variation within species (chapter 13, p 225). The taxonomy, being primarily Australian, favours central Indo-Pacific species names.

5   J Maragos has long provided the author with collections of Hawaiian corals for this study: taxonomic decisions about them are joint ones.

6   S Blake, G Paulay and J Pandolfi have contributed substantial collections from Henderson, Ducie and Pitcairn Islands to this study.

7   P Glynn and his colleagues have long contributed collections from the far eastern Pacific to this study.

8   Unpublished study with P Glynn.

9   These are superficial observations only and are unsupported by publications.

10 The three species of *Mussismilia* contain an unusual amount of inter-specific variation, suggesting a need of taxonomic revision. Two of the three species could be placed in the genus *Acanthastrea*.

11 Satellite data used in this study were from the Mundocart dataset, divided into 251 grid squares (5° latitude × 5° longitude). Data compilation was undertaken courtesy of the World Conservation Monitoring Centre, Cambridge.

Chapter 10

1   The fourth sequence occurs along the east coast of Africa. Other sequences (the east coast of Yemen and Oman and the west coast of India) are smaller, with latitudinal distribution limited by bathymetry.

2   Thirty-seven per cent of the world's reefs, as determined from satellite data (p 169), are contained within the 70 genera contour of figure 48, p 158.

3   Branches of dendrograms can be rotated at their nodes. These dendrograms are measures of dissimilarity between branches, not geographic sequences among them.

4   Many computer packages allow deeper analyses of distribution data and correspondence between distribution and environmental data than those presented in this book. Exploratory analyses, the outcome of which could be partly dependent on particular algorithms, have not been included here, but may nevertheless be valid and useful.

5   For these reasons, dendrograms, rather than two-dimensional ordinations (as is figure 59, p 182), are used to indicate geographic relationships.

6   A major *Acanthaster* outbreak was observed by the author in 1981, a population that is genetically distinct from that of Lord Howe Island (Benzie 1992).

7   A reasonable hypothesis might predict that, in time, all localities within a subsequence will have the same species complement, provided all species have adequate dispersion capability.

8   Presence/absence differences between offshore and onshore locations within the Great Barrier Reef have not been determined and thus are not separated in figures 55 and 58. By subjective evaluation, differences between species of the offshore reefs of the Great Barrier Reef and those of the North-west Shelf are even less than indicated by figure 58.

9   Of the fifteen groups of *Acropora* defined by Veron and Wallace (1984) (figure 77, p 244), the *A. palifera* and *A. echinata* groups of western Australia are largely restricted

to North-west Shelf reefs where colonies are similar in abundance and appearance to those of the Great Barrier Reef. Three common members of the *A. humilis* group (*A. humilis, A. gemmifera* and *A. monticulosa*) are similarly restricted or absent.

10 Torres Strait and the Arafura Sea to the west have either been emergent or have had a sedimentary environment hostile to corals since the Plio-Pleistocene at least. Torres Strait has probably never acted as a significant dispersion corridor.

11 Vanuatu is the only location inside the Indo-Pacific centre of diversity (but outside the three latitudinal diversity sequences), for which presence/absence data are comprehensive (Veron 1990b).

12 Figure 55 also shows a difference (of nineteen species) between the central and northern Great Barrier Reef: this may be sampling error, as the central region has received much more study than the remote north.

13 For example, *Turbinaria peltata* and *T. patula* are common in high-latitude locations and inshore tropical locations, but uncommon offshore in the tropics. All *Acanthastrea* except *A. echinata* are very rare on reefs, but all are found in inshore tropical locations and are relatively common in high latitudes. The same applies to *Acropora glauca, A. lovelli, A. solitaryensis* and *A. tortuosa.*

14 All but seven species recorded at Lord Howe Island have been recorded from Elizabeth/Middleton Reef.

15 *Fungia* has been recorded from a single specimen, almost certainly a recent immigrant.

Chapter 11

1 Molecular methods should greatly enhance our understanding of biogeographic variation within the province, but they will not provide final solutions. As with morphological taxonomy, results will be matters of interpretation and assessment in the light of other data (eg about reproductive compatibilities). Both morphological and molecular methods can only indicate delineations within continuous biogeographic and taxonomic variation (figure 74, p 228).

2 Issues that arise when taxonomic complexities are *not* set aside are discussed in the next chapter.

3 The bimodal pattern of figure 62 is primarily determined by bathymetry, not biogeographic attributes of species.

4 This has not been statistically tested.

5 Regions where contours of species richness are concentrated (figure 50, p 160) indicate common distribution boundaries, but not dissimilarities between regions.

6 *Madracis decactis, Stephanocoenia michelini, Agaricia agaricites, A. fragilis, Porites astreoides, Montastrea cavernosa, Scolymia* sp.

7 For example, *Stylocoeniella cocosensis, Australomussa rowleyensis, Echinopora ashmorensis* and *Boninastrea boninensis* were all named after places they were thought (by the author in the first three cases) to be endemic, but they were later discovered in another country.

8 *Ctenella, Erythrastrea, Indophyllia, Mussismilia, Parasimplastrea* and *Simplastrea.*

9 *Coscinaraea mcneilli, C. marshae, Symphyllia wilsoni* and an undescribed *Turbinaria.* To these may be added an undescribed *Montipora* of Lord Howe Island and three undescribed *Montipora* and an undescribed *Acropora* of the Houtman Abrolhos Islands region. Of these undescribed species, it is possible to be confident about the endemicity and/or taxonomy only of the first three *Montipora.*

10 *Acropora tanegashimensis, A. pruinosa, A. tumida,* an undescribed *Acropora, Goniopora cellulosa, Alveopora japonica, Hyndophora bonsai, Goniastrea deformis* and *Euphyllia paraglabrescens.* Of these species, it is possible to be taxonomically confident about the described *Acropora,* the *Alveopora* and the *Euphyllia.* Two of these species (the first and last) are known from a very restricted area (Tanegashima, Japan); the others have relatively wide ranges.

11 An undescribed *Montipora,* five undescribed *Acropora* as well as *A. azurea* and *A. cardinae,* an undescribed *Goniopora* and *Porites myrmidonensis.* All except the last are rare.

12 *Montigyra kenti, Turbinaria conspicua* and undescribed species of *Porites, Goniopora* (three species), *Favia, Favites, Goniastrea, Leptastrea* and *Cyphastrea.*

13 *Montipora setosa, M. confusa, M. orientalis, M. florida, Acropora magnifica, Porites cumulatus, Pachyseris foliosa, Galaxea alta, Oxypora crassispinosa, Euphyllia paradivisa, Plerogyra turbida* and *Physogyra exerta.* The number increases as the geographic area increases: for example; four species are added if the Yaeyama Islands are included.

14 An undescribed *Acropora, A. sekiseiensis, Favites stylifera* and *Platygyra yaeyamensis.*

15 A *Montipora,* two *Acropora,* a *Porites* and a *Pavona,* all apparently undescribed. This represents 5 per cent of the atoll's total species complement.

16 Of *Agaricia, Cladocora, Diploria, Favia,* zooxanthellate *Madracis, Montastrea, Porites* and *Siderastrea.*

17 Bernard (1896–1905) provides the best examples (p 19).

18 For example, *Acropora grandis, A. microclados, Montipora incrassata, M. efflorescens* and *Astreopora explanata* all display major geographic variations in parts of their central Indo-Pacific range. *Acropora grandis* and *A. florida* have subtle variations in some tropical regions and major ones in others. *Acropora aspera* has subtle variations in the tropics which are distinct from more major ones in high latitudes. Minor geographic variations within the central Indo-Pacific, from which the above examples are drawn, may become major over greater ranges. Veron (1990a) illustrates this in *Acanthastrea ishigakiensis.*

19 For example, *Montipora capitata* consistently has thicker branches at Vanuatu than in the Philippines or the Ryukyu Islands; *Acropora danai* forms more compact colonies in the northern hemisphere than in the southern.

20 For example, *Caulastrea tumida* has growth forms ranging from phaceloid to plocoid and *Blastomussa wellsi* has growth forms ranging from phaceloid to cerioid depending on both region and environment. Similar variations occur in *Psammocora contigua, Leptoseris mycetoseroides* and *Pavona maldivensis.*

21 For example, *Seriatopora hystrix* has an ecomorph found only on the reef flats of North-west Shelf reefs of western Australia. Similar occurrences occur in *Montipora angulata, Acropora digitifera, A. anthocercis* and *Pavona decussata. Pocillopora damicornis* has ecomorphs associated with shallow turbid habitats on the Great Barrier Reef, but these are absent in similar environments of western Australia.

22 For example, *Pocillopora verrucosa* shows no unusual morphological characteristics at Tanegashima (Japan), the northern limit of its distribution range. Similar observations are applicable to Philippine species that occur as far north as the Yaeyama Islands

23 For example, there are few intra-specific differences between species of *Montipora* common to the North-west Shelf reefs of western Australia and the Great Barrier Reef. The same applies to comparisons between the *Montipora* of the Philippines and southern Ryukyu Islands. Similar examples can be found with most major genera, especially where comparisons are restricted to the tropics.

24 For example, *Montipora foveolata* is less easily distinguished from *M. venosa* in Japan than it is on the Great Barrier Reef. *Acropora dendrum* is readily confused with *A. valida* in mainland Japan whereas elsewhere these species are usually clearly distinct. *Leptastrea transversa* and *L. purpurea* are clearly distinct in some regions and readily confused in others. Likewise, as originally observed by Chevalier (1975), *Echinopora* species are generally more distinct in some parts of their range, for example, western Indian Ocean, than in others, for example, the south Pacific. Conversely, differences between *Pocillopora verrucosa* and *P. meandrina* appear to be constant throughout their overlapping ranges.

25 For example, *Cyphastrea decadia* and some species with disjunct populations (p 193) have non-contiguous geographic subspecies.

26 For example, *Anacropora puertogalerae, Acropora divaricata, A. valenciennesi, A. horrida, A. paniculata, A. grandis, A. acuminata, A. florida, Pavona clavus, Scapophyllia cylindrica, Goniastrea australensis* and *Turbinaria heronensis* (as well as the species listed

immediately above) are distinct species over most of their ranges within the central Indo-Pacific, but in some regions this identification becomes arbitrary.

27 For example, *Acropora aspera* in mainland Japan forms a morphologically distinct geographic subspecies that also has distinct colours; *Acropora palifera* commonly develops incipient axial corallites at Lord Howe Island, but not elsewhere; *Stylophora pistillata* has similar latitude-correlated variations in morphology in both south-east and south-west Australia.

28 Many species, but for example, *Leptoseris mycetoseroides, Pachyseris rugosa, Galaxea fascicularis* and *Caulastrea tumida,* display extreme environment-correlated variations. *Diaseris distorta, Hydnophora rigida* and *H. exesa* have specific habitat-correlated variations in particular locations.

29 For example, *Montastrea valenciennesi* has larger calices in most northern regions than in the south.

30 For example, *Montipora capricornis* from Australia and Vanuatu can be distinguished by details of skeletal microstructure, but not by gross morphology. Skeletal details of *Palauastrea ramosa* do not vary geographically, but some characters normally found in Philippine and Ryukyu coralla are rare in Great Barrier Reef ones. This is partly, but not completely, correlated with habitat differences.

31 For example, most species at the Houtman Abrolhos Islands and in some regions of the Philippines have very low levels of skeletal density, probably due to high nutrient levels and unusually rapid growth.

32 For example, *Montipora hispida* has an unusually low skeletal density in areas of mainland Japan where most species have a high density.

33 For example, *Acanthastrea lordhowensis* (illustrated Veron 1986, pp 410–11).

34 For example, *Pocillopora verrucosa* and *Oulastrea crispata* both have stable skeletal pigments throughout very wide geographic ranges; *Acropora acuminata* also has stable skeletal pigments throughout the distribution range except in Japan (where additional characters indicate a distinct geographic subspecies).

35 For example, *Turbinaria reniformis* can be immediately distinguished *in situ* from other very similar species of *Turbinaria* by its distinctive yellow-green coloration. *Acropora valida* and *A. secale* both have a distinctive colour morph (illustrated Veron 1986 pp 185 and 186, respectively) that is identical and that occurs widely throughout the distribution ranges of both species. *Catalaphyllia jardeni* has a very distinctive, complex colour pattern throughout its distribution range, even in a very isolated population in mainland Japan. Most other caryophylliids have similarly conserved colours.

36 For example, *Mycedium elephantotus* characteristically has a very wide range of colours, but is usually dark green in Tanegashima and mainland Japan; *Australomussa rowleyensis* has a wide range of colours in Thailand, but is always green and/or grey in western Australia; *Acanthastrea hillae* has a wide range of colours over most of its distribution range, but is creamy green or brown in western Australia.

37 *Blastomussa wellsi* is greenish or red wherever it occurs; *Acropora pruinosa* is always greenish or brown at the Amakusa Islands, but has additional colours elsewhere.

38 The '*echinata* group' provides the best examples. *Acropora aculeus* is usually yellow on the Great Barrier Reef and in the Philippines, pale brown in western Australia; *A. echinata* is greenish grey in the Ryukyu Islands, cream with blue or purple branchlet tips on the Great Barrier Reef and in Vanuatu; *A. rosaria* is pale brown or pinkish cream in the Ryukyu Islands, deep blue on the Great Barrier Reef and in southern Papua New Guinea. In other *Acropora, A. microclados* is pink on the Great Barrier Reef, usually grey in Vanuatu, cream or yellow in the Philippines, grey or brown in the Ryukyu Islands; *Acropora nana* is commonly bright blue in the Ryukyu Islands, brown, blue-grey or cream at Tanegashima, Japan.

39 Examples of region-specific colours: *Psammocora profundacella* is usually grey or creamy yellow, sometimes pink or blue in the Ryukyu Islands, commonly pink at Tanegashima (Japan), mostly dark brown or green in mainland Japan, green or pale

pink at Cocos (Keeling) Atoll; *Fungia scutaria* is commonly bright green in the Ryukyu Islands, uniform pale brown at Tanegashima (Japan), brown on the Great Barrier Reef, cream with white tentacular lobes or occasionally pink at Cocos (Keeling) Atoll; *Blastomussa merleti* is usually greenish in Australia, red, greenish or brown in Japan.

40 For example, *Alveopora spongiosa* is usually chocolate brown, but often green in the Ryukyu Islands.

41 For example, *Pocillopora damicornis* is pale to dark brown or green throughout its range, is sometimes pink in the Ryukyu Islands and Tanegashima (Japan), but has dark colours in mainland Japan; *Pavona decussata* is usually yellowish brown, but is chocolate brown with green tentacles in mainland Japan; *Acanthastrea hemprichii* is orange-yellow in the Ryukyu Islands, but is brownish grey, dark green or dark red further north; *Oxypora lacera* is usually pale grey or brown in the Ryukyu Islands, yellowish-brown to creamy grey at Tanegashima (Japan), but bright green, sometimes red, in mainland Japan; *Hyndophora exesa* is usually pale coloured in the Ryukyu Islands, but may be green, red or brown in mainland Japan.

42 For example, *Favia danae* is usually mottled in the tropics but dark in high latitudes of both Australia and Japan; *Favia lizardensis* is pinkish brown with cream or green oral discs on the Great Barrier Reef and in Japan, commonly uniform grey in tropical western Australia, brown at the Houtman Abrolhos Islands; *Favites abdita* is usually honey-coloured or cream in the tropics, but dark grey, green or brown in high-latitude eastern and western Australia and Japan; *Goniastrea pectinata* is usually pinkish brown in the tropics, but may have dark colours in mainland Japan.

43 *Goniastrea retiformis* is usually pale orange-brown throughout the tropics, but in mainland Japan may become bright green, specifically where light appears to be limiting.

44 For example, *Acropora florida* in the Houtman Abrolhos Islands is morphologically distinct, and has distinct colours, compared with colonies of this species further north; *Symphyllia radians* occurs at Tanegashima (Japan) as small, encrusting colonies, which are a distinctive dark red colour; *S. valenciennesi* forms a geographic subspecies in mainland Japan which has a wider range of colours than elsewhere; *Scapophyllia cylindrica* forms a non-columnar geographic subspecies that is a distinctive pale grey colour in the Ryukyu Islands; *Favia rotundata* from the Ryukyu Islands has slightly smaller corallites than those of the Great Barrier Reef and this difference coincides with a minor but constant colour difference; *Favites pentagona* has a wide range of mottled colours throughout the tropics, but is particularly colourful in mainland Japan where it has a relatively common encrusting growth form.

45 For example, *Montipora mollis, M. caliculata, M. hispida, Porites aranetai, P. vaughani, Coscinaraea wellsi, Podabacia motuporensis, Symphyllia agaricia* and *Platygyra ryukyuensis*.

46 For example, *Montipora peltiformis, M. nodosa, Psammocora profundacella* and *Leptastrea transversa*.

47 For example, *Montipora floweri, M. undata, Acropora acuminata, A. kirstyae, A. aculeus, Alveopora catali, A. fenestrata, Leptoseris scabra, Echinophyllia echinata, E. echinoporoides, Goniastrea favulus, Leptastrea bewickensis, Cyphastrea ocellina* and *Euphyllia divisa*.

48 At least fifty-two species are cited by Veron (1993), but this appears to be true, to varying extents, for about 20 per cent of all species.

49 For example, *Montipora verrucosa, M. efflorescens, M. grisea, Acropora gemmifera, A. formosa, Psammocora contigua, P. superficialis, P. digitata, Pavona cactus, Leptoseris mycetoseroides* and *Goniastrea retiformis*.

50 *Acropora longicyathus* and *A. subglabra* are limited to protected lagoons of North-west Shelf reefs of western Australia, but are common in many environments elsewhere.

51 The abundance of *Porites compressa* at Kaneohe Bay, Hawaii, which dominates all shallow reef flats, has few equals of species dominance or abundance. At Cocos (Keeling) Atoll, *Pachyseris speciosa* and *Turbinaria reniformis* both occur in great abundance in one locality and nowhere else.

52 For example, *Acropora aspera* and *A. hyacinthus* are both common on the western Australian coast, but are uncommon at Cocos (Keeling) Atoll.

53 Of thirty-eight species that occur in the Yaeyama Group of the Ryukyu Islands but not further north, three are common, ten are uncommon and twenty-five are rare. The ratio of these abundance categories throughout all Japan is 129:151:119 (Veron 1992c).

54 For example, *Montipora capricornis, Acropora abrolhosensis, A. lovelli, A. glauca, A. tortuosa, A. stoddarti, A. striata* and *Turbinaria patula.*

55 For example, *Acropora humilis, Leptoseris yabei, Fungia repanda* and *Favia pallida.*

56 For example, *Acropora danai, A. horrida, A. pulchra* and *Oxypora glabra,* but there are many others involving different levels of taxonomic uncertainty.

57 Large colonies of some species have been recorded in regions where the species is relatively rare: for example, *Astreopora moretonensis* at Lord Howe Island; *Porites lutea* at the Houtman Abrolhos Islands; *Pavona clavus* at Vanuatu and Ryukyu Islands; *Echinopora mammiformis* at the Ryukyu Islands.

58 For example, *Acropora yongei* and *Pocillopora damicornis* both form monospecific stands at Lord Howe Island and Rottnest Island on the south-east and south-west Australian coasts (respectively).

59 For example, exceptionally large colonies of *Astreopora moretonensis* (1.5 m diameter), *Alveopora allingi* (8 m diameter), *Acanthastrea hillae* (1.5 m diameter) and *Turbinaria peltata* (3.5 m diameter) occur at Lord Howe Island and exceptionally large colonies of *Turbinaria mesenterina* (3 m diameter) occur in far south-west Australia. A *Plesiastrea versipora* colony 3.1 m in diameter has been recorded at St Vincents Gulf, far southern Australia (Howchin 1909)

60 All but approximately ten species (of which *Acrhelia horrescens* is the only conspicuous example) of the Great Barrier Reef have been recorded on both offshore platform reefs and inshore fringing reefs; nearly one-third also occupy coastal habitats with very high turbidity (figure60, p 182).

61 This high number, dominated by *Acropora* species, includes non-reefal high-latitude occurrences and all inter-reef and soft-bottom records; the remainder (33 per cent of species) have never been recorded in the literature, or observed by the author, in non-reef habitats.

62 Only a few offshore *Acropora* (eg *A. nana*) and inshore *Montipora* (eg *M. digitata*) are always found in shallow habitats. Many more species (eg the *Acropora humilis* group and *Goniastrea* species) are *usually* found in shallow habitats.

63 Some species, including most *Goniastrea* and some *Acropora* are particularly abundant in inter-tidal habitats and may, as a result, have enhanced distributions over areas with high tidal ranges. Conversely, some species may be limited geographically by habitat availability. *Coscinaraea mcneilli* and *C. marshae* are probably ecologically limited to high-latitude non-reef communities of Australia; *Symphyllia wilsoni* is similarly limited to shallow exposed rock flats. *Alveopora japonica* is limited to partly protected sub-tidal habitats, and the same may apply to other high-latitude endemics. *Acropora nana* is limited to the exposed fronts of tropical reefs, *A. cardinae* to deep waters where other *Acropora* seldom occur.

64 For example, *Acrhelia horrescens* in Australia is virtually limited to clear waters, but occurs in turbid waters elsewhere in the central Indo-Pacific.

65 For example, *Leptoseris papyracea* is usually limited to deep waters where light availability is low, but at Vanuatu and Ashmore Reef (western Australia) it occurs in shallow waters exposed to full sunlight. *Plerogyra sinuosa* and *Physogyra lichtensteini* are both usually limited to protected waters where light availability is low, but both occur in shallow lagoons of North-west Shelf reefs of western Australia. *Diaseris fragilis* is also usually limited to deep waters, but in Cebu, the Philippines, it occurs in inter-tidal sea-grass beds in great abundance (p 250).

66 For example, *Acropora longicyathus* and *A. subglabra* are both found in a wide range of habitats throughout the central Indo-Pacific, but are limited to protected lagoons of the North-west Shelf reefs of western Australia.

67 Possible examples, such as the aforementioned *Leptoseris papyracea* on North-west Shelf reefs of western Australia and Vanuatu, or two recorded geographic subspecies of *Cyphastrea decadia*, have environment-correlated variations that mask geographic ones.

68 For example, *Polyphyllia talpina, Heliofungia actiniformis, Catalaphyllia jardinei* and several *Euphyllia* species.

69 Most species of *Goniopora* and *Alveopora* have middle-sized semi-permanently extended polyps that retract in a few seconds after stimulation. *Porites evermanni* and *Stephanocoenia michelinii* have very small polyps which do the same. In the latter species, polyp retraction causing blanching can be used for field identification of the species.

70 For example, colonies of *Acropora horrida* in Australian locations usually have polyps partly extended during the day while those in the northern hemisphere seldom do. This behavioural difference is correlated with colour and some morphological differences. Colonies of *Acropora paniculata* in Vanuatu commonly have polyps extended during the day; in this case, there is no indication of a geographic subspecies.

71 For example, colonies of *Symphyllia valenciennesi* in mainland Japan have fleshy polyps extended throughout the day, whereas they are not extended in Ryukyu Island colonies. This behaviour may be associated with a geographic subspecies. *Pavona decussata* and *Hydnophora exesa* also have distinctive behaviours in high latitudes, but here there is no association with a geographic subspecies. *Physogyra lichtensteini* and *Plerogyra* colonies always have a conspicuous surface layer of nematocyst-loaded vesicles which, in the case of the former species, retract after stimulation. The shape of undisturbed vesicles varies from semi-tubular to egg-shaped according to degree of inflation. There is some geographic variation in this behaviour as vesicles of Japanese colonies are seldom fully inflated, whereas those of the Great Barrier Reef almost always are.

72 For example, colonies of many species of *Acropora* in the Houtman Abrolhos Islands, including *A. pulchra, A. abrolhosensis, A. tenuis, A. hyacinthus* and *A. spicifera,* as well as those of other genera, for example, *Leptastrea purpurea* have, unlike colonies of the same species further north, polyps and/or one or a few tentacles extended most of the day as well as night. The same observation applies to a lesser extent to colonies at Lord Howe Island and some Japanese mainland locations.

73 For example, *Stylocoeniella armata, Seriatopora hystrix, Montipora foliosa, Porites rus, Gardineroseris planulata* and *Diploastrea heliopora.*

74 Although accurately identified, *Sylocoeniella guentheri, Pavona clavus, Leptoseris explanata, Coeloseris mayeri, Podabacia crustacea,* and approximately seven species of *Acropora* recorded by Riegl (1993) from South Africa, are sufficiently dissimilar from Great Barrier Reef equivalents that they can be considered distinct geographic subspecies.

75 Several Hawaiian species (eg *Montipora capitata* and *Cyphastrea agassizi*) are indistinguishable from equivalents in Japan.

76 For example, most *Pocillopora* species in French Polynesia and all *Porites* species in Hawaii are readily separable from their central Indo-Pacific equivalents.

77 Fourteen of the sixty-one species recorded from Pitcairn Islands (p 165) can reasonably be distinguished as subspecies. All seven species from Clipperton Atoll have that level of distinction.

78 Some very common species are examples: *Porites lobata* can be considered one species that spans the Indo-Pacific; it can also be considered a distinct species, subspecies, or a sibling species in almost every region where it occurs. Thus, in the far eastern Pacific it forms a single discrete species that has *P. lutea*-like septal characters. Specimens are similar to those from the Pitcairn Islands where *P. lobata* is readily distinguishable from co-occurring *P. lutea.* Whether or not far eastern Pacific *P. lobata* has, at some past time, hybridised with *P. lutea* cannot be determined morphologically. The same can be said for taxonomically isolated species, such as

Stylocoeniella armata as well as species complexes, such as *Favia pallida* and *Cyphastrea serailia*.

79 For example, *Acropora hyacinthus* and *A. cytherea* are very easily separated on the Great Barrier Reef, but not so in the central Pacific. *Leptastrea transversa* and *L. purpurea* are distinctive in the Red Sea, but may be inseparable in the western Pacific. *Echinopora mammiformis* and *Lobophyllia corymbosa* appear to be the only species of their respective genera that retain their individual identity over wide geographic ranges.

Chapter 12

1 Two simple hypothetical examples of surface circulation vicariance in different spatial and temporal contexts are as follows. The first is that increase in global surface circulation vectors will cause *decreased* isolation between most regions (eg the central Pacific), but at the same time will cause increased isolation in other regions (eg 'downstream' locations under the influence of continental boundary currents (chapter 10, p 171). The changed pattern would presumably need to be maintained for thousands of years before it significantly influenced genetic patterns. The second is that the Leeuwin Current is an aberrant boundary current in that it flows poleward on the western side of a continental land mass (Australia) (figure 56, p 174); it may thus be relatively unstable in geological time. Reversal of the current, if only for a single spawning season, would be likely to have substantial genetic consequences for upstream locations on the west Australian coast.

2 This assumes that natural selection produces evolutionary change through increased reproductive success.

3 The term 'introgression' is most commonly used in plants. It was originally restricted to gene flow between 'species', but usage is now wider, essentially because 'species' in plants involve similar issues to 'species' in corals.

4 Genetic aspects of hybrid zones are complex and poorly understood. In most literature, hybrid zones are clines maintained by environmental, reproductive, and/or genetic mechanisms.

5 Hybridisation increases homoplasy (which is, in any case, overwhelmingly important in corals, p 24); it increases the number of parsimonious cladograms that will collapse to consensus trees; and it will distort patterns of relationships among non-hybrids.

6 Inevitably with a subject as complicated as evolution, many ifs and buts are associated with this conclusion.

Chapter 13

1 The terms 'metaspecies' and 'syngameon' are not well established in the literature: the former has previously been used in publications about species concepts and is a corollary of the term 'metapopulation'; the latter has previously been used in plant genetics.

2 Syngameons: a familiar example. I can recognise the species of *Eucalyptus* trees that surround my house at a glance, for most *Eucalyptus* species are readily separable where they co-occur. However, a few hundred kilometres away, the supposedly same species appear rather unfamiliar; further afield again, I do not know if they are the same species or not. Different botanists have divided *Eucalyptus* into between 500 and 850 species with equally variable geographic ranges. Only a few of these species are always distinct; most intergrade (hybridise) with other species in a reticulate mosaic across the Australian continent. Thus, although an individual species may be clearly identifiable at a single location (eg around my house), it may have no clear geographic boundaries because it has no clear taxonomic boundaries. In concept, the genus can be divided into species that do not frequently hybridise but which intergrade in space and time into syngameons.

3 Species complexes: some species may have many taxonomically useful characters, while others have few; there is little equality from one taxon to the next. A

'species complex' may thus be a complex because it contains discontinuous reticulate patterns or, equally, because it simply lacks good (conservative) diagnostic characters.

4   Subspecies names are sometimes added to species names by some authors. These may be intended to apply to geographic subspecies, ecomorphs or races; in most cases they have no taxonomic status or meaning.

5   The alternative strategy, as exploited by insects, is to have large numbers of species and little variation within each of those species.

6   This involves the doubtful assumption that reticulate patterns form in the western Atlantic.

7   The point is illustrated by the question: 'why are there about 700 species of coral in the Indo-Pacific and not 70 or 7000?'

8   This is not a proposal for Darwinian centres of origin: there is no displacement of species from a centre of evolution, but there *is* a centre of evolution.

Appendix

1   There may be some discrepancies between the ages of genera recorded here and those given in (or used quantitatively in) chapter 8, owing to a range of identification and dating issues. If so, and unless otherwise stated (see next endnote), this chapter has the more conservative record.

2   Untraced records of Wells (1956) are indicated by reference to that study. Other generic records, which are older than those indicated here, are to be found in the literature, but have not been subsequently verified, or may belong to a different genus.

3   The two species of the *Acropora palifera* group (making up the sub-genus *Isopora*) are exceptions to these observations, yet are the overwhelmingly dominant frame builders of the outer Great Barrier Reef.

4   There are many records of *Agaricia* in the Indo-Pacific; all are incorrect.

5   *Leptoseris* is commonly known as *Helioseris* in the Caribbean.

6   The single specimen on which this observation is based appears to have been lost.

7   *Astrangia* is not included in Veron (1993) as it has no species which are obligatory zooxanthellates.

8   *Indophyllia* was also formerly believed extinct.

9   *Archohelia,* found alive on the Great Barrier Reef, but formerly believed extinct (Wells and Alderslade 1979), is azooxanthellate.

10 The full range of depth-correlated intra-specific variation in *Favites russelli* is illustrated in Veron et al (1977, pp 74–5). This variation is readily observed on most exposed reef slopes in the tropical western Pacific.

11 *Favites* and *Acanthastrea* are very similar in high-latitude regions, making identification problematic.

12 Illustrated in Veron et al (1977 p 199). This variation is readily observed in protected lagoons throughout the range of the species.

LITERATURE
CITED

Achituv, Y. and Dubinsky, Z. 1990. Evolution and zoogeography of coral reefs **in** Dubinsky Z. (ed), *Coral reefs, ecosystems of the world*. 25. Elsevier, Amsterdam: 1–9.

Adams, C.G. 1981. An outline of Tertiary paleogeography **in** Cocks L.R.M (ed), *The evolving earth*. British Museum (Natural History) Cambridge University Press 14:221–35.

Adams, C.G. et al 1977. The Messinian salinity crisis and evidence of late Miocene eustatic changes in the world ocean. *Nature* 269:383–6.

Adams, C.G. et al 1990. Conflicting isotopic and biotic evidence for tropical sea-surface temperatures during the Tertiary. *Palaeogeogr. Palaeoclimat. Palaeoecol.* 77:289–313.

Agassiz, L. 1865. **In** Winsor M.P. 1991 (ed), *Reading the shape of nature*. Univ. Chicago Press, Chicago, 1991.

Ahron, P. and Chappell, J. 1986. Oxygen isotopes, sea level changes and the temperature history of a coral reef environment in New Guinea over the last 10^5 years. *Palaeogeogr. Palaeoclimat. Palaeoecol.* 56:337–79.

Allan, R.J. et al 1991. A further extension of the Tahiti–Darwin SOI, early ENSO events and Darwin pressure. *J. Climate* 4(7):743–9.

Alvarez, W. et al 1984. The end of the Cretaceous: sharp boundary or gradual transition? *Science* 223:1183–1186.

Arnold, M.L. 1992. Natural hybridization as an evolutionary process. *Annu. Rev. Ecol. Syst.* 23:237–61.

Arnold, M.L. et al 1990. Natural hybridisation between *Iris fulva* and *I. hexagona*: pattern of ribosomal DNA variation. *Evolution* 44: 1512–21.

Arthur, M.A. and Garrison, R.E. 1986. Cyclicity in the Milankovity band through geological time: an introduction. *Paleoceanography* 1:369–72.

Arthur, M.A. et al 1985. Variations in the global carbon cycle during the Cretaceous related to climate, volcanism and changes in atmospheric CO_2 in Sundquist E.T. and Broecker W.S. (eds), The carbon cycle and atmospheric CO_2: natural variations archean to present. *Geophys. Monogr.* 32:504–29.

Avise, J.C. et al 1987. Intraspecific phylogeography: The mitochondrial DNA bridge between population genetics and systematics. *Annu. Rev. Ecol. Syst.* 18:489–522.

Ayre, D.J. and Dufty, S. in press. Evidence for restricted gene flow in the viviparous coral *Seriatopora hystrix* on Australia's Great Barrier Reef. *Evolution*.

Ayre, D.J. and Resing, J.M. 1986. Sexual and asexual production of planulae in reef corals. *Mar. Biol.* 90:187–90.

Ayre, D.J. and Willis, B.L. 1988. Population structure in the coral *Pavona cactus:* clonal genotypes show little phenotypic plasticity. *Mar. Biol.* 99:495–505.

Ayre, D.J. et al 1991. The corals *Acropora palifera* and *Acropora cuneata* are genetically and ecologically distinct. *Coral Reefs* 10:13–18.

Babcock, R.C. et al 1986. Synchronous spawnings of 105 scleractinian coral species on the Great Barrier Reef. *Mar. Biol.* 90:379–94.

Babcock, R.C. et al in press. Mass spawning of corals on a high latitude coral reef. *Coral Reefs.*

Barnes, D.J. and Chalker, B.E. 1990. Calcification and photosynthesis in reef-building corals and algae **in** Dubinsky Z. (ed), *Coral reefs, ecosystems of the world.* 25. Elsevier, Amsterdam: 109–31.

Barron, E.J. 1987. Eocene equator-to-pole surface ocean temperatures: A significant climate problem? *Paleoceanography* 2:729–39.

Barron, E.J. and Peterson, W.H. 1989. Model simulation of the Cretaceous ocean circulation. *Science* 244:684–6.

Barton, N.H. and Hewitt, G.M. 1985. Analysis of hybrid zones. *Annu. Rev. Ecol. Syst.* 16:113–48.

Bartlein, P.J. and Prentice, I.C. 1989. Orbital variations, climate and paleoecology. *Trends Ecol. & Evol.* 4:195–9.

Beauvais, L. 1980. Sur la taxinomie des Madréporaires Mésozoïques. *Acta Palaeontol. Pol.* 25:345–60.

Beauvais, L. 1982. Paléobiogéographie des Madréporaires du Trias. *Bull. Soc. Geol. Fr.* XXIV(5–6):963–70.

Beauvais, L. 1984. Evolution and diversification of Jurassic Scleractinia Palaeontographica Americana, *4th International Symposium on Fossil Cnidaria,* Washington: 219–24.

Beauvais, L. 1986. Evolution paléobiogéographique des formations à Scléractiniaires du Bassin téthysien au cours du Mésozoïque. *Bull. Soc. Geol. Fr.* 8:499–509.

Beauvais, L. 1989. Jurassic corals from the circum Pacific area. *Mem. Assoc. Australasian Palaeontol.* 8:291–302.

Beauvais, L. 1992. Palaeobiogeography of the Early Cretaceous corals. *Palaeogeogr. Palaeoclimatol. Palaeoecol.* 92:233–47.

Belasky, P. 1992. Assessment of sampling bias in biogeography by means of a probabilistic estimate of taxonomic diversity: application to modern Indo-Pacific reef corals. *Palaeogeogr. Palaeoclimatol. Palaeoecol.* 99:243–70.

Belasky, P. and Runnegar, B. 1993. Biogeographic constraints for tectonic reconstructions of the Pacific region. *Geology* 21:979–82.

Belbin, L. 1987. PATN: *Pattern analysis package, reference manual,* Part 1. CSIRO Division of Wildlife and Rangelands Research, Melbourne.

Belém, M.J.C. et al 1986. S.O.S. Corais. *Ciencia Hoje,* Rio de Janeiro 4(26):34–42.

Bennett, K.D. 1990. Milankovitch cycles and their effect on species in ecological and evolutionary time. *Paleobiology* 16:11–21.

Benzie, J.A.H. 1987. The biogeography of Australian *Daphnia:* clues of an ancient (>70 m.y.) origin of the genus. *Hydrobiologica* 145:51–65.

Benzie, J.A.H. 1992. Review of the genetics, dispersal and recruitment of Crown-of-thorns starfish. (*Acanthaster planci) Aust. J. Mar. Freshwater Res.* 43:597–610.

Benzie, J.A.H. et al in press. Variation in the genetic composition of coral (*Pocillopora damicornis* and *Acropora palifera*) populations from different reef habitats. *Mar. Biol.*

Bernard, H.M. 1896–1905. *Catalogue Madreporarian corals. British Museum (Natural History),* vols 1–5, British Mus. London.

Blank, R.J. and Trench, R.K. 1986. Nomenclature of endosymbiotic dinoflagellates. *Taxon* 35:286–94.

Boekschoten, G.J. and Wijsman-Best, M. 1981. *Pocillopora* in the Miocene reef at Baixo, Porto Santo (Eastern Atlantic). *Palaeontology* B84:13–20.

Borel Best, M. and Hoeksema, B.W. 1987. New observations on scleractinian corals from Indonesia. 1. Free living species belonging to the Faviina. *Zool. Meded. (Leiden)* 61:387–403.

Borel Best, M. et al 1984. Species concept and ecomorph variation in living and fossil Scleractinia. *Palaentol. Am.* 54:70–79.

Borel Best, M. et al 1989. Recent scleractinian coral species collected during the Snellius–II expedition in eastern Indonesia. *Neth. J. Sea Res.* 23:107–15.

Bradley, R.S. 1985. *Quaternary paleoclimatology. Methods of paleoclimatic reconstruction.* Allen & Unwin, Boston.

Briggs, D.E.G. and Crowther, P.R. 1990. *Palaeobiology: a synthesis.* Blackwell Scientific Publications, Oxford.

Briggs, J.C. 1974. *Marine Zoogeography.* McGraw-Hill, New York.

Briggs, J.C. 1984. *Centers of origin in biogeography.* Biogeography Monograph 1, University of Leeds, Leeds.

Briggs, J.C. 1987. Antitropicality and vicariance. *Syst. Zool.* 36:206–7.

Briggs, J.C. 1987. Antitropical distribution and evolution in the Indo–west Pacific Ocean. *Syst. Zool.* 36:237–47.

Briggs, J.C. 1987. *Biogeography and plate tectonics.* Elsevier, Amsterdam.

Brooks, D.R. 1990. Parsimony analysis in historical biogeography and coevolution: methodological and theoretical update. *Syst. Zool.* 39:14–30.

Brown, B.E. and Suharsono 1990. Damage and recovery of coral reefs affected by El Niño related seawater warming in the Thousand Islands, Indonesia. *Coral Reefs* 8:163–70.

Brundin, L.Z. 1981. Croizat's panbiogeography versus phylogenetic biogeography **in** Nelson G. and Rosen D.E. (eds), *Vicariance Biogeography. A Critique.* Columbia University Press, New York: 95–138.

Bryan, J.R. 1991. A Palaeocene coral-algal-sponge reef from southwestern Alabama and the ecology of the early Tertiary reefs. *Lethaia* 24:423–38.

Budd, A.F. 1979. Phenotypic plasticity in the reef corals *Montastrea annularis* (Ellis and Solander) and *Siderastrea siderea* (Ellis and Solander). *J. Exp. Mar. Biol. Ecol.* 39: 25–54.

Budd, A.F. 1983. The relationship between corallite morphology and colony shape in some massive reef corals. *Coral Reefs* 2:19–25.

Budd, A.F. 1984. The species concept in fossil hermatypic corals: a statistical approach. *Palaeontogr. Am.* 54:58–69.

Budd, A.F. 1985. Variation within coral colonies and its importance for interpreting fossil species. *J. Paleontol.* 59:1359–81.

Budd, A.F. 1986. Neogene paleontology in the northern Dominican Republic 3. The Family Poritidae (Anthozoa: Scleractinia). *Bull. Am. Paleontol.* 90:47–123.

Budd, A.F. 1987. Neogene paleontology in the northern Dominican Republic 4. The genus *Stephanocoenia* (Anthozoa: Scleractinia: Astrocoencidae). *Bull. Am. Paleontol.* 93:5–74.

Budd, A.F. 1993. Variation within and among morphospecies of *Montastrea. Cour. Forsch.–Inst.* Senckenberg 164:241–54.

Budd, A. and Coates, A.G. 1992. Nonprogressive evolution in a clade of Cretaceous *Montastrea*-like corals. *Paleobiology* 18:425–46.

Budd, A. F. and Guzmán, H.M. in press. *Siderastrea glynni*, a new species of scleractinian coral (Cnidaria: Anthozoa) from the eastern Pacific. *Proc. Biol. Soc. Washington.*

Budd, A.F. et al 1992. Eocene Caribbean reef corals: a unique fauna from the Gatuncillo formation of Panama. *J. Paleontol.* 66:570–94.

Budd, A.F. et al in press. Stratigraphic distribution of genera and species of Neogene to Recent Caribbean reef corals. *Paleontology.*

Buddemeier, R.W. and Fautin, D.G. 1993. Coral bleaching as an adaptive mechanism. *BioScience* 43:320–6.

Buddemeier, R.W. and Fautin, D.G. in press. Geochemistry and the evolution of hexa-corals. *Proc 7th Biomineralisation Symp.* Monaco.

Burger, W.C. 1975. The species concept in *Quercus. Taxon* 24:45–50.

Burns, T.P. 1985. Hard-coral distribution and cold-water disturbances in South Florida: Variation with depth and location. *Coral Reefs* 4:117–24.

Burrage, D. 1993. Coral Sea currents. *Corella* 17:135–45.

Campbell, J.W. and Aarup, T. 1989. Photosynthetically available radiation at high latitudes. *Limnol. Oceanogr.* 34:1490–9.

Campbell, K.S.W. and Day, M.F. 1987. *Rates of evolution.* Allen and Unwin, London.

Carson, H.L. 1971. Speciation and the founder principle. *Stadler Genetics Symposium,* University Missouri 3:51–70.

Carson, H.L. and Templeton, A.R. 1984. Genetic revolutions in relation to speciation phenomena: the founding of new populations. *Annu. Rev. Ecol. Syst.* 15:97–131.

Chalker, B.E. et al 1984. Seasonal changes in primary production and photoadaptation by the reef-building coral *Acropora granulosa* on the Great Barrier Reef **in** Holm-Hansen O. et al (eds), *Marine Phytoplankton and Productivity.* Springer-Verlag, Berlin: 73–87.

Chalker, B.E. et al 1988. Light and reef-building corals. *Interdisciplinary Sci. Rev.* 13(3):222–37.

Chandler, R.C. and Gromko, M.H. 1989. On the relationship between species concepts and speciation processes. *Syst. Zool.* 38:116–25.

Chappell, J. 1974. Geology of coral terraces, Huon Peninsula, New Guinea: a study of Quaternary tectonic movements and sea-level changes. *Bull. Am. Geol. Soc.* 85:553–70.

Chappell, J. and Shackleton, N.J. 1986. Oxygen isotopes and sea level. *Nature* 32(13):137–40.

Charlesworth, B. et al 1982. A neo-Darwinian commentary on macroevolution. *Evolution* 36:474–98.

Chen, C.A. et al in press. Systematic relationships within the Anthozoa (Cnidaria: Anthozoa) using the 5'-end of the 28S rDNA. *Molecular Phylogenetics Evol.*

Chesson, P.L. and Chase, T.J. 1986. Overview: nonequilibrium community theories: chance, variability, history and coexistence **in** Diamond J. and Case T.J. (eds), *Community ecology.* Harper and Row, New York: 229–39.

Chevalier, J.P. 1961. Recherches sur les Madreporaires et les formations recifales miocenes de la Mediterranee occidentale. *Mem. Soc. Geol. Fr.* 93:1–562.

Chevalier, J.P. 1966. Contribution a l'etude des Madreporaires des cotes occidentales d'Afrique tropicale. *Bull. Inst. Fr. Afr. Noire Ser. A,* 28:912–75; 1356–405.

Chevalier, J.P. 1972. Les Scléractiniaires du Miocene de Porto Santo (Archipel de Madere). Etude Paleontologique. *Ann. Paleontol. Invertebr.* LVIII:141–60.

Chevalier, J.P. 1975. Les Scléractiniaires de la Mélanésie française (Nouvelle Calédonie, Iles Chesterfield, Iles Loyauté, Nouvelles Hébrides). 2éme Partie, *Expéd. Française récifs Nouvelle Calédonie,* Edn. Fond. Singer-Polignac, Paris, 7:5–407.

Chevalier, J.P. and Beauvais, L. 1987. Classification en sous-orderes des Scléractiniaires **in** Grassé P.P. (ed) *Traité de Zoologie* 3:679–764.

Chornesky, E.A. 1989. Repeated reversals during spatial competition between corals. *Ecology* 70:843–55.

Chornesky, E.A. and Peters, 1987. Sexual reproduction and colony growth in the scleractinian coral *Porites astreoides. Biol. Bull.* 172:161–77.

Clarke, A. 1992. Is there a latitudinal species diversity cline in the sea? *Trends Ecol. & Evol.* 7:286–7.

Clausen, C.D. 1971. Effects of temperature on the rate of 45-Calcium uptake by *Pocillopora damicornis* in Lenhoff H.M. et al (eds), *Experimental coelenterate biology.* Hawaii University Press, Honolulu: 246–59.

Clausen, C.D. and Roth, A.A. 1975. Effect of temperature and temperature adaptation on calcification rate in the hermatypic coral *Pocillopora damicornis. Mar. Biol.* 33:93–100.

CLIMAP Project Members 1976. The surface of the ice-age earth. *Science* 191:1131–7.

CLIMAP Project Members 1981. Seasonal reconstruction of the Earth's surface at the last interglacial maximum. *Geol. Soc. Am. Map Chart* Ser. MC-36.

Coates, A.G. and Jackson, J.B.C. 1987. Clonal growth, algal symbiosis and reef formation by corals. *Paleobiology* 13:363–78.

Coates, A.G. et al 1992. Closure of the Isthmus of Panama: the near shore marine record of Costa Rica and western Panama. *Geol. Soc. Am. Bull.* 104:814–28.

Cock, A.G. 1977. Bernard's Symposium — the species concept in 1900. *Biol. J. Linn. Soc.* 9:1–30.

COHMAP Members 1988. Climatic changes of the last 18,000 years: observations and model simulations. *Science* 241:1043–52.

Coles, S.L. 1988. Limitations on reef coral development in the Arabian Gulf: temperature or algal competition. *Proc. 6th Int. Coral Reef Symp.,* Australia 3:211–16.

Coles, S.L. and Jokiel, P.L. 1977. Effects of temperature on photosynthesis and respiration in hermatypic corals. *Mar. Biol.* 43:209–16.

Coles, S.L. and Jokiel, P.L. 1978. Synergistic effects of temperature, salinity and light on the hermatypic coral. *Montipora verrucosa. Mar. Biol.* 49:187–95.

Coles, S.L. and Fadlallah, Y.H. 1991. Reef coral survival and mortality at low temperatures in the Arabian Gulf: a new species-specific lower temperature limits. *Coral Reefs* 9:231–7.

Coles, S.L. et al 1976. Thermal tolerance in tropical *versus* subtropical Pacific reef corals. *Pac. Sci.* 30:159–66.

Connell, J.H. 1978. Diversity in tropical rain forests and coral reefs. *Science* 199:1302–09.

Connell, J.H. 1987. Maintenance of species diversity in biotic communities **in** Kawano S. (ed), *Evolution and co-adaptation in biotic communities.* University of Tokyo Press: 201–18.

Connell, J.H. and Slatyer, R.O. 1977. Mechanisms of succession in natural communities and their role in community stability and organisation. *Am. Nat.* 111:1119–41.

Connell, J.H. and Keough, M.J. 1985. Disturbance and patch dynamics of subtidal marine animals on hard substrata **in** Pickett S.T.A. and White P.S. (eds), *The Ecology of Natural Disturbance and Patch Dynamics.* Academic Press, Orlando: 125–51.

Cook, C.B. et al 1990. Elevated temperatures and bleaching on a high latitude coral reef: the 1988 Bermuda event. *Coral Reefs* 9:45–9.

Cortés, J.N. 1986. Biogeografia de corales hermatípicos: el istmo centro Americano. *An. Inst. Cienc. del mar y Limnol. Univ. Nal. Autón. México* 13:297–304.

Cortés, J.N. et al 1994. Holocene growth history of an eastern Pacific fringing reef, Punta Islotes, Costa Rica. *Coral Reefs* 13:65–73.

Coudray, J. and Montaggioni, L. 1982. Coraux et recifs coralliens de la province Indo-Pacifique: repartition geographique et altitudinale en relation avec la tectonique globale. *Bull. Soc. Geol. Fr.* 24:981–93.

Cowen, R. 1988. The role of algal symbiosis in reefs through time. *PALAIOS* 3:221–7.

Cracraft, J. 1984. The terminology of allopatric speciation. *Syst. Zool.* 33:115–6.

Cracraft, J. 1987. Species concepts and the ontology of evolution. *Biol. & Philos.* 2:63–80.

Cracraft, J. 1989. Species and its ontogeny: the empirical consequences of alternative species concepts for understanding patterns and processes of differentiation **in** Otte D. and Endler J.A. (eds), *Speciation and its consequences.* Sinauer, Massachusetts.

Craw, R. 1988a. Continuing the synthesis between panbiogeography, phylogenetic systematics and geology as illustrated by empirical studies on the biogeography of New Zealand and the Chatham Islands. *Syst. Zool.* 37:291–310.

Craw, R. 1988b. Panbiogeography: method and synthesis in biogeography **in** Myers A.A. and Giller P.S. (eds), *Analytical biogeography.* Chapman and Hall: 405–35.

Craw, R.C. and Weston, P. 1984. Panbiogeography: a progressive research program? *Syst. Zool.* 33:1–13.

Croizat, L. 1974. Centres of origin and related concepts. *Syst. Zool.* 23:265–87.

Croizat, L. 1981. Biogeography: past, present and future **in** Nelson G. and Rosen D.E. (eds), *Vicariance Biogeography: A Critique.* Columbia University Press, New York: 501–23.

Croizat, L. 1982. Vicariance/vicariism, panbiogeography, 'vicariance biogeography', etc.: a clarification. *Syst. Zool.* 31:291–304.

Crossland, C.J. 1981. Seasonal growth of *Acropora cf. formosa* and *Pocillopora damicornis* on the high latitude reef (Houtman Abrolhos, Western Australia). *Proc. 4th Coral Reef Symp.*, Manila 1:663–8.

Crossland, C.J. 1984. Seasonal variation in the rates of calcification and productivity in the coral *Acropora formosa* on a high-latitude reef. *Mar. Ecol. Prog. Ser.* 15:135–40.

Crowley, T.J. 1991. Past CO_2 changes and tropical sea surface temperatures. *Paleoceanography* 6:387–94.

Cuffey, R.J. and Pachut, J.F. 1991. Clinal morphological variation along a depth gradient in the living Scleractinian reef coral *Favia pallida*: effects on perceived evolutionary tempos in the fossil record. *Soc. Sedimentary Geol. Res. Reports*: 580–8.

Cuif, J.–P. 1980. Microstructure versus morphology in the skeleton of Triassic scleractinian corals. *Acta Palaeontol. Pol.* 25:361–74.

Daly, R.A. 1915. The glacial-control theory of coral reefs. *Proc. Am. Acad. Arts Sci.* Boston. 51:155–251.

Dana, J.D. 1843. On the temperature limiting the distribution of corals. *Am. J. Sci.* 45:130–1.

Dana, J.D. 1857. **In** M.P. Winsor (ed), *Reading the shape of nature.* University of Chicago Press, Chicago.

Dana, T.F. 1975. Development of contemporary eastern Pacific coral reefs. *Mar. Biol.* 33:355–74.

Darlington, P.J. 1965. *Biogeography of the southern end of the world.* Harvard University Press, Cambridge, Mass.

Darwin, C. 1842. *The structure and distribution of coral reefs.* Smith, Elder.

Darwin, C. 1859. *The origin of species by means of natural selection, or the preservation of favoured races in the struggle for life.* Murray.

Davies, P.J. et al 1989. Facies models in exploration — the carbonate platforms of northeastern Australia **in** Crevello P.D. et al (eds), Controls on carbonate platform and basin development. *Soc. Econ. Paleont. Min. Spec. Publ.* 44:233–58.

Davis, G.E. 1982 A century of natural change in coral distribution in the dry Tortugas: a comparison of reef maps from 1881 and 1976. *Bull. Mar. Sci.* 32:608–23.

Davis, W.M. 1928. *The coral reef problem.* Special Publication American Geographic Society 9. New York.

Davis, J.I. and Nixon, K.C. 1992. Populations, genetic variation, and the delimitation of phylogenetic species. *Syst. Biol.* 41:421–35.

de Queiroz, K. and Donoghue, M.J. 1988. Phylogenetic systematics and the species problem. *Cladistics* 4:317–38.

Debiche, 1987. The motion of allochthonous terranes across the north Pacific basin. *Geol. Soc. Am. Spec.* Paper 207:1–47.

deMenocal, P.B. 1993. Influences of high- and low-latitude processes on African terrestrial climate: Pleistocene eolian records from equatorial Atlantic drilling program site 663. *Paleoceanography* 8:209–42.

Dinesen, Z.D. 1977. The coral fauna of the Chagos Archipelago. *Proc. 3rd Int. Coral Reef Symp.*, Miami 1:155–61.

Dobzhansky, Th. 1935. A critique of species concept in biology. *Philos. Sci.* 2:344–55.

Dobzhansky, Th. 1937. *Genetics and the origin of species.* Columbia University Press, New York.

Dodge, R.E. and Lang, J.C. 1983. Environmental correlates of hermatypic coral (*Montastrea annularis*) growth on the East Flower Garden Bank, northwest Gulf of Mexico. *Limnol. Oceanogr.* 28:228–40.

Done, T.J. 1983. Coral zonation: its nature and significance **in** D.J. Barnes (ed), *Perspectives on coral reefs.* Clouston, Manuka: 107–47.

Done, T.J. 1988. Simulation of recovery of pre-disturbance size structure in populations of *Porites* spp. damaged by the crown of thorns starfish *Acanthaster planci. Mar. Biol.* 100:51–61.

Done, T.J. 1992. Constancy and change in some Great Barrier Reef coral communities: 1980–1990. *Amer. Zool.* 32:655–62.

Downing, N. 1985. Coral reef communities in an extreme environment: the northwestern Arabian Gulf. *Proc. 5th Int. Coral Reef Congr.*, Tahiti 6:343–8.

Doyle, J.J. 1992. Gene trees and species trees: molecular systematics as one-character taxonomy. *Syst. Bot.* 17:144–63.

Durham, J.W. 1947. Corals from the Gulf of California and the North Pacific coast of America. *Geol. Soc. Amer.* 20:1–46.

Eldredge, N. and Gould, S.J. 1972. Punctuated equilibria: an alternative to phyletic gradualism **in** T.J.M. Schopf (ed), *Models in Paleobiology*. Freeman Cooper, San Francisco: 82–115.

Esteban, M. 1980. Significance of the upper Miocene coral reefs of the Western Mediterranean. *Palaeogeogr. Palaeoclimatol. Palaeoecol.* 29:169–88.

Fadlallah, Y.H. 1983. Sexual reproduction, development and larval biology in Scleractinian corals. *Coral Reefs* 2:129–50.

Fagerstrom, J.A. 1987. *The evolution of reef communities*. John Wiley & Sons.

Fautin, D.G. and Lowenstein, J.M. 1994. Phylogenetic relationships among scleractinians, actinians and corallimorpharians (Coelenterata: Anthozoa). *Proc. 7th Int. Coral Reef Symp.*, Guam: 665–71.

Fedorowski, J. 1989. Intraspecific variation in Carboniferous and Permian rugosa. *Fossil Cnidaria* 5:7–12.

Fenner, D.P. 1993. Species distinctions among several Caribbean stony corals. *Bull. Mar. Sci.* 53:1099–116.

Fischer, A.G. 1986. Climatic rhythms recorded in Strata. *Annu. Rev. Earth Planet Sci.* 14:351–76.

Fisk, D.A. and Done, T.J. 1977. Taxonomic and bathymetric patterns of bleaching in corals, Myrmidon Reef (Queensland). *Proc. 3rd Int. Coral Reef Symp.*, Miami 6:149–54.

Flügel, E. 1982. Evolution of Triassic reefs: current concepts and problems. *Facies* 6:297–328.

Flügel, E. and Flügel-Kahler, E. 1992. Phanerozoic reef evolution: basic questions and data base. *Facies* 26:167–278.

Frakes, L.A. 1979. *Climate throughout geologic time*. Elsevier, Amsterdam.

Frakes, L.A. et al 1992. *Climate modes of the Phanerozoic*. Cambridge University Press.

Franzisket, L. 1969. Riffkorallen konnen autotroph leben. *Naturwissenschaften* 56:144.

Fritt, W.K. et al 1993. Recovery of the coral *Montastrea annularis* in the Florida Bay after the 1987 Caribbean 'bleaching event'. *Coral Reefs* 12:57–64.

Frost, S.H. 1972. Evolution of Cenozoic Caribbean coral faunas. *Memoirs 6th Caribbean Geological Conference*, Margarita, Venezuela: 461–4.

Frost, S.H. 1977a. Oligocene reef coral biogeography, Caribbean and western Tethys. *Bull. Bur. Rech. Geol. Minieres Mem.* 89:342–52.

Frost, S.H. 1977b. Miocene to Holocene evolution of Caribbean province reef-building corals. *Proc. 3rd Int. Coral Reef Symp.*, Miami 2:354–9.

Frost, S.H. 1977c. Ecologic controls of Caribbean and Mediterranean Oligocene reef coral communities. *Proc. 3rd Int. Coral Reef Symp.*, Miami 2:367–73.

Frost, S.H. 1981. Oligocene reef coral biofacies of the Vicentin, northeast Italy. *Soc. Econ. Paleontol. Mineral. Spec. Pub.* 30:483–539.

Frost, S.H. and Langenheim, R.L. 1974. *Cenozoic reef biofacies: Tertiary larger foraminifera and scleractinian corals from Chipas, Mexico*. Northern Illinois University Press, DeKalb.

Fukuda, T. 1981. The growth of coral transplanted to big aquaria. *Mar. Pavilion* 10(1):4 (in Japanese).

Fukuda, T. 1984a. March low water temperature and the dynamics of fish fauna around the underwater observatory. *Mar. Pavilion* 13(4):20–3 (in Japanese).

Fukuda, T. 1984b. *Acropora* species survive low water temperature at Tenjin–Saki. *Mar. Pavilion* 13(7):1–3 (in Japanese).

Fulthorpe, C.S. and Schlanger, S.O. 1989. Paleo-oceanographic and tectonic settings of early Miocene reefs and associated carbonates of offshore Southeast Asia. *Am. Assoc. Pet. Geol. Bull.* 73:729–56.

Funnell, B.M. 1990. Global and European Cretaceous shorelines, stage by stage **in** Ginsberg R.N. and Beaudoin B. (eds), *Cretaceous resources, events and rhythms. Background and plans for research.* NATO Series C: Mathematical and physical sciences. 304:221–35.

Futuyma, D.J. 1986. *Evolutionary Biology 2nd. ed.* Sinauer, Sunderland.

Futuyma, D.J. and Mayer, G.C. 1980. Non-allopatric speciation in animals. *Syst. Zool.* 29:254–71.

Gardiner, J.S. 1905. The Madreporarian corals. I. The Family Fungiidae, with a revision of its genera and species and an account of their geographical distribution. *Trans. Linn. Soc. Lond. Zool. Ser.* 2, 12:257–90.

Gardiner, J.S. 1909. Percy Sladen Trust Expedition to the Indian Ocean in 1905. The Madreporarian corals. 1 The Family Fungiidae, with a revision of its genera and species and an account of their geographical distribution. *Trans. Linn. Soc. Lond. Zool. Ser.* 2, 12:257–90.

Garthwaite, R.L. et al 1994. Electrophoretic identification of poritid species (Anthozoa: Scleractinia). *Coral Reefs* 13:49–56.

Gates, R. 1990. Seawater temperature and sublethal coral bleaching in Jamaica. *Coral Reefs* 8:193–7.

Gattuso, J.P. et al 1991. Physiology and taxonomy of scleractinian corals: a case study in the genus *Stylophora. Coral Reefs* 9:173–82.

Gaudian, G. 1988. *Taxonomic and ecological studies on Red Sea corals.* PhD Thesis, University of York.

Gayon, J. 1990. Critics and criticisms of the Modern Synthesis. The viewpoint of a Philosopher **in** Hecht M.K. et al (eds) *Evolutionary Biology.* Plenum Press, New York 24:1–50.

Geister, J. 1977. Occurrence of *Pocillopora* in late Pleistocene Caribbean coral reefs. *Mem. Bur. Rech. Geol. Min.* 89:378–88.

Ghiselin, M.T. 1974. A radical solution to the species problem. *Syst. Zool.* 23:536–44.

Ginsberg, R.N. and Beaudoin, B. 1990. *Cretaceous resources, events and rhythms. Background and plans for research.* NATO Series C: Mathematic and Physical Sciences 304.

Glynn, P.W. 1976. Some physical and biological determinants of coral community structure in the eastern Pacific. *Ecol. Monogr.* 46:431–56.

Glynn, P.W. 1984. Widespread coral mortality and the 1982–83 El Niño warming event. *Environ. Conserv.* 11:133–46.

Glynn, P.W. 1990. *Global ecological consequences of the 1982–83 El Niño-southern oscillation.* Elsevier Oceanography Series 52.

Glynn, P.W. and Wellington, G.M. 1983. *Corals and Coral Reefs of the Galapagos Islands.* University of California Press, Berkeley.

Glynn, P.W. et al 1991. Reef coral reproduction in the eastern Pacific: Costa Rica, Panama, and Galapagos Islands (Ecuador). *Mar. Biol.* 109:355–68.

Glynn, P.W. et al 1994. Experimental responses of Okinawan (Ryukyu Islands, Japan) reef corals to high sea temperature and UV radiation. *Proc. 7th Int. Coral Reef Symp.,* Guam: 27–8.

Goreau, T.F. 1961. On the relation of calcification to primary production in reef-building organisms **in** Lenhoff H.M. and Loomis W.F. (eds), *Biology of Hydra and some other coelenterates.* University of Miami Press, Florida.

Goreau, T.F. 1969. Post Pleistocene urban renewal in coral reefs. *Micronesica* 5:323–6.

Gower, J.C. and Ross, G.J.S. 1969. Minimum spanning trees and single linkage cluster analysis. *Appl. Stat.* 18:65–74.

Grant, V. 1981. *Plant speciation.* Columbia University Press, New York.

Grigg, R.W. 1981. *Acropora* in Hawaii. Part 2. Zoogeography. *Pac. Sci.* 35:15–24.

Grigg, R.W. 1988. Palaeoceanography of coral reefs in the Hawaiian-Emperor chain. *Science* 240:1737–43.

Grigg, R.W. and Epp, D. 1989. Critical depth for the survival of coral islands: effects on the Hawaiian Archipelago. *Science* 243:638–41.

Grigg, R.W. and Hey, R. 1992. Paleoceanography of the tropical eastern Pacific Ocean. *Science* 255:172–8.

Guzmán, H.M. and Cortés, J. 1989. Coral reef community structure at Caño Island, Pacific Costa Rica. *Mar. Ecol.* 10:23–41.

Guzmán, H.M. and Cortés, J. 1992. Cocos Island (Pacific coast of Costa Rica) coral reefs after the 1982–83 El Niño disturbance. *Rev. Biol. Trop.* 40:309–24.

Hallam, A. 1986. Evidence of displaced terrains from Permian to Jurassic faunas around the Pacific margins. *J. Geol. Soc.* (Lond.) 143:209–16.

Hallam, A. 1989. The case for sea-level change as a dominant causal factor in mass extinction in marine invertebrates. *Phil. Trans. Roy. Soc.* B 325:437–55.

Hallam, A. 1992. Phanerozoic sea-level changes. Columbia University Press, New York.

Haq, B.U. and van Eysinga, F.W.B. 1987. *Geological time table.* Elsevier, New York.

Haq, B.U. et al 1987. Chronology of fluctuating sea levels since the Triassic (250 million years ago to present). *Science* 235:1156–67.

Haq, B.U. et al 1988. Mesozoic and Cenozoic chronostratigraphy and eustatic cycles **in** Wilgus C.K. et al (eds), Sea level change: an integrated approach. *Soc. Econ. Paleontol. Mineral. Spec. Publ.* 42:71–108.

Harriott, V.J. 1992. Recruitment patterns of scleractinian corals in an isolated sub-tropical reef system. *Coral Reefs* 11:215–19.

Harris, P.Y. and Davies, P.J. 1989. Submerged reefs and terraces on the shelf edge of the Great Barrier Reef, Australia. Morphology, occurrence and implications for reef evolution. *Coral Reefs* 8:87–98.

Harrison, P.L. 1985. Sexual characteristics of scleractinian coral: systematic and evolutionary implementations. *Proc. 5th Int. Coral Reef Congr.*, Tahiti, 2:337–42.

Harrison, P.L. 1988. Pseudo-gynodioecy: an unusual breeding system in the scleractinian coral. *Galaxea fascicularis Proc. 6th Int. Coral Reef Symp.*, Australia, 2:699–705.

Harrison, P.L. and Wallace, C.C. 1990. Reproduction, dispersal and recruitment of Scleractinian corals **in** Dubinsky Z. (ed), *Coral reefs, ecosystems of the world.* 25. Elsevier, Amsterdam: 133–207.

Harrison, P.L. et al 1984. Mass spawning in tropical reef corals. *Science* 223:1186–9.

Harrison, R.G. 1993. Hybrids and hybrid zones: historical perspective **in** Harrison R.G. (ed), *Hybrid zones and the evolutionary process.* Oxford University Press, New York.

Hatcher, B.G. 1991. Coral reefs in the Leeuwin Current — a ecological perspective **in** Pearce A.F. and Walker D.I. (eds), The Leeuwin Current. *J. Roy. Soc. Western Australia* 74:115–27.

Hayashibara, T. et al 1993. Patterns of coral spawning at Akajima island, Okinawa, Japan. *Mar. Ecol. Prog. Ser.* 101:253–62.

Hays, J.D. and Pittman, W.C. 1973. Lithospheric plate motion, sea level changes, and climatic and ecological consequences. *Nature* 246:18–22.

Hayward, B.W. 1977. Lower Miocene corals from the Waitakere Ranges, North Auckland, New Zealand. *J. Roy. Soc. N.Z.* 7:99–111.

Hecht, M.K. and Hoffman, A. 1986. Why not neo-Darwinism? *Oxf. Surv. Evol. Biol.* 3:1–47.

Heck, K.L. and McCoy, E.D. 1978. Long-distance dispersal and the reef-building corals of the eastern Pacific. *Mar. Biol.* 48:348–56.

Hennig, W. 1966. *Phylogenetic systematics.* University of Illinois Press, Urbana.

Heyward, A.J. 1985. Chromosomes of the coral *Goniopora lobata* (Anthozoa: Scleractinia). *Heredity* 55:269–71.

Heyward, A.J. and Babcock, R.C. 1986. Self- and cross-fertilization in scleractinian corals. *Mar. Biol.* 90:191–5.

Heyward, A.J. and Stoddart, J.A. 1985. Genetic structures of two species of *Montipora* on a patch reef: conflicting results from electrophoresis and histocompatibility. *Mar. Biol.* 85:117–21.

Heyward, A. et al 1987. Sexual reproduction of corals in Okinawa. *Galaxea* 6:331–43.

Hillis, D.M. 1987. Molecular versus morphological approaches to systematics. *Annu. Rev. Ecol. Syst.* 18:23–42.

Hillis, D.M. and Dixon, M.T. 1991. Ribosomal DNA: molecular evolution and phylogenetic inheritance. *Quart. Rev. Biol.* 66:411–53.

Hillis, D.M. and Moritz, C. 1990. Molecular systematics. Sinauer, Sunderland.

Hodell, D.A. et al 1986. Late Miocene benthic $\delta^{18}O$ changes, global ice volume, sea level and the 'Messinian salinity crisis'. *Nature* 320:411–14.

Hodgson, G. in press. Scleractinian corals of Kuwait. *Pac. Sci.*

Hoegh-Guldberg, O. and Smith, G.J. 1989. The effect of sudden changes in temperature, light and salinity on the population density and export of zooxanthellae from the reef corals *Stylophora pistillata* Esper and *Seriatopora hystrix* Dana. *J. Exp. Mar. Biol. Ecol.* 129:279–303.

Hoeksema, B.W. 1989. Systematics and ecology of mushroom corals (Scleractinia: Fungiidae). *Zool. Verh. (Leiden)* 254.

Hoeksema, B.W. 1994. The position of northern New Guinea in the centre of maximum shallow-water diversity: a reef coral perspective. *Proc. 7th Int. Coral Reef Symp.*, Guam: 710–18.

Hoffman, A. 1989. *Arguments on evolution. A paleontologist's perspective.* Oxford University Press, Oxford.

Howarth, M.K. 1981. Palaeogeography of the Mesozoic **in** Cocks R.M. (ed), *The evolving earth.* Cambridge University Press, New York: 197–220.

Howard, W.R. and Prell, W. 1992. Late Quaternary surface circulation of the southern Indian Ocean and its relationship to orbital variations. *Paleoceanography* 7:79–117.

Howchin, W. 1909. Notes on the discovery of a large mass of living coral in Gulf St. Vincent. *Trans. Proc. Roy. Soc. South Australia* 33:242–52.

Hsu, K.J. 1986. Cretaceous/Tertiary boundary event **in** Hsu K.J. (ed), *Mesozoic and Cenozoic oceans.* Am. Geophysical Union Geodynamics Ser.15:75–84.

Hughes, T.P. 1989. Community structure and diversity of coral reefs: the role of history. *Ecology* 70:275–9.

Hughes, T.P. et al 1992. The evolutionary ecology of corals. *Trends Ecol. & Evol.* 7:292–5.

Hughes, R.D. et al in press. A western boundary current along the coast of Papua New Guinea. *Deep Sea Res.*

Hull, D.L. 1970. Contemporary systematic philosophies. *Annu. Rev. Ecol. Syst.* 1:19–54.

Hull, D.L 1976. Are species really individuals? *Syst. Zool.* 25:174–91.

Hull, D.L. 1979. The limits of cladism. *Syst. Zool.* 28:416–40.

Humphries, C.J. and Parenti, L.R. 1986. *Cladistic biogeography.* Oxford Monographs on Biogeography 2, Clarendon Press, Oxford.

Hunter, C.L. 1985. Assessment of clonal diversity and population structure of *Porites compressa* (Cnidaria, Scleractinia). *Proc. 5th Int. Coral Reef Congr.*, Tahiti, 6:69–74.

Huston, M.A. 1985. Patterns of species diversity on coral reefs. *Annu. Rev. Ecol. Syst.* 16:149–77.

Imbrie, J. and Imbrie, K.P. 1979. *Ice ages: solving the mystery.* Macmillan, London.

Imbrie, J. et al 1992. On the structure and origin of major glaciation cycles 1. Linear responses to Milankovitch forcing. *Paleoceanography* 7:701–38.

Irie, M. 1980. Growth of corals. *Mar. Pavilion* 9(3):5–8 (in Japanese).

Jackson, J.B.C. 1986. Modes of dispersal of clonal benthic invertebrates: consequences for species distributions and genetic structure of local populations. *Bull. Mar. Sci.* 39:588–606.

Jackson, J.B.C. 1991. Adaptation and diversity of reef corals. *Bioscience* 41:475–82.

Jackson, J.B.C. and Buss, L.W. 1975. Allelopathy and spatial competition among coral reef invertebrates. *Proc. Nat. Acad. Sci. U.S.A.* 72:5160–3.

Jacques, T.G. et al 1983. Experimental ecology of the temperate scleractinian coral *Astrangia danae.* II. Effects of temperature, light intensity and symbiosis with zooxanthellae on metabolic rate and calcification. *Mar. Biol.* 76:135–48.

Johannes, R.E. et al 1983. Latitudinal limits of coral reef growth. *Mar. Ecol. Prog. Ser.* 11:105–11.

Jokiel, P.L. 1990a. Long-distance dispersal by rafting: re-emergence of an old hypothesis. *Endeavour* 14:66–73.

Jokiel, P.L. 1990b. Transport of reef corals into the Great Barrier Reef. *Nature* 347:665–7.

Jokiel, P.L. and Coles, S.L. 1977. Effects of temperature on the mortality and growth of Hawaiian reef corals. *Mar. Biol.* 43:201–8.

Jokiel, P.L. and Coles, S.L. 1990. Response of Hawaiian and other Indo-Pacific reef corals to elevated temperature. *Coral Reefs* 8:155–62.

Jokiel, P. and Martinelli, F.J. 1992. The vortex model of coral reef biogeography. *J. Biogeogr.* 19:449–58.

Karlson, R.H. and Hurd, L.E. 1993. Disturbance, coral reef communities, and changing ecological paradigms. *Coral Reefs* 12: 117–25.

Kauffman, E.G. 1984. The fabric of Cretaceous marine extinctions **in** Berggren W.A. and Van Couvering J.A. (eds), *Catastrophes and Earth History*. Princeton University Press, Princeton: 151–246.

Kauffman, E.G. and Fagerstrom, J.A. 1993. The phanerozoic evolution of reef diversity **in** Ricklefs R.E. and Schluter D. (eds), *Species diversity in ecological communities*. University Chicago Press, Chicago: 315–29.

Kauffman, E.G. and Johnson, C.C. 1988. The morphological and ecological evolution of middle and upper Cretaceous reef-building rudistids. *PALAIOS* 3:194–216.

Kawabe, M. 1985. El Niño effects in the Kuroshio and western north Pacific **in** Wooster W.S. and Fluhardy D.L. (eds), *El Niño North: Niño effects in the eastern subarctic Pacific Ocean*. Sea Grant Project, University of Washington: 31–43.

Kay, E.A. 1984. Patterns of speciation in the Indo–west Pacific **in** Radovsky F.J. et al (eds), *Biogeography of the tropical Pacific*. Bishop Mus. Spec. Publ. 72:15–30.

Kay, E.A. and Palumbi, S.R. 1987. Endemism and evolution in Hawaiian marine invertebrates. *Trends Ecol. & Evol.* 2:183–6.

Kendrick, G.W. et al 1991. Plio-Pleistocene coastal events and history along the western margin of Australia. *Quat. Sci. Rev.* 10:419–30.

Kennett, J.P. 1982. Sea-level change and the coastal zone *Marine Geology*. Prentice-Hall, Englewood Cliffs, New Jersey.

Kenyon, J.C. 1992. Sexual reproduction in Hawaiian *Acropora*. *Coral Reefs* 11:37–43.

Kenyon, J.C. 1994. Chromosome number in ten species of the coral genus *Acropora*. *Proc. 7th Int. Coral Reef Symp.*, Guam: 471–6.

Kimura, M. 1991. Quaternary land bridges of the Ryukyu Arc detected on seismic reflecting profiles. Spec. vol. Univ. Tohoku 35:109–17 (in Japanese).

Kinsey, D.W. 1977. Seasonality and zonation in coral reef productivity and calcification. *Proc. Third Intern. Coral Reef Symp.* 2:383–388.

Kinsey, D.W. 1985. The functional role of back reef and lagoonal systems in the central Great Barrier reef. *Proc. 5th Int. Coral Reef Congr.*, Tahiti 6:223–8.

Kluge, A.G. 1988. Parsimony in vicariance biogeography: a quantitative method and a greater Antillean example. 137:315–28.

Knowlton, N. et al 1992. Sibling species in *Montastraea annularis*, coral bleaching, and the coral climate record. *Science* 255:330–3.

Kruskal, J.B. 1964. Multidimensional scaling by optimizing goodness of fit to a nonmetric hypothesis. *Psychometrika* 29:1–27.

Laborel, J. 1974. West African reef corals: an hypothesis on their origin. *Proc. 2nd Int. Coral Reef Symp.*, Brisbane 1:425–43.

Lang, J.C. 1984. Whatever works: the variable importance of skeletal and of non-skeletal characters in scleractinian taxonomy. *Palaeontogr. Am.* 54:18–44.

Lasaga, A.C. et al 1985. An improved geochemical model of atmospheric CO_2 fluctuations over the past 100 million years **in** Sundquist E.T. and Broecker W.S. (eds), The carbon cycle and atmospheric CO_2: natural variations archean to present. *Geophys. Monogr.* 32:397–411.

Lathuillière, B. 1988. Analyse de populations d'Isastrees bajociennes (Scleractiniaires Jurassiques de France). Consequences taxonomiques, strategraphiques et paleoecologiques. *Geobios* 21:269–305.

Lessios, H.A. et al 1983. Mass mortalities of coral reef organisms. *Science* 222:715.

Levinton, J. 1988. *Genetics, paleontology, and macroevolution.* Cambridge University Press, Cambridge.

Lewontin, R.C. 1974. *The genetic basis of evolutionary change.* Columbia University Press, New York.

Lincoln, J.M. and Schlanger, S.O. 1987. Miocene sea-level falls related to the geologic history of Midway Atoll. *Geology* 15:454–7.

Lincoln, R.J. et al 1982. *A dictionary of ecology, evolution and systematics.* Cambridge University Press, Cambridge.

Linden, M. 1990. Replicators, hierarchy, and the species problem. *Cladistics* 6:183–6.

Lowenstein, J.M. 1985. Molecular approaches to the identification of species. *Am. Sci.* 73:541–7.

Lowenstein, J.M. 1986. Molecular phylogenetics. *Annu. Rev. Earth Planet. Sci.* 14:71–83.

Loya,Y. 1972. Community structure and species diversity of hermatypic corals at Eilat, Red Sea. *Mar. Biol.* 13:100–23.

MacArthur, R.H. and Wilson, E.O. 1963. An equilibrium theory of insular biogeography. *Evolution* 17:373–87.

MacArthur, R.H. and Wilson, E.O. 1967. *The theory of island biogeography.* Princeton University Press, Princeton.

MacIntyre, I.G. and Pilkey, O.H. 1969. Tropical reef corals: tolerance of low temperatures in the North Carolina continental shelf. *Science* 166:374–5.

Maier-Reimer, E. and Mikolajewicz, U. 1990. Ocean general circulation model sensitivity experiment with an open Central American Isthmus. *Paleoceanography* 5:349–66.

Mann, K.H. and Lazier, J.R.N. 1991. *Dynamics of marine ecosystems: biological–physical interactions in the oceans.* Blackwell Scientific Publications, Boston.

Maragos, J.E. and Jokiel, P.L. 1978. Reef corals of Canton Atoll. I. Zoogeography. *Atoll Res. Bull.* 221:55–70.

Maragos, J.E. and Jokiel, P.L. 1986. Reef corals of Johnston Atoll: one of the world's most isolated reefs. *Coral Reefs* 4:141–50.

Marsh, L.M. 1994. The occurrences and growth of *Acropora* in extra-tropical waters off Fremantle, Western Australia. *Proc. 7th Int. Coral Reef Symp.*, Guam: 1233–9.

Matthews, R.K. and Poore, R.Z. 1980. Tertiary $\delta^{18}O$ record and glacio-eustatic sea level fluctuations. *Geology* 8:501–4.

Mayr, E. 1942. *Systematics and the origin of species.* Columbia University Press, New York.

Mayr, E. 1963. *Animal species and evolution.* Harvard University Press, Cambridge, Mass.

Mayr, E. 1970. *Populations, species and evolution.* Harvard University Press, Cambridge, Mass.

Mayr, E. 1988. The why and how of species.. *Biol. & Philos.* 3:431–41.

McCoy, E.D. and Heck, K.L. 1976. Biogeography of corals, seagrasses and mangroves: an alternative to the centre of origin concept. *Syst. Zool.* 25:201–10.

McCreary, J.P and Anderson, D.L.T. 1991. An overview of couples ocean–atmosphere models of El Niño and the southern ocean. *J. Geophys. Res.* 96:3125–50.

McDade, L.A. 1992. Hybrids and phylogenetic systematics II. the impact of hybrids on cladistic analysis. *Evolution* 46:1329–46.

McManus, J.W. 1985. Marine speciation, tectonics and sea-level changes in southeast Asia. *Proc. 5th Int. Coral Reef Congr.*, Tahiti 4:133–8.

McMillan, D. and Miller, D.J. 1990. Highly repeated DNA sequences in the scleractinian coral genus *Acropora:* evaluation of cloned repeats as taxonomic probes. *Mar. Biol.* 104:483–7.

Mercer, J.H. 1987. The Antarctic ice sheet during the late Neogene. *Paleoecology of Africa* 18:21–32.

Miller, D. and Veron, J.E.N. 1990. Biochemistry of a special relationship. *New Sci.* June 1990: 44–9.

Miller, K.J. 1994. *The Platygyra species complex: implications for coral taxonomy and evolution.* PhD thesis, James Cook University of North Queensland, Australia.

Misaki, H. 1984. On the mortality of scleractinian corals due to cold winter. *Mar. Pavilion* 13(12):1–3 (in Japanese).

Misaki, H. 1985. On the survival of corals against low water temperature in winter. *Mar. Parks J.* 68:15–19 (in Japanese).

Mishler, B.D. and Donoghue, M.J. 1982. Species concepts: a case for pluralism. *Syst. Zool.* 31:491–503.

Moore, W.S. 1982. Late Pleistocene sea-level history **in** Ivanovitch M. and Harmon R.S. (eds), *Uranium series disequilibrium: applications to environmental problems.* Clarendon Press, Oxford.

Moritz, C. 1992. Molecular systematics. *Australian Biol.* 5:40–7.

Moritz, C. and Hillis, D. 1990. Molecular systematics: context and controversies **in** Hillis D. and Moritz C. (eds), *Molecular systematics.* Sinauer, Sunderland.

Muscatine, L. 1990. The role of symbiotic algae in carbon and energy flux in reef corals **in** Dubinsky Z. (ed), *Coral reefs, ecosystems of the world.* 25. Elsevier, Amsterdam: 75–87.

Muscatine, L. and Porter, J.W. 1977. Reef corals: mutualistic symbiosis adapted to nutrient-poor environments. *Bioscience* 27: 454–60.

Myers, A.A. and Giller, P.S. 1988. *Analytical biogeography: an integrated approach to the study of animal and plant distributions.* Chapman and Hall, London.

Nei, M. 1987. *Molecular evolutionary genetics.* Columbia University Press, New York.

Neigel, J.E. and Avise, J.C. 1986. Phylogenetic relationships of mitochondrial DNA under various demographic models of speciation **in** Nevo, E. and Karlin, S. (eds), *Evolutionary processes and theory.* Academic Press, New York.

Nelson, G. 1989. Species and taxa: systematics and evolution **in** Otte, D. and Endler, J.A. (eds), *Speciation and its consequences.* Sinauer, Massachusetts.

Nelson, G. and Platnick, N.I. 1981. *Systematics and Biogeography: cladistics and vicariance.* Columbia University Press, New York.

Newell, N.D. 1971. An outline history of tropical organic reefs. *Am. Mus. Novit.* 2465: 1–37.

Newman, W.A. 1986. Origin of the Hawaiian marine fauna and vicariance as indicated by barnacles and other organisms, in crustacean biogeography **in** Gore R.H. and Heck K.L. (eds), *Crustacean biogeography.* Balkema, Rotterdam: 21–49.

Newton, C.R. 1988. Significance of 'Tethyan' fossils in the American Cordillera. *Science* 242:385–91.

Nicolis, C. and Keppenne, C.L. 1989. Climate predictability: a dynamical view **in** Berger A. et al (eds), *Climate and Geo-sciences; a challenge for science and society in the 21st century.* Kluwer Academic Publ., Dordrecht.

Nixon, K.C. and Wheeler, Q.D. 1990. An amplification on the phylogenetic species concept. *Cladistics* 6:211–23.

Nomura, K. 1986. Stony corals of Kuroshima which did not recover (2) Death of stony corals due to low water temperature. *Mar. Pavilion* 15(7), 3–4 (in Japanese).

Odum, E.P. 1969. The strategy of ecosystem development. *Science* 164:262–70.

Officer, C.B. et al 1987. Late Cretaceous and paroxysmal Cretaceous/Tertiary extinctions. *Nature* 326:143–9.

Oliver, J. et al 1983. Bathymetric adaptations of reef building corals at Davies Reef, Great Barrier Reef. I. Long term growth responses of *Acropora formosa* (Dana 1846). *J. Exp. Mar. Biol. Ecol.* 73:11–35.

Oliver, J. et al 1988. The geographic extent of mass coral spawning: clues to ultimate causal factors. *Proc. 6th Int. Coral Reef Symp.,* Australia 2:803–10.

Oliver, W.A. 1980. The relationship of the scleractinian corals to the rugose corals. *Paleobiology* 6:146–60.

Oliver, W.A., 1989. Intraspecific variation in pre-Carboniferous rugose corals: a subjective review. *Fossil Cnidaria* 5:1–6.

Oosterbaan, A.F.F. 1990. Notes on a collection of Badenian (Middle Miocene) corals from Hungary in the National Museum of Natural History at Leiden (The Netherlands). *Contrib. Tert. Quat. Geol.* 27:3–15.

Page, R.D.M. 1988 Quantitative cladistic biogeography: constructing and comparing area cladograms. *Syst. Zool.* 37:254–70.

Page, R.D.M. 1990a. Component analysis: a valiant failure? *Cladistics* 6:119–36.

Page, R.D.M. 1990b. Temporal congruence and cladistic analysis of biogeography and cospeciation. *Syst. Zool.* 39:205–26.

Pandolfi, J.M. 1992. Successive isolation rather than evolutionary centres for the origination of Indo-Pacific reef corals. *J. Biogeogr.* 19:593–609.

Pandolfi, J.M. 1994. Tectonic history of New Guinea and its biogeographic significance. *Proc. 7th Int. Coral Reef Symp.*, Guam: 718–29.

Park, J. and Herbert, T.D. 1987. Hunting for paleoclimatic periodicities in a geologic time series with an uncertain time scale. *J. Geophys. Res.* 92:14027–40.

Paterson, H.E.H. 1981. The continuing search for the unknown and unknowable: A critique of contemporary ideas on speciation. *S. Afr. J. Sci.* 77:113–19.

Paterson, H.E.H. 1985. The recognition concept of species **in** Vrba E.S. (ed), *Species and speciation*. Transvaal Museum Monographs 4, Pretoria.

Patterson, C. 1981. Significance of fossils in determining evolutionary relationships. *Annu. Rev. Ecol. Syst.* 12:195–223.

Paulay, G. 1990. Late Cenozoic sea level fluctuations and the diversity and species composition of insular shallow water marine faunas **in** Dudley E.C. (ed), *The unity of evolutionary biology. Proceedings 4th International Congress of Systematic and Evolutionary Biology*. Dioscorides Press, Portland, Oregon: 184–93.

Paulay G. and Spencer, T. 1988. Geomorphology, paleoenvironments and faunal turnover, Henderson Is, S.E. Polynesia. *Proc. 6th Int. Coral Reef Symp.*, Australia 3:461–6.

Pearce, A.F. and Walker, D.I. (eds.) 1991. The Leeuwin Current. *J. Roy. Soc. Western Australia* 74.

Pestiaux, P. et al 1988. Paleoclimatic variability at frequencies ranging from 1 cycle per 10,000 years to 1 cycle per 1,000 years: evidence for nonlinear behaviour of the climate system. *Climatic change* 12:9–37.

Pfister, T. 1977. Das problem der variationsbreite von korallen aus beispiel der oligozanen *Antiguastraea lucasiana* (Defrance). *Eclogae Geol. Helv.* 70:825–43.

Pickard, G.L. et al 1977. *A review of the physical oceanography of the Great Barrier Reef and the western Coral Sea*. Australian Institute of Marine *Science* Monograph Series 2.

Pielou, E.C. 1979. *Biogeography*. John Wiley & Sons, New York.

Platnick, N.I. 1987. Cladistics and phylogenetic analysis today **in** Fernholm B. et al (eds), *The hierarchy of life. Molecules and morphology in phylogenetic analysis*. Excerpta Medica, Amsterdam: 17–24.

Platnick, N.I. and Funk, V.A. 1983. Advances in cladistics *Proceedings of the second meeting of the Willi Hennig Society 2*. Columbia University Press, New York.

Plaziat, J.C. and Perrin, C. 1992. Multikilometer sized reefs built by foraminiferans (Solenomeris) from the early Eocene of the Pyrenean domain (S. France, N. Spain): paleoecologic relations with coral reefs. *Palaeogeogr. Palaeoclimatol. Palaeoecol.* 96:195–232.

Pomar, L. 1991. Reef geometries, erosion surfaces and high-frequency sea-level changes, upper Miocene reef complex, Mallorca, Spain. *Sedimentology* 38:243–69.

Porter, J.W. 1974. Zooplankton feeding by the Caribbean reef-building coral. *Montastrea cavernosa Proc. 2nd Int. Coral Reef Symp.*, Brisbane 1:111–25.

Porter, J.W. et al 1982. Perturbation and change in coral reef communities. *Proc. Nat. Acad. Sci. USA* 79:1678–81.

Porter, J.W. et al 1984. Primary production and photoadaptation in light and shade-adapted colonies of the symbiotic coral *Stylophora pistillata*. *Proc. R. Soc. Lond. Ser. B Biol. Sci.* 222:161–80.

Potts, B.M. and Reid, J.B. 1985. Analysis of a hybrid swarm between *Eucalyptus risdonii* Hook and *E. amygdalina* Labill. *Aust. J. Bot.* 33:543–62.

Potts, D.C. 1983. Evolutionary disequilibrium among Indo-Pacific corals. *Bull. Mar. Sci.* 33:619–32.

Potts, D.C. 1984a. Natural selection in experimental population of reef-building corals (Scleractinia). *Evolution* 38:1059–78.

Potts, D.C. 1984b. Generation times and the Quaternary evolution of reef-building corals. *Paleobiology* 19:48–58.

Potts, D.C 1985. Sea level fluctuation and speciation in Scleractinia. *Proc. 5th Int. Coral Reef Congr.*, Tahiti 4:127–32.

Potts, D.C. and Garthwaite, R.L. 1990. Evolution of reef-building corals during periods of rapid global change **in** Dudley E.C. (ed), *The unity of evolutionary biology. Proceedings 4th International Congress of Systematic and Evolutionary Biology.* Dioscorides Press, Portland, Oregon: 170–8.

Potts, D.C. et al 1985. Dominance of a coral community of the genus Porites *(Scleractinia). Mar. Ecol. Prog. Ser.* 23:79–84.

Potts, D.C. et al in press. Soft tissue vs. skeletal approaches to species recognition and phylogeny reconstruction in corals. *6th Fossil Cnidaria Symp. Proc.*

Prahl, H. von and Mejia, A. 1985. Primer informe de un coral acroporida, *Acropora valida* (Dana, 1846) (Scleractinia: Astrocoeniida: Acroporidae) para el Pacifico americana. *Rev. Biol. Trop.* 33:39–43.

Prell, W.I. and Kutzbach, J.E. 1987. Monsoon variability over the last 15,000 years. *J. Geophys. Res.* 92:8411–25.

Preston, F.W. 1962. The canonical distribution of commonness and rarity. *Ecology* 43:185–215.

Purdy, E.G. 1974. Reef configurations: cause and effect reefs in time and space: selected examples from the recent and ancient. *Soc. Econ. Paleontol. Mineral. Spec. Publ.* 18.

Randall, J.E. 1981. Examples of antitropical and antiequatorial distribution of Indo–west Pacific fishes. *Pac. Sci.* 35:197–209.

Raup, D.M. 1979. Biases in the fossil record of species and genera. *Bull. Carnegie Mus. Nat. Hist.* 13:85–91.

Raup, D.M. and Boyajian, G.E. 1988. Patterns of generic extinction in the fossil record. *Paleobiology* 14:109–25.

Raup, D.M. and Sepkoski, J. 1984. Periodicity of extinctions in the geologic past. *Proc. Nat. Acad. Sci.* 81:801–5.

Raup, D.M. and Stanley, S.M. 1978. *Principles of paleontology.* W.H. Freeman & Co., San Francisco.

Reising, J.M. and Ayre, D.J. 1985. The usefulness of tissue-grafting bioassay as an indicator of clonal identity in scleractinian corals (Great Barrier Reef, Australia). *Proc. 5th Intern. Coral Reef Congr.* 6:75–81.

Reys Bonilla, H. 1992. New records for hermatypic corals (Anthozoa: Scleractinia) in the Gulf of California, Mexico, with an historical and biogeographical discussion. *J. Nat. Hist.* 26:1163–75.

Rhode, K. 1992. Latitudinal gradients in species diversity: the search for the primary cause. *Oikos* 65:514–27.

Richmond, R.H. 1985. Reversible metamorphosis in coral planula larvae. *Mar. Ecol. Prog. Ser.* 22:181–5.

Richmond, R.H. 1987a. Energetics, competency, and long-distance dispersal of planula larvae of the coral *Pocillopora damicornis. Mar. Biol.* 93:527–33.

Richmond, R.H. 1987b. Energetic relationships and biogeographic differences among fecundity, growth and reproduction in the reef coral *Pocillopora damicornis. Bull. Mar. Sci.* 41:594–604.

Richmond, R.H. 1990. The effects of the El Niño/southern oscillation on the dispersal of corals and other marine organisms **in** P.W. Glynn (ed), *Global ecological consequences of the 1982–83 El Niño southern oscillation.* Elsevier, Amsterdam.

Richmond, R.H. and Hunter, C.L. 1990. Reproduction and recruitment of corals: comparisons among the Caribbean, the tropical Pacific, and the Red Sea. *Mar. Ecol. Prog. Ser.* 60:185–203.

Richmond, R.H. and Jokiel, P.L. 1984. Lunar periodicity in larva release in the reef coral *Pocillopora damicornis* at Enewetak and Hawaii. *Bull. Mar. Sci.* 34: 280–7.

Ricklefs, R.E. 1987. Community diversity: relative roles of local and regional processes. *Science* 235:167–71.

Riegl, B. 1993. Ecology and taxonomy of South African reef corals. PhD thesis. Univ. Cape Town.

Rieppel, O. 1986. Species are individuals. A review and critique of the argument. *Evol. Biol.* 20:283–317.

Riseberg, L.H. and Wendel, J.F. 1993. Introgression and its consequences in plants **in** Harrison R.G. (ed), *Hybrid zones and the evolutionary process.* Oxford University Press, New York.

Rinkevich, B. and Loya, Y. 1979. The reproduction of the Red Sea coral *Stylophora pistillata.* II. Synchronization in breeding and seasonality of planular shedding. *Mar. Ecol. Prog. Ser.* 1:145–52.

Roberts, H.H. et al 1982. Cold water stress in Florida Bay and northern Bahamas: a product of winter air outbreaks. *J. Sediment. Petrol.* 52:145–55.

Roemmich, D. and Cornuelle, B. 1990. Observing the fluctuations of gyre-scale ocean circulation: a study of the subtropical South Pacific. *J. Phys. Oceanogr.* 20:191–4.

Rögl, F and Steininger, F.F. 1984. Neogene Paratethys, Mediterranean and Indo-Pacific seaways **in** Brenchley P. (ed), *Fossils and Climate.* John Wiley and Sons: 171–200.

Romine, K and Lombari, G. 1985. Evolution of Pacific circulation in the Miocene: radiolarian evidence from DSDP site 289. *Geol. Soc. Am. Mem.* 163:273–90.

Roniewicz, E. 1984. Aragonitic Jurassic corals from erratic boulders on the south Baltic coast. *Ann. Soc. Geol. Poloniae* 54:65–77.

Roniewicz, E. 1989. Triassic scleractinian corals of the Zlambach Beds, northern Calcareous Alps, Austria. *Osterreichische Akademie Wissenschaften.* Springer-Verlag, Wien: 1–152.

Roniewicz, E. and Morycowa, E. 1989. Triassic Scleractinia and the Triassic/Liassic boundary. *Mem. Assoc. Aust. Palaeontol.* 8:347–54.

Rosen, B.R. 1971a. The distribution of reef coral genera in the Indian Ocean **in** Stoddart D.R. and Yonge C.M. (eds), Regional variation in Indian Ocean coral reefs. *Symp. Zool. Soc. Lond.* 28:263–99.

Rosen, B.R. 1971b. Annotated check list and bibliography of corals of the Chagos Archipelago (including the recent collection from Diego Garcia) with remarks on their distribution. *Atoll Res. Bull.* 149:67–88.

Rosen, B.R. 1975. The distribution of reef corals. *Report Underwater Association* 1:1–16.

Rosen, B.R. 1981. The tropical high diversity enigma — the corals' eye view **in** P.L. Foley (ed), *Chance, Change and Challenge. The Evolving Biosphere.* London and Cambridge: 103–29.

Rosen, B.R. 1984. Reef coral biogeography and climate through the late Cainozoic: just islands in the sun or a critical pattern of islands **in** Brenchley P. (ed), *Fossils and Climate.* John Wiley & Sons: 201–62.

Rosen, B.R. 1988a. Biogeographic patterns: a perceptual overview **in** Myers A.A. and Giller P.S. (eds), *Analytical Biogeography: an integrated approach to the study of animal and plant distributions.* Chapman and Hall, London: 23–55.

Rosen, B.R. 1988b. Progress, problems and patterns in the biogeography of reef corals and other tropical marine organisms. *Helgol. Meeresunters.* 40:269–301.

Rosen, B.R. and Turnsek, D. 1989. Extinction patterns and biogeography of scleractinian corals across the Cretaceous/Tertiary boundary. *Mem. Assoc. Aust. Palaeontol.* 5:355–70.

Rotondo, G.M. et al 1981. Plate movement and island integration — a possible mechanism in the formation of endemic biotas, with special reference to the Hawaiian Islands. *Syst. Zool.* 30:12–21.

Rouchy, J.M.. et al 1986. Evolution and antagonism of coral and microbial communities at the end of the Miocene in the western Mediterranean: biology and sedimentology. *Bull. Cent. Rech. Explor. Prod. Elf-Aquitaine* 10:333–48.

Rowan, R. and Knowlton, N. (in prep). Intraspecific diversity and ecological zonation in coral-algal symbiosis.

Rowan, R. and Powers, D.A. 1992. Ribosonial RNA sequences and the diversity of symbiotic dinoflagellates (zooxanthellae). *Proc. Nat. Acad. Sci.* USA 89:3639–43.

Sammarco, P.W. 1986. Polyp bail-out: an escape response to environmental stress and a new means of reproduction in corals. *Mar. Ecol. Prog. Ser.* 10:57–65.

Saville-Kent, W. 1900. *The Great Barrier Reef of Australia: its products and potentialities.* Allen and Co., London.

Scheer, G. 1984. The distribution of reef corals in the Indian Ocean with a historical review of its investigation. *Deep-Sea Res.* A 31:885–900.

Schlanger, S.O. 1986. High-frequency sea-level fluctuations in Cretaceous time: an emerging geophysical problem **in** Hsu K.J. (ed), *Mesozoic and Cenozoic oceans.* Am. Geophysical Union Geodynamics Ser.15:61–74.

Schuhmacher, H. and Zibrowius, H. 1985. What is hermatypic? a redefinition of ecological groups in corals and other organisms. *Coral Reefs* 4:1–9.

Scott, R.W. 1988. Evolution of late Jurassic and early Cretaceous reef biotas. *PALAIOS* 3:184–93.

Scrutton, C.T. 1989. Intracolonial and intraspecific variation in tabulate corals. *Fossil Cnidaria* 5:33–43.

Scrutton, C.T. and Clarkson, N.K. 1991. A new scleractinian-like coral from the Ordovician of the southern uplands, Scotland. *Palaeontology* 34:179–94.

Seberg, O. 1986. A critique of the theory and methods of panbiogeography. *Syst. Zool.* 35:369–80.

Semter, A.J and Chervin, R.M. 1992. Ocean general circulation from global eddy-resolving model. *J. Geophys. Res.* 97:5493–500.

Shackleton, N.J. and Kennett, J.P. 1975. Paleotemperature history of the Cenozoic and the initiation of Antarctic glaciation: oxygen and carbon analysis in DSDP sites 277, 279 and 281. *Init. Repts.* DSDP 29:743–5.

Sheppard, C.R.C. 1985. Reefs and coral assemblages of Saudi Arabia. Fringing reefs in the southern region, Jeddah to Jizan. *Fauna Saudi Arabia* 7:37–58.

Sheppard, C.R.C. 1987. Coral species of the Indian Ocean and adjecent seas: a synonymised compilation of some regional distribution patterns. *Atoll Res. Bull.* 307: 1–32.

Sheppard, C.R.C. and Sheppard, A.L.S. 1991. *Corals and coral communities of Saudi Arabia.* Fauna of Saudi Arabia 12:1–170.

Shinn, E.A. 1976. Coral reef recovery in Florida and the Persian Gulf, *Environ. Geol.* 1:241–54.

Sikes, E.L. et al 1991. Pliocene paleoceanography: circulation and oceanographic changes associated with the 2.4ma glacial event. *Palaeoceanography* 6:245–57.

Simberloff, D. et al 1981. There have been no statistical tests of cladistic biogeographical hypotheses **in** Nelson G. and D.E. Rosen (eds), *Vicariance Biogeography. A Critique.* Columbia University Press, New York: 40–63.

Simpson, C.J. 1985. *Mass spawning of scleractinian corals in the Dampier Archipelago and the implications for management of coral reefs in Western Australia.* Western Australian Department of Conservation and Environment, Perth, Western Australia. Bulletin.

Simpson, C.J. 1991. Mass spawning of corals on Western Australian reefs and comparisons with the Great Barrier Reef. *J. Roy. Soc. Western Australia* 74:85–92.

Simpson, G.G. 1944. *Tempo and mode in evolution.* Columbia University Press, New York.

Slater, R.A. and Phipps, C.V.G. 1977. A preliminary report on the coral reefs of Lord Howe Island and Elizabeth Reef, Australia. *Proc. 3rd Int. Coral Reef Symp.,* Miami 2:314–18.

Smith, A.G. et al 1981. *Phanerozoic paleocontinental world maps.* Cambridge University Press.

Smith, S.D.A. and Simpson, R.D. 1991. Nearshore corals of the Coffs Harbour region, mid north coast, New South Wales. *Wetlands (Australia)* 11:111–19.

Smith, S.V. 1981. The Houtman Abrolhos Islands: carbon metabolism of coral reefs at high latitude. *Limnol. Oceanogr.* 26:612–21.

Smith, S.V. and Buddemeier, R.W. 1992. Global change and coral reef ecosystems. *Annu. Rev. Ecol. Syst.* 23:89–118.

Sober, E. 1988. The conceptual relationship of cladistic phylogenetics and vicariance biogeography. *Syst. Zool.* 37:245–53.

Sorauf, J.E. and Mackey, S.D. 1989. Variation and biometrics in rugose corals. *Fossil Cnidaria* 5:23–31.

Springer, V.G. 1982. Pacific Plate biogeography, with special reference to shorefishes. *Smithson. Contrib. Zool.* 367:1–181.

Springer, V.G. and Williams, J.T. 1990. Widely distributed Pacific plate endemics and lowered sea-level. *Bull. Mar. Sci.* 47:631–40.

Stanley, G.D. 1981. Early history of scleractinian corals and its geological consequences. *Geology* 9:507–11.

Stanley, G.D. 1987. Travels of an ancient reef: the odyssey of a coral island is just part of the epic of continental drift. *Nat. Hist.* 11:36–44.

Stanley, G.D. 1988. The history of early Mesozoic reef communities: a three-step process. *PALAIOS* 3:170–83.

Stanley, G.D. 1991. Marine faunal and floral similarities between the Tethys and western North America: tectonic and pantropical models. *Saito Hon-on Kai Spec. Publ.* 3 Rendai: 51–2.

Stanley, G.D. and Swart, P.K. 1984. A geochemical method for distinguishing zooxanthellate and non-zooxanthellate corals in the fossil record **in** Ginsberg R.N. (ed), *Advances in reef science*. Univ. Miami Rosenstiel School. Mar. Atmospheric Sci., Florida:118–19.

Stanley, S.M. 1978. Chronospecies' longevities, the origin of genera, and the punctuational model of evolution. *Paleobiology* 4:26–40.

Stanley, S.M. 1979. *Macroevolution — pattern and process*. W.H. Freeman, San Francisco.

Stanley, S.M. 1982. Macroevolution and the fossil record. *Evolution* 36:460–73.

Stanley, S.M. 1984a. Marine mass extinctions: a dominant role for temperature **in** Nitecki M.H. (ed), *Extinctions*. University of Chicago Press: 79–117.

Stanley, S.M. 1984b. Temperature and biotic crises in the marine realm. *Geology* 12:205–8.

Stanley, S.M. 1988. Climatic cooling and mass extinction of Paleozoic reef communities. *PALAIOS* 3:228–32.

Stanley, S.M. 1990. The general correlation between rate of speciation and rate of extinction: fortuitous causal linkages **in** Ross R.M. and Allmon W.D. (eds), *Causes of evolution. A paleontological perspective*. University of Chicago Press, Chicago: 103–27.

Stehli, F.G. and Wells, J.W. 1971. Diversity and age patterns in hermatypic corals. *Syst. Zool.* 20:115–26.

Stenseth, N.C. and Maynard Smith, J. 1984. Coevolution in ecosystems: Red Queen evolution or stasis? *Evolution* 38:870–80.

Stevens, G.C. 1989. The latitudinal gradient in geographical range: how so many species coexist in the tropics. *Am. Nat.* 133:240–56.

Stobart, B. and Benzie, J.A.H. 1994. Allozyme electrophoresis demonstrates that the scleractinian coral *Montipora digitata* is two species. *Mar. Biol.* 118:183–90.

Stobart, B. et al 1994. Biannual spawning of three species of scleractinian coral from the Great Barrier Reef. *Proc. 7th Int. Coral Reef Symp.*, Guam: 494–500.

Stoddart, D.R. 1969. Ecology and morphology of recent coral reefs. *Biol. Rev.* 44:433–498.

Stoddart, J.A. 1983. Asexual production of planulae in the coral *Pocillopora damicornis*. *Mar. Biol.* 76: 279–84.

Stoddart, J.A. 1984. Genetic structure within populations of the coral *Pocillopora damicornis*. *Mar. Biol.* 81:19–30.

Stoddart, J.A. et al 1985. Self-recognition in sponges and corals? *Evolution* 39:461–3.

Surlyk, F. 1990. Cretaceous–Tertiary (Marine) in Briggs D.E.G. and Crowther P.R. (eds), *Palaeobiology: a synthesis*. Blackwell Scientific: 198–203.

Szmant, A.M.. and Gassman, N.J. 1990. The effects of prolonged bleaching on the tissue biomass and reproduction of the reef coral *Montastrea annularis*. *Coral Reefs* 8:217–24.

Templeton, A.R. 1980. Modes of speciation and inferences based on genetic distances. *Evolution* 34:719–29.

Templeton, A.R. 1981. Mechanisms of speciation a population genetic approach. *Annu. Rev. Ecol. Syst.* 12:23–48.

Templeton, A.R. 1989. The meaning of species and speciation: a genetic perspective **in**

Otte D. and Endler J.A. (eds), *Speciation and its consequences.* Sinauer, Massachusetts.

Thiel, M.E. 1928. Beitrage zur Kenntnis der Meeresfauna Westafrikas. *Madreporaria,* Hamburg 3:253–350.

Tozer, E.T. 1982. Marine Triassic faunas of North America: their significance in assessing plate and terrane movements. *Geologische Rundschau* 71:1077–1104.

Trench, R.K. 1992. Microalgal-invertebrate symbiosis, current trends. *Encycl. Microbiol.* 3:129–42.

Tribble, G.W. and Randall, R.H. 1986. A description of the high-latitude shallow water coral communities of Miyake-jima, Japan. *Coral Reefs* 4:151–9.

Utinomi, H. 1970. *Corals of Ehime Prefecture.* National Museum of Ehime Prefecture Publication (in Japanese).

Vail, P.R. and Hardenbol, J. 1979. Sea-level changes during the Tertiary. *Oceanus* 22:71–80.

Valentine, 1973. *Evolutionary paleoecology of the marine biosphere.* Prentice-Hall, New Jersey.

Valentine, J.W. 1984a. Climate and evolution in the shallow sea **in** Brenchley P. (ed), *Fossils and Climate.* John Wiley and Sons: 265–77.

Valentine, J.W. 1984b. Neogene marine climate trends: implications for biogeography and evolution of the shallow-sea biota. *Geology* 12:647–50.

Valentine, J.W., and Jablonski, D. 1991. Biotic effects of sea level change: the Pleistocene test. *J. Geophys. Res.* 96(B4):6873–8.

Van Valen, L.M. 1973. A new evolutionary law. *Evol. Theory* 1:1–30.

Van Veghel, M.L.J. and Bak, R.P.M. 1993. Intraspecific variation of a dominant Caribbean reef building coral, *Montastrea annularis:* genetic, behavioral and morphometric aspects. *Mar. Ecol. Prog. Ser.* 92:255–65.

Vaughan, T.W. 1907. Recent Madreporaria of the Hawaiian Islands and Laysan. *U.S. Nat. Mus. Bull.* 59(9):1–427.

Vaughan, T.W. 1918. Temperature of the Florida coral-reef tract. *Carnegie Inst. Washington Publ.* 213:321–39.

Vaughan, T.W. 1919. Corals and the formation of coral reefs. *Smithson. Inst. Annu. Rep.:* 189–238.

Vaughan, T.W. and Wells, J.W. 1943. Revision of the suborders, families and genera of the Scleractinia. *Geol. Soc. Am. Spec. Pap.* 44.

Veevers. J.J. and Ettriem, S.L. 1988. Reconstruction of Antartica and Australia at breakup (95±5 Ma) and before rifting (160 Ma). *Aust. J. Earth Sci.* 35:355–62.

Vermeij, G.J. 1978. *Biogeography and adaptation, patterns of marine life.* Harvard University Press, Cambridge, Mass.

Veron, J.E.N. 1974. Southern geographic limits to the distribution of Great Barrier Reef hermatypic corals. *Proc. 2nd Int. Coral Reef Symp.,* Brisbane 1:465–73.

Veron, J.E.N. 1985. Aspects of the biogeography of hermatypic corals. *Proc. 5th Int. Coral Reef Congr.,* Tahiti 4:83–8.

Veron, J.E.N. 1986. *Corals of Australia and the Indo-Pacific.* Angus & Robertson, Sydney.

Veron, J.E.N. 1988. Comparison between the hermatypic corals of the Southern Ryukyu Islands of Japan and the Great Barrier Reef of Australia. *Galaxea* 7:211–31.

Veron, J.E.N. 1990a. New scleractinia from Japan and other Indo-west Pacific countries. *Galaxea* 9:95–173.

Veron, J.E.N. 1990b. Checklist of the hermatypic corals of Vanuatu. *Pac. Sci.* 44:51–70.

Veron, J.E.N. 1990c. Re-examination of the reef corals of Cocos (Keeling) Atoll. *Rec. W. Aust. Mus.* 14:553–81.

Veron, J.E.N. 1992a. Environmental control of Holocene changes to the world's most northern hermatypic coral outcrop. *Pac. Sci.* 46:405–25.

Veron, J.E.N. 1992b. *Hermatypic corals of Japan.* Australian Institute of Marine Science Monograph Series 8.

Veron, J.E.N. 1992c. Conservation of biodiversity: a critical time for the corals of Japan. *Coral Reefs* 11:13–21.

Veron, J.E.N. 1993. *A biogeographic database of hermatypic corals; species of the central Indo-*

Pacific, genera of the world. Australian Institute of Marine Science Monograph Series 9.

Veron, J.E.N. and Done, T.J. 1979. Corals and coral communities of Lord Howe Island. *Aust. J. Mar. Freshwater Res.* 30(2):1–34.

Veron, J.E.N. and Hodgson, G. 1989. Annotated checklist of the hermatypic corals of the Philippines. *Pac. Sci.* 43:234–87.

Veron, J.E.N. and Kelley, R. 1988. Species stability in reef corals of Papua New Guinea and the Indo-Pacific. *Assoc. Aust. Palaeontol. Memoirs* 6:1–69.

Veron, J.E.N. and Marsh, L.M. 1988. Records and annotated check list of the hermatypic corals of Western Australia. *Rec. Western Australia Mus. Suppl.* 29.

Veron, J.E.N. and Minchin, P.R. 1992. Correlations between sea surface temperature, circulation patterns and the distribution of hermatypic corals of Japan. *Continental Shelf Res.* 12:835–57.

Veron, J.E.N. and Pichon, M. 1976. *Scleractinia of eastern Australia. I. Families Thamnasteriidae, Astrocoeniidae, Pocilloporidae.* Australian Institute of Marine Science Monograph Series 1.

Veron, J.E.N. and Pichon, M. 1979. *Scleractinia of eastern Australia. III. Families Agariciidae, Siderastreidae, Fungiidae, Oculinidae, Merulinidae, Mussidae, Pectiniidae, Caryophylliidae, Dendrophylliidae.* Australian Institute of Marine Science Monograph Series 4.

Veron, J.E.N. and Pichon, M. 1982. *Scleractinia of eastern Australia. IV. Family Poritidae.* Australian Institute of Marine Science Monograph Series 5.

Veron, J.E.N. and Wallace, C. 1984. *Scleractinia of eastern Australia. V. Family Acroporidae.* Australian Institue of Marine Science Monograph Series 6.

Veron, J.E.N. et al 1974. Corals of the Solitary islands, central New South Wales. *Aust. J. Mar. Freshwater Res.* 25:193–208.

Veron, J.E.N. et al 1977. *Scleractinia of eastern Australia. II. Families Faviidae, Trachyphylliidae.* Australian Institute of Marine Science Monograph Series 3.

Vrba, E.S. 1985. Species and speciation. *Transvaal Mus. Monogr.* 4, Pretoria.

Wafar, M.V.M. 1986. Corals and coral reefs of India. *Proc. Indian Acad. Sci. (Anim. Sci. Plant Sci.) Suppl.*:19–43.

Walker, N.D. et al 1982. Thermal history of reef-associated environments during a record cold-air outbreak event. *Coral Reefs* 1:83–7.

Wallace, C.C. et al 1991. Indo-Pacific coral biogeography: a case study for the *Acropora selago* group. *Aust. Syst. Bot.* 4:199–210.

Ward, S. 1992. Evidence for broadcast spawning as well as brooding in the scleractinian coral *Pocillopora damicornis. Mar. Biol.* 112:641–6.

Wells, J.W. 1954a. Recent corals of the Marshall Islands. *U.S. Geol. Survey Prof. Pap.* 260:385–486.

Wells, J.W. 1954b. Fossil Corals from Bikini Atoll. *U.S. Geol. Survey Prof. Pap.* 260: 609–15.

Wells, J.W. 1955. A survey of the distribution of reef coral genera in the Great Barrier Reef region. *Report Great Barrier Reef Committee* 6:1–9.

Wells, J.W. 1956. Scleractinia **in** Moore R.C. (ed), *Treatise on Invertebrate Paleontology. Coelenterata.* Geological Society of America and University of Kansas Press: 328–440.

Wells, J.W. 1957. Corals. *Mem. Geol. Soc. Am.* 67:1087–104.

Wells, J.W. 1966. Evolutionary development in the scleractinian family Fungiidae. *Symp. Zool. Soc. Lond.* 16:223–416.

Wells, J.W. 1969. Aspects of Pacific coral reefs. *Micronesica* 5:317–26.

Wells, J.W. 1972. Notes on Indo-Pacific scleractinian corals. Part 8, scleractinian corals from Easter Island. *Pac. Sci.* 26:183–90

Wells, J.W. 1982. Fossil corals from Midway Atoll. *U.S. Geol. Survey Prof. Pap.* 680:1–7.

Wells, J.W. 1986. A list of scleractinian generic and subgeneric taxa, 1758–1985. *Fossil Cnidaria* 15(11).

Wells, J.W. and Alderslade P.N. 1979. The scleractinian coral *Archohelia* living on the coastal shores of Queensland, Australia. *Rec. Aust. Mus.* 32:211–16.

Wells, S.M. (ed) 1988. *Coral reefs of the world.* International Union of Conservation, Nature and Natural Resources vols. 1–3.

Wheeler, C.W. and Ahron, P. 1991. Mid-oceanic carbonate platforms as oceanic dipsticks: examples from the Pacific. *Coral Reefs* 10:101–14.

White, B.N. 1986. The Isthmian Link, antitropicality and American biogeography: distributional history of the Atherinopsinae (Pisces: Atherinidae). *Syst. Zool.* 35:176–94.

White, M.J.D. 1968. Models of speciation. *Science* 159:1065–70.

White, M.J.D. 1978. *Models of speciation.* Freeman, San Francisco.

Whitham, T.G. and Slobodchikoff, C.N. 1981. Evolution by individuals, plant-herbivore interactions, and mosaics of genetic variability: the somatic significance of somatic mutations in plants. *Oecologia* 49:287–92.

Wijsman, H.J.W. and Wijsman-Best, M. 1973. A note on the chromosomes of some madreporarian corals from the Mediterranean (Cnidaria, Anthozoa, Scleractinia: Dendrophyllidae, Caryophyllidae, Faviidae). *Genen Phaenen* 16:61–4.

Wijsman-Best, M. 1974. Habitat-induced modification of reef corals (Faviidae) and its consequences for taxonomy. *Proc. 2nd Int. Coral Reef Symp.*, Brisbane 2:217–28.

Wiley, E.O. 1978. The evolutionary species concept reconsidered. *Syst. Zool.* 27:17–26.

Wilkinson, C.R. and Buddemeier, R.W. 1994. *Global climatic change and coral reefs: implications for people and reefs. Report of the UNEP–IOC–ASPEI–IUCN global task team on the implications of climate change on coral reefs.* IUCN, Gland.

Williams, E.H. and Bunkley-Williams, L. 1990. The world-wide coral reef bleaching cycle and related sources of coral mortality. *Atoll Res. Bull.* 3335:1–71.

Willis, B. 1985. Phenotypic plasticity versus phenotypic stability in the reef corals *Turbinaria mesenterina* and *Pavona cactus*. *Proc. 5th Int. Coral Reef Congr.*, Tahiti 4:107–12.

Willis, B.L. 1990. Species concepts in extant scleractinian corals: considerations based on reproductive biology and genotypic population structures. *Syst. Bot.* 15:136–49.

Willis, B. and Ayre, D.J. 1985. Asexual reproduction and genetic determination of growth form in the coral *Pavona cactus*: biochemical genetic and immunogenic evidence. *Oecologia* 65:516–25.

Willis, B. et al 1985. Patterns in the mass spawning of corals on the Great Barrier Reef from 1981 to 1984. *Proc. 5th Int. Coral Reef Congr.*, Tahiti 4:343–8.

Willis, B. et al 1994. Experimental evidence of hybridization in reef corals involved in mass spawning events. *Proc. 7th Int. Coral Reef Symp.*, Guam (abstract):504.

Wilson, E.C. 1990. Mass mortality of the reef coral *Pocillopora* on the south coast of Baja California Sur, Mexico. *Bull. S. Cal. Acad. Sci.* 89:39–41.

Wilson, E.C. 1991. Composition, origin and succession of Pliocene to Recent hermatypic coral faunas of the Baja California peninsula area. *Résumé 1st International Meeting of the Geology of Baja California Peninsula*:78–9.

Wilson, J.K. 1975. *Carbonate facies in geologic history.* Springer Verlag, Berlin

Wolfe, J.A. 1985. Distribution of major vegetational types during the Tertiary **in** Sundquist E.T. and Broecker, W.S. (eds), The Carbon Cycle and Atmospheric CO₂: Natural Variations Archean to Present. *Geophys. Monogr.* 3.

Yabe, H. and Sugiyama, T. 1935. Revised lists of the reef corals from the Japanese seas and of the fossil reef corals of the raised reefs and the Ryukyu limestone of Japan. *J. Geol. Soc. Japan.* 42(502):279–403.

Yajima, T. et al 1986. Ecological distribution of the reef coral *Oulastrea crispata* (Lamarck) at the shore region in the vicinity of Tsukumo Bay. *Bull. Japan Sea Research Institute, Kanazawa University* 18:21–36 (in Japanese).

Yamaguchi, M. 1987. Occurrences and persistency of *Acanthaster planci* pseudo-population in relation to oceanographic conditions along the Pacific Coast of Japan. *Galaxea* 6:277–88.

Yeemin, T. et al 1990. Sexual reproduction of the scleractinian coral, *Montastrea valenciennesi*, from a high-latitude coral community, southwest Japan. *Publication Amakusa Marine Biological Laboratory* 10:105–21.

Yellowlees, D. (ed) 1991. *Land use patterns and nutrient loading of the Great Barrier Reef region.* Proc. Workshop held at James Cook University of North Queensland, 17–18 November, 1990. James Cook University of North Queensland.

Yonge, C.M. 1931. The significance of the relationship between corals and zooxanthellae. *Nature* 128:309–10.

Yonge, C.M. 1940. The biology of reef building corals. *Scientific Report, Great Barrier Reef Expedition 1928–29*. British Museum (Natural History) 1(13):353–891.

Zabala, M. and Ballesteros, E. 1989. Surface-dependent strategies and energy flux in benthic marine communities or, why corals do not exist in the Mediterranean. *Selent. Mar.* 53:3–17.

Zachos, J.C. and Arthur, M.A. 1986. Paleoceanography of the Cretaceous/Tertiary boundary event: inferences from stable isotopic and other data. *Paleoceanography* 1:5–26.

INDEX